Ergebnisse der Mathematik und ihrer Grenzgebiete

Volume 45

3. Folge

A Series of Modern Surveys in Mathematics

Editorial Board

S. Feferman, Stanford M. Gromov, Bures-sur-Yvette
J. Jost, Leipzig J. Kollár, Princeton
H.W. Lenstra, Jr., Berkeley P.-L. Lions, Paris
M. Rapoport, Köln J.Tits, Paris D. B. Zagier, Bonn
G. M. Ziegler, Berlin

Managing Editor R. Remmert, Münster

Springer
Berlin
Heidelberg
New York
Hong Kong
London
Milan
Paris
Tokyo

Thomas Kappeler
Jürgen Pöschel

KdV & KAM

Springer

Thomas Kappeler
Institut für Mathematik
Universität Zürich
Winterthurerstr. 190
8057 Zürich, Switzerland
e-mail: tk@math.unizh.ch

Jürgen Pöschel
Fakultät Mathematik und Physik
Universität Stuttgart
Pfaffenwaldring 57
70569 Stuttgart, Germany
e-mail: j@poschel.de

Cataloging-in-Publication Data applied for

Bibliographic information published by Die Deutsche Bibliothek

Die Deutsche Bibliothek lists this publication in the Deutsche Nationalbibliografie; detailed bibliographic data is available in the Internet at <http://dnb.ddb.de>.

Mathematics Subject Classification (2000):
34B30, 35B20, 37K10, 37K35, 37J40

ISSN 0071-1136
ISBN 3-540-02234-1 Springer-Verlag Berlin Heidelberg New York

Springer-Verlag Berlin Heidelberg New York
a member of BertelsmannSpringer Science+Business Media GmbH

Printed in Germany

Typeset by the authors using a Springer TEX macro package.
Printed on acid-free paper 44/3142LK - 5 4 3 2 1 0

In
memory
of

JÜRGEN MOSER

teacher
mentor
friend

Preface

This book is concerned with two aspects of the theory of integrable partial differential equations. The first aspect is a *normal form theory* for such equations, which we exemplify by the periodic *K*orteweg *d*e *V*ries equation – undoubtedly one of the most important nonlinear, integrable pdes. This makes for the 'KdV' part of the title of the book.

The second aspect is a theory for Hamiltonian perturbations of such pdes. Its prototype is the so called *KAM theory*, developed for finite dimensional systems by *K*olmogorov, *A*rnold and *M*oser. This makes for the 'KAM' part of the title of the book.

To be more specific, our starting point is the periodic KdV equation considered as an infinite dimensional, integrable Hamiltonian system admitting a complete set of independent integrals in involution. We show that this leads to a single, global, real analytic system of Birkhoff coordinates – the cartesian version of action-angle coordinates –, such that the KdV Hamiltonian becomes a function of the actions alone. In fact, these coordinates work simultaneously for all Hamiltonians in the KdV hierarchy.

While the existence of *global* Birkhoff coordinates is a special feature of KdV, *local* Birkhoff coordinates may be constructed via our approach for many integrable pdes anywhere in phase space. Specifically this holds true for the defocusing nonlinear Schrödinger equation, for which parallel results were developed in [51].

The global coordinates make it evident that all solutions of the periodic KdV equation are periodic, quasi-periodic, or almost-periodic in time. It also provides a convenient handle to study small Hamiltonian perturbations, by applying a suitable generalization of the KAM theory to partial differential operators. To check the pertinent nondegeneracy conditions, we construct Birkhoff normal forms up to order six to gain sufficient control over the KdV frequencies as functions of the actions. In fact, these Birkhoff normal forms are just the first terms in the power series expansion of the KdV Hamiltonian in Birkhoff coordinates.

Finally, we describe the set up, assumptions and conclusions of a general infinite dimensional KAM theorem, that is applicable here and goes back to Kuksin. The situation differs from more conventional applications of KAM to pdes in that the perturbations are given by unbounded operators. This is only partially compensated by a smoothing effect of the small divisors. In addition, one has to modify the iteration scheme and use normal forms which also depend on angular variables.

Only recently, monographs on KAM theory for integrable pdes appeared, by Bourgain [17], Craig [29], and Kuksin [75]. Of these, the first two choose a different approach, setting up a functional equation and applying a Lyapunov-Schmidt decomposition scheme pioneered by Craig & Wayne [28]. The latter employs a normal form theory for Lax-integrable pdes near finite dimensional tori, which is based on the Its-Matveev formula. In contrast, the normal form theory presented in this book with its global features goes much further. It allows us to obtain a perturbation theory for KdV from an abstract KAM theorem of a particularly simple form, and to use Birkhoff normal forms to check the relevant nondegeneracy conditions. Moreover, this normal form might turn out to be useful for other long time stability results for perturbed integrable pdes such as Nekhoroshev estimates.

This book is not only intended for the handful of specialists working at the intersection of integrable partial differential equations and Hamiltonian perturbation theory, but also researchers farther away from these fields. In fact, it is our intention to reach out to graduate students as well. It is for this reason that first of all, we have included a chapter on the classical theory, describing the finite dimensional background of integrable Hamiltonian systems and their perturbation theory according to the theory initiated by Kolmogorov, Arnold and Moser.

Secondly, we made the book self-contained, omitting only those proofs which can be found in well known textbooks. We therefore included numerous appendices – some of them, we hope, of independent interest – on topics from complex analysis on Hilbert spaces, spectral theory of Schrödinger operators, Riemann surface theory, representation of holomorphic differentials, and certain aspects of the KdV equation such as the KdV hierarchy and new formulas for the KdV frequencies.

Thirdly, we wrote the book in a modular manner, where each of its five main chapters – chapters II to VI – as well as its appendices may be read independently of each other. Every chapter has its own introduction, and the notation is explained. As a result, there is some natural repetition and overlap among them. Moreover, the results of these chapters are summarized in the very first chapter, titled "The Beginning", and here too we took the liberty to quote from the introductions to the later chapters. We consider these repetitions a benefit for the reader rather than a nuisance, since it allows him, or her, to peruse the material in a nonlinear manner.

This book took many years to complete, and during this long time we benefitted from discussions and collaborations with many friends and colleagues. We would like to thank all of them, in particular Benoît Grébert, with whom we developped parallel results for the defocusing nonlinear Schrödinger equation in [51], and Jürg Kramer, for his contribution to the nondegeneracy result for the first KdV Hamiltonian. Most of all we are indebted to Jürgen Moser, who initiated this joint effort and never failed to encourage us as long as he was able to do so. We dedicate this book to him.

The second author also gratefully acknowledges the hospitality of the Forschungsinstitut at the ETH Zürich and the Institute of Mathematics at the University of Zürich during many periods of our collaborative efforts, as well as the support of the Deutsche Forschungsgemeinschaft, while the first author gratefully acknowledges the support of the Swiss National Science Foundaton and of the European Research Training Network HPRN-CT-1999-00118.

Finally we would like to thank Jules Hobbes for his never tiring TEXpertise from the very first lines through many, many revisions up to the final, press-ready output, and Jürgen Jost and Springer Verlag for their pleasant cooperation to make this book happen.

Last, but not least we thank our families for their patience and support during these many years.

Zürich/Stuttgart
February 14/16, 2003 *TK/JP*

Contents

List of Figures

I

The Beginning

1 Overview

In this book we consider the Korteweg-de Vries (KdV) equation

$$u_t = -u_{xxx} + 6uu_x.$$

The KdV equation is an evolution equation in one space dimension which is named after the two Dutch mathematicians Korteweg and de Vries [66] – see also Boussinesq [18] and Rayleigh [113]. It was proposed as a model equation for long surface waves of water in a narrow and shallow channel. Their aim was to obtain as solutions solitary waves of the type discovered in nature by Russell [114] in 1834. Later it became clear that this equation also models waves in other homogeneous, weakly nonlinear and weakly dispersive media. Since the mid-sixties the KdV equation received a lot of attention in the aftermath of the computational experiments of Kruskal and Zabusky [69], which lead to the discovery of the interaction properties of the solitary wave solutions and in turn to the understanding of KdV as an infinite dimensional integrable Hamiltonian system.

Our purpose here is to study small Hamiltonian perturbations of the KdV equation with periodic boundary conditions. In the unperturbed system *all* solutions are periodic, quasi-periodic, or almost-periodic in time. The aim is to show that large families of periodic and quasi-periodic solutions persist under such perturbations. This is true not only for the KdV equation itself, but in principle for all equations in the KdV hierarchy. As an example, the second KdV equation will also be considered.

The KdV Equation

Let us recall those features of the KdV equation that are essential for our purposes. It was observed by Gardner [46], see also Faddeev & Zakharov [40], that the KdV equation can be written in the Hamiltonian form

$$\frac{\partial u}{\partial t} = \frac{\mathrm{d}}{\mathrm{d}x}\frac{\partial H}{\partial u}$$

with the Hamiltonian

$$H(u) = \int_{S^1} \left(\tfrac{1}{2}u_x^2 + u^3\right) \mathrm{d}x,$$

where $\partial H/\partial u$ denotes the L^2-gradient of H, representing the Fréchet derivative of H with respect to the standard scalar product on L^2. Since we are interested in *spatially periodic solutions*, we take as the underlying phase space the Sobolev space

$$\mathcal{H}^N = H^N(S^1; \mathbb{R}), \qquad S^1 = \mathbb{R}/\mathbb{Z},$$

of real valued functions with period 1, where $N \geq 1$ is an integer, and endow it with the Poisson bracket proposed by Gardner,

$$\{F, G\} = \int_{S^1} \frac{\partial F}{\partial u(x)} \frac{\mathrm{d}}{\mathrm{d}x} \frac{\partial G}{\partial u(x)} \,\mathrm{d}x.$$

Here, F and G are differentiable functions on $\mathcal{H}^N$ with L^2-gradients in $\mathcal{H}^1$. This makes $\mathcal{H}^N$ a Poisson manifold, on which the KdV equation may also be represented in the form $u_t = \{u, H\}$ familiar from classical mechanics.

We note that the initial value problem for the KdV equation on the circle S^1 is well posed on every Sobolev space $\mathcal{H}^N$ with $N \geq 1$: for initial data $u^o \in \mathcal{H}^N$ it has been shown by Temam for $N = 1, 2$ [128] and by Saut & Temam for any real $N \geq 2$ [121] that there exists a unique solution evolving in $\mathcal{H}^N$ and defined globally in time. For further results on the initial value problem see for instance [78, 88, 126] as well as the more recent results [14, 15, 64].

The KdV equation admits infinitely many conserved quantities, or *integrals*, in involution, and there are many ways to construct such integrals [46, 94, 95]. Lax [77] obtained a set of Poisson commuting integrals in a particularly elegant way by considering the spectrum of an associated Schrödinger operator. For

$$u \in \mathcal{H}^0 = L^2 = L^2(S^1, \mathbb{R})$$

consider the differential operator

$$L = -\frac{\mathrm{d}^2}{\mathrm{d}x^2} + u$$

on the interval $[0, 2]$ of *twice* the length of the period of u with periodic boundary conditions. It is well known [80, 82, 84] that its spectrum, denoted $\operatorname{spec}(u)$, is pure point and consists of an unbounded sequence of *periodic eigenvalues*

$$\lambda_0(u) < \lambda_1(u) \leq \lambda_2(u) < \lambda_3(u) \leq \lambda_4(u) < \dots .$$

Equality or inequality may occur in every place with a '$\leq$'-sign, and one speaks of the *gaps* $(\lambda_{2n-1}(u), \lambda_{2n}(u))$ of the *potential* u and its *gap length*

$$\gamma_n(u) = \lambda_{2n}(u) - \lambda_{2n-1}(u), \qquad n \geq 1.$$

If some gap length is zero, one speaks of a *collapsed gap*, otherwise of an *open gap*.

For $u = u(t, \cdot)$ depending also on t define the corresponding operator

$$L(t) = -\frac{\mathrm{d}^2}{\mathrm{d}x^2} + u(t, \cdot).$$

Lax observed that u is a solution of the KdV equation if and only if

$$\frac{\mathrm{d}}{\mathrm{d}t} L = [B, L],$$

where $[B, L] = BL - LB$ denotes the commutator of L with the *anti-symmetric* operator

$$B = -4\frac{\mathrm{d}^3}{\mathrm{d}x^3} + 3u\frac{\mathrm{d}}{\mathrm{d}x} + 3\frac{\mathrm{d}}{\mathrm{d}x}u.$$

It follows by an elementary calculation that the solution of

$$\frac{\mathrm{d}}{\mathrm{d}t} U = BU, \qquad U(0) = I,$$

defines a family of *unitary* operators $U(t)$ such that $U^*(t)L(t)U(t) = L(0)$. Consequently, the spectrum of $L(t)$ is independent of t, and so the periodic eigenvalues $\lambda_n = \lambda_n(u)$ are conserved quantities under the evolution of the KdV equation, a fact first observed by Gardner, Greene, Kruskal & Miura [47]. Thus, the flow of the KdV equation defines an *isospectral deformation* on the space of all potentials in $\mathcal{H}^N$.

From an analytical point of view, however, the periodic eigenvalues are not satisfactory as integrals, as λ_n is not a smooth function of u whenever the corresponding gap collapses. But McKean & Trubowitz [89] showed that the *squared gap lengths*

$$\gamma_n^2(u), \qquad n \geq 1,$$

together with the *average*

$$[u] = \int_{S^1} u(x)\,\mathrm{d}x$$

form another set of integrals, which are *real analytic* on all of L^2 and Poisson commute with each other. Moreover, the squared gap lengths together with the average determine uniquely the periodic spectrum of a potential [48].

The space L^2 thus decomposes into the *isospectral sets*

$$\mathrm{Iso}(u) = \left\{ v \in L^2 \colon \ \mathrm{spec}(v) = \mathrm{spec}(u) \right\},$$

which are invariant under the KdV flow and may also be characterized as

$$\mathrm{Iso}(u) = \left\{ v \in L^2 \colon \ \text{gap lengths}(v) = \text{gap lengths}(u), \ \ [v] = [u] \right\}.$$

As shown by McKean & Trubowitz [89] these are compact connected *tori*, whose dimension equals the number of positive gap lengths and is infinite generically. Moreover, as the asymptotic behavior of the gap lengths characterizes the regularity of a potential in exactly the same way as its Fourier coefficients do [84], we have

$$u \in \mathcal{H}^N \quad \Leftrightarrow \quad \mathrm{Iso}(u) \subset \mathcal{H}^N$$

for each $N \geq 1$. Hence also the phase space $\mathcal{H}^N$ decomposes into a collection of tori of varying dimension which are *invariant* under the KdV flow.

Angle-Action and Birkhoff Coordinates

In classical mechanics the existence of a foliation of the phase space into Lagrangian invariant tori is tantamount, at least locally, to the existence of angle-action coordinates. This is the content of the Liouville-Arnold-Jost theorem. In the infinite dimensional setting of the KdV equation, however, the existence of such coordinates is far less clear as the dimension of the foliation is *nowhere* locally constant. Invariant tori of infinite and finite dimension each form dense subsets of the foliation. Nevertheless, angle-action coordinates *can* be introduced *globally* in the form of Birkhoff coordinates as we describe now. They will form the basis of our study of perturbations of the KdV equation.

To formulate the statement we define the phase spaces more precisely. For any integer $N \geq 0$, let

$$\mathcal{H}^N = \left\{ u \in L^2(S^1, \mathbb{R}) \colon \; \|u\|_N < \infty \right\},$$

where

$$\|u\|_N^2 = \left|\hat{u}(0)\right|^2 + \sum_{k \in \mathbb{Z}} |k|^{2N} \left|\hat{u}(k)\right|^2$$

is defined in terms of the discrete Fourier transform $\hat{u}$ of u. The Poisson structure $\{\cdot, \cdot\}$ is degenerate on $\mathcal{H}^N$ and admits the average $[\,\cdot\,]$ as a Casimir function. The leaves of the corresponding symplectic foliation are given by $[u] = \text{const}$. Instead of restricting the KdV Hamiltonian to each leaf, it is more convenient to fix one such leaf, namely

$$\mathcal{H}_0^N = \left\{ u \in \mathcal{H}^N \colon \; [u] = 0 \right\},$$

which is symplectomorphic to each other leaf by a simple translation, and consider the mean value as a parameter. On $\mathcal{H}_0^N$ the Poisson structure is nondegenerate and induces a symplectic structure. Writing $u = v + c$ with $[v] = 0$ and $c = [u]$, the Hamiltonian then takes the form

$$H(u) = H_c(v) + c^3$$

with

$$H_c(v) = \int_{S^1} \left(\tfrac{1}{2} v_x^2 + v^3\right) \mathrm{d}x + 6c \int_{S^1} \tfrac{1}{2} v^2 \,\mathrm{d}x.$$

We consider H_c as a 1-parameter family of Hamiltonians on $\mathcal{H}_0^N$.

We remark that

$$H^0 = \frac{1}{2} \int_{S^1} v^2 \,\mathrm{d}x$$

corresponds to translation and is the zero-th Hamiltonian of the KdV hierarchy, as described in appendix C.

To describe the angle-action variables on $\mathcal{H}_0^N$ we introduce the model space

$$\mathfrak{h}_r = \ell_r^2 \times \ell_r^2$$

with elements (x, y), where

$$\ell_r^2 = \Big\{ x \in \ell^2(\mathbb{N}, \mathbb{R}) \colon \; \|x\|_r^2 = \sum_{n \ge 1} n^{2r} \, |x_n|^2 < \infty \Big\}.$$

We endow $\mathfrak{h}_r$ with the standard Poisson structure, for which $\{x_n, y_m\} = \delta_{nm}$, while all other brackets vanish.

The following theorem was first proven in [5] and [6]. A quite different approach for this result – and the one we expand on here – was first presented in [60]. For a related result for the nonlinear Schrödinger equation see [51].

Theorem 1.1. *There exists a diffeomorphism*

$$\Psi \colon \; \mathfrak{h}_{1/2} \to \mathcal{H}_0^0$$

with the following properties.

(i) *Ψ is one-to-one, onto, bi-analytic, and preserves the Poisson bracket.*

(ii) *For each $N \ge 0$, the restriction of Ψ to $\mathfrak{h}_{N+1/2}$, denoted by the same symbol, is a map*

$$\Psi \colon \; \mathfrak{h}_{N+1/2} \to \mathcal{H}_0^N,$$

which is one-to-one, onto, and bi-analytic as well.

(iii) *The coordinates (x, y) in $\mathfrak{h}_{3/2}$ are global Birkhoff coordinates for KdV. That is, for any $c \in \mathbb{R}$, the transformed Hamiltonian $H_c \circ \Psi$ depends only on $x_n^2 + y_n^2$, $n \ge 1$, with (x, y) being canonical coordinates.*

Thus, in the coordinates (x, y) the KdV Hamiltonian is a real analytic function of the actions alone:

$$H_c = H_c(I_1, I_2, \dots), \qquad I_n = \frac{1}{2}(x_n^2 + y_n^2),$$

with equations of motion

$$\dot{x}_n = \omega_n(I) y_n, \qquad \dot{y}_n = -\omega_n(I) x_n,$$

where

$$\omega_n = \omega_{c,n} = \frac{\partial H_c}{\partial I_n}(I), \qquad I = (I_n)_{n \ge 1}.$$

The whole system appears now as an infinite chain of anharmonic oscillators, whose frequencies depend on their amplitudes in a nonlinear and real analytic fashion.

These results are not restricted to the KdV Hamiltonian. They simultaneously apply to *every* real analytic Hamiltonian in the Poisson algebra of all Hamiltonians which Poisson commute with all actions $I_1, I_2, \ldots$. In particular, one obtains Birkhoff coordinates for every Hamiltonian in the KdV hierarchy defined in appendix C. As an example, we will later also consider the second KdV Hamiltonian.

The existence of Birkhoff coordinates makes it evident that every solution of the KdV equation is *almost-periodic* in time. In the coordinates of the model space every solution is given by

$$\begin{aligned} x_n(t) &= \sqrt{2I_n^{\mathrm{o}}}\,\sin(\theta_n^{\mathrm{o}} + \omega_n(I^{\mathrm{o}})t), \\ y_n(t) &= \sqrt{2I_n^{\mathrm{o}}}\,\cos(\theta_n^{\mathrm{o}} + \omega_n(I^{\mathrm{o}})t), \end{aligned}$$

where $(\theta^{\mathrm{o}}, I^{\mathrm{o}})$ corresponds to the initial data u^{o}. Hence, it winds around the underlying invariant torus

$$T_{I^{\mathrm{o}}} = \left\{ (x, y) \colon x_n^2 + y_n^2 = 2I_n^{\mathrm{o}},\ \ n \geq 1 \right\},$$

The solution in the original space $\mathcal{H}_0^N$ is thus winding around the embedded torus $\Psi(T_{I^{\mathrm{o}}})$, and expanding Ψ into its Taylor series, it is of the form

$$\begin{aligned} u(t) &= \Psi(x(t), y(t)) \\ &= \sum_{k \in \mathbb{Z}^\infty, |k| < \infty} \Psi_k(I^{\mathrm{o}}, \theta^{\mathrm{o}})\, \mathrm{e}^{\mathrm{i}\langle k, \omega(I^{\mathrm{o}})\rangle t}. \end{aligned}$$

Here, $\langle k, \omega\rangle = \sum_n k_n \omega_n$, and each $\Psi_k(I^{\mathrm{o}}, \theta^{\mathrm{o}})$ is an element of $\mathcal{H}_0^N$. Thus, every solution is almost-periodic in time.

We remark that the solution above can also be represented in terms of the Riemann theta function. The corresponding formula is due to Its & Matveev [33].

Among all almost-periodic solutions there is a dense subset of quasi-periodic solutions, which are characterized by a *finite* number of frequencies and correspond to finite gap potentials. To describe them more precisely, let $A \subset \mathbb{N}$ be a *finite* index set, and consider the set of *A-gap potentials*

$$\mathcal{G}_A = \left\{ u \in \mathcal{H}_0^0 \colon \gamma_n(u) > 0 \Leftrightarrow n \in A \right\}.$$

That is, $u \in \mathcal{G}_A$ if and only if precisely the gaps $(\lambda_{2n-1}(u), \lambda_{2n}(u))$ with $n \in A$ are open. Clearly,

$$u \in \mathcal{G}_A \quad \Leftrightarrow \quad \mathrm{Iso}(u) \subset \mathcal{G}_A,$$

and all finite gap potentials are smooth, in fact real analytic, as almost all gap lengths are zero.

As might be expected there is a close connection between the set $\mathcal{G}_A$ and the subspace

$$\mathscr{h}_A = \left\{ (x, y) \in \mathscr{h}_0 \colon x_n^2 + y_n^2 > 0 \Leftrightarrow n \in A \right\}.$$

Addendum to Theorem 1.1. *The canonical transformation* Ψ *also has the following property.*

(iv) *For every finite index set* $A \subset \mathbb{N}$, *the restriction* Ψ_A *of* Ψ *to* $\mathscr{h}_A$ *is a map*

$$\Psi_A\colon\ \mathscr{h}_A \to \mathscr{G}_A,$$

which is one-to-one, onto, and bi-analytic.

It follows that $\mathscr{G}_A$ is a real analytic, invariant submanifold of $\mathscr{H}_0^0$, which is completely foliated into invariant tori $\mathrm{Iso}(u)$ of the same dimension $|A|$. The KdV flow consists of a quasi-periodic winding around each such torus which is characterized by the frequencies ω_n with $n \in A$. The above representation reduces to a convergent, real analytic representation

$$u(t) = \sum_{k \in \mathbb{Z}^A} \Psi_k(I_A^{\text{o}}, \theta_A^{\text{o}})\, \mathrm{e}^{\mathrm{i}\langle k, \omega_A(I_A^{\text{o}})\rangle t},$$

with $I_A = (I_n)_{n \in A}$, and similarly defined θ_A and ω_A.

Outline of the Proof of Theorem 1.1

The proof of the theorem splits into four parts. First we define actions I_n and angles θ_n for a potential q following a procedure introduced for finite dimensional integrable systems. The formula for the actions I_n, due to Flaschka & McLaughlin [43], is given entirely in terms of the periodic spectrum of the potential. The angles θ_n, which linearize the KdV equation, were introduced even earlier by a number of authors, namely Dubrovin, Its, Krichever, Matveev, Novikov [32, 33, 35, 36, 57] (see also [34]), McKean & van Moerbeke [88], and McKean & Trubowitz [89, 90]. They are defined in terms of the Riemann surface $\Sigma(q)$ associated with the periodic spectrum of a potential as explained in section 6. We show that each I_n is real analytic on L_0^2, while each θ_n, taken modulo 2π, is real analytic on the dense open domain $L_0^2 \setminus D_n$, where D_n denotes the subvariety of potentials with collapsed n-th gap.

Next, we define the cartesian coordinates x_n and y_n canonically associated to I_n and θ_n. Although defined originally only on $L_0^2 \setminus D_n$, we show that they extend real analytically to a complex neighbourhood W of L_0^2. Surely, the angle θ_n blows up when γ_n collapses, but this blow up is compensated by the rate at which I_n vanishes in the process. In particular, for real q the resulting limit will vanish.

Then we show that the thus defined map $\Omega\colon q \mapsto (x, y)$ is a diffeomorphism between L_0^2 and $\mathscr{h}_{1/2}$. The main problem here is to verify that $d_q\Omega$ is a linear isomorphism at *every* point q. This is done with the help of orthogonality relations among the coordinates, which are in fact their Poisson brackets. For the nonlinear Schrödinger equation the corresponding orthogonality relations have first been established by McKean & Vaninsky [91, 92]. It turned out that many of their ideas can also be used in the case of KdV.

It is immediate that each Hamiltonian in the KdV hierarchy becomes a function of the actions alone, using their characterization in terms of the asymptotic expansion of one of the Floquet multipliers $w(\lambda)$ and hence as spectral invariants.

Finally, we verify that Ω preserves the Poisson bracket. As it happens, it is more convenient to look at the associated symplectic structures. This way, we only need to establish the regularity of the gradient of θ_n at special points, not everywhere. Thus, we equivalently show that Ω is a symplectomorphism. This will complete the proof of the main results of Theorem 1.1.

Perturbations of the KdV Equation

Our aim is to investigate whether sufficiently small Hamiltonian perturbations of the KdV equation, namely

$$\frac{\partial u}{\partial t} = \frac{\mathrm{d}}{\mathrm{d}x}\left(\frac{\partial H_c}{\partial u} + \varepsilon \frac{\partial K}{\partial u}\right),$$

admit almost-periodic solutions as well, winding around invariant tori in phase space.

In the classical setting of integrable Hamiltonian systems of finitely many degrees of freedom this question is answered by the theory of Kolmogorov, Arnold and Moser, known as *KAM theory*. It states that with respect to Lebesgue measure the majority of the invariant tori of a real analytic, nondegenerate integrable system persist under sufficiently small, real analytic Hamiltonian perturbations. They are only slightly deformed and still completely filled with quasi-periodic motions. The base of this partial foliation of the phase space into invariant tori, however, is no longer open, but has the structure of a *Cantor set*: it is a nowhere dense, closed set with no isolated points.

Till now, there is no general infinite dimensional KAM theory to establish the persistence of infinite dimensional tori with almost-periodic solutions for Hamiltonian systems arising from partial differential equations, such as the KdV equation. There does exist a KAM theory to this effect, but it is restricted to systems, in which the coupling between the different modes of oscillations is of short range type. See [108] and the references therein. Such a theory does not apply here, and we can not make any statement about the persistence of almost-periodic solutions.

But as noted above, there are also families of finite dimensional tori on real analytic submanifolds, corresponding to finite gap solutions and filling the space densely. In the classical setting a KAM theorem about the persistence of such lower dimensional tori was first formulated by Melnikov [93], and proven later by Eliasson [38]. It was independently extended to the infinite dimensional setting of partial differential equations by Kuksin [70, 72]. In the following we describe this kind of perturbation theory for finite gap solutions, following [107, 109].

Again, let $A \subset \mathbb{N}$ be a *finite* index, and let $\Gamma \subset \mathbb{R}^A_+$ be a compact set of positive Lebesgue measure. We then set

$$\mathcal{T}_\Gamma = \bigcup_{I \in \Gamma} \mathcal{T}_I \subset \mathcal{G}_A, \qquad \mathcal{T}_I = \Psi_A(T_I),$$

where

$$T_I = \left\{(x, y) \in h_A : x_n^2 + y_n^2 = 2I_n,\ \ n \in A\right\} \cong \mathbb{T}^A \times \{I\},$$

with $\mathbb{T} = \mathbb{R}/2\pi\mathbb{Z}$ the circle of length 2π. Notice that $\mathcal{T}_\Gamma \subset \bigcap_{N\geq 0} \mathcal{H}_0^N$ in view of the Addendum to Theorem 1.1.

We show that under sufficiently small real analytic Hamiltonian perturbations of the KdV equation the majority of these tori persists together with their translational flows, the tori being only slightly deformed. The following theorem was first proven by Kuksin for the case $c = 0$.

Theorem 1.2. *Let $A \subset \mathbb{N}$ be a finite index set, $\Gamma \subset \mathbb{R}_+^A$ a compact subset of positive Lebesgue measure, and $N \geq 1$. Assume that the Hamiltonian K is real analytic in a complex neighbourhood U of $\mathcal{T}_\Gamma$ in $\mathcal{H}_{0,\mathbb{C}}^N$ and satisfies the regularity condition*

$$\frac{\partial K}{\partial u}\colon \ U \to \mathcal{H}_{0,\mathbb{C}}^N, \qquad \left\| \frac{\partial K}{\partial u} \right\|_{N;U}^{\sup} = \sup_{u\in U} \left\| \frac{\partial K}{\partial u} \right\|_N \leq 1.$$

Then, for any real c, there exists an $\varepsilon_0 > 0$ depending only on A, N, c and the size of U such that for $|\varepsilon| < \varepsilon_0$ the following holds. There exist

(i) *a nonempty Cantor set $\Gamma_\varepsilon \subset \Gamma$ with* $\operatorname{meas}(\Gamma - \Gamma_\varepsilon) \to 0$ *as $\varepsilon \to 0$,*

(ii) *a Lipschitz family of real analytic torus embeddings*

$$\Xi\colon \ \mathbb{T}^n \times \Gamma_\varepsilon \to U \cap \mathcal{H}_0^N,$$

(iii) *a Lipschitz map $\chi : \Gamma_\varepsilon \to \mathbb{R}^n$,*

such that for each $(\theta, I) \in \mathbb{T}^n \times \Gamma_\varepsilon$, the curve $u(t) = \Xi(\theta + \chi(I)t, I)$ is a quasi-periodic solution of

$$\frac{\partial u}{\partial t} = \frac{\mathrm{d}}{\mathrm{d}x}\left(\frac{\partial H_c}{\partial u} + \varepsilon \frac{\partial K}{\partial u}\right)$$

winding around the invariant torus $\Xi(\mathbb{T}^n \times \{I\})$. Moreover, each such torus is linearly stable.

Remark 1. Note that the L^2-gradient of a function on $\mathcal{H}_0^N$ has mean value zero by the definition of the gradient. On the other hand, the L^2-gradient of a function on the larger space $\mathcal{H}^N$ usually has mean value different from zero, and the gradient of its restriction to $\mathcal{H}_0^N$ is the *projection* of the former onto $\mathcal{H}_0^N$:

$$\nabla\big(K|_{H_0^N}\big) = \operatorname{Proj}_{\mathcal{H}_0^N} \nabla K = \nabla K - [\nabla K],$$

with $\nabla = \partial/\partial u$. This, however, does not affect their Hamiltonian equations, since the derivative of the constant function $[\nabla K]$ vanishes. Therefore, we will not explicitly distinguish between these two gradients.

Remark 2. We already mentioned that

$$\mathcal{G}_A \subset \bigcap_{N\geq 0} \mathcal{H}_0^N.$$

Thus, the quasi-periodic solutions remain in $\mathcal{H}_0^N$ if the gradient of K is in $\mathcal{H}_0^N$.

Remark 3. The regularity assumption on K entails that K depends only on u, but not on its derivatives. So the perturbation effected by K is of *lower order* than the unperturbed KdV equation.

In view of the derivation of the KdV equation as a model equation for surface waves of water in a certain regime – see for example [136] – it would be interesting to obtain perturbation results which also include terms of higher order, at least in the region where the KdV approximation is valid. However, results of this type are still out of reach, if true at all.

Remark 4. We point out that the perturbing term $\varepsilon\partial K/\partial u$ need not be a differential operator. For example,

$$K(u) = \left(\int_{S^1} u^3 \,\mathrm{d}x\right)^2$$

has an L^2-gradient

$$\frac{\partial K}{\partial u} = 6u^2 \int_{S^1} u^3 \,\mathrm{d}x$$

to which the theorem applies as well.

Remark 5. The invariant embedded tori are linearly stable in the sense that the variational, or linearized, equations of motion along such a torus are reducible to constant coefficient form whose spectrum is located on the imaginary axis. Hence, all Lyapunov exponents of such a torus vanish.

Similar results hold for any equation in the KdV hierarchy. Consequently, for any given finite index set A, the manifold of A-gap potentials is foliated into the same family of invariant tori. The difference is only in the frequencies of the quasi-periodic motions on each of these tori. Therefore, similar results should also hold for the higher order KdV equations, once we can establish the corresponding nonresonance conditions.

As an example we consider the second KdV equation, which reads

$$\partial_t u = \partial_x^5 u - 10u\partial_x^3 u - 20\partial_x u\partial_x^2 u + 30u^2\partial_x u.$$

Its Hamiltonian is

$$H^2(u) = \int_{S^1} \left(\tfrac{1}{2}u_{xx}^2 + 5uu_x^2 + \tfrac{5}{2}u^4\right) \mathrm{d}x,$$

which is defined on $\mathcal{H}^2$. Again, with $u = v + c$, where $[v] = 0$, we get

$$H^2(u) = H_c^2(v) + \frac{5}{2}c^4$$

with

$$H_c^2(v) = H^2(v) + 10cH^1(v) + 30c^2H^0(v).$$

Here,

$$H^1(v) = \int_{S^1} \left(\frac{1}{2}v_x^2 + v^3\right) \mathrm{d}x, \qquad H^0(v) = \frac{1}{2}\int_{S^1} v^2 \,\mathrm{d}x$$

are the familiar KdV Hamiltonian and the zero-th Hamiltonian of translation, respectively. We study this Hamiltonian on the space $\mathcal{H}_0^N$ with $N \geq 2$, considering c as a real parameter.

Theorem 1.3. *Let $A \subset \mathbb{N}$ be a finite index set, $\Gamma \subset \mathbb{R}_+^A$ a compact subset of positive Lebesgue measure, and $N \geq 3$. Assume that the Hamiltonian K is real analytic in a complex neighbourhood U of $\mathcal{T}_\Gamma$ in $\mathcal{H}_{0,\mathbb{C}}^N$ and satisfies the regularity condition*

$$\frac{\partial K}{\partial u}\colon\ U \to \mathcal{H}_{0,\mathbb{C}}^{N-2}, \qquad \left\|\frac{\partial K}{\partial u}\right\|_{N-2;U}^{\sup} \leq 1.$$

If $c \notin \mathcal{E}_A^2$, where the exceptional set $\mathcal{E}_A^2$ is an at most countable subset of the real line not containing 0 and with at most $|A|$ accumulation points, then the same conclusions as in Theorem 1.2 hold for the system with Hamiltonian $H_c^2 + \varepsilon K$.

Remark 1. The gradient $\partial K/\partial u$ is only required to be in $\mathcal{H}_0^{N-2}$. Still, the regularity assumption ensures that the perturbation is of *lower order* than the unperturbed equation.

Remark 2. A more detailed description of the set $\mathcal{E}_A^2$ is given in appendix J. It is likely that the theorem is true for *all* $c \in \mathbb{R}$.

We now give two simple examples of perturbations to which the preceding theorems apply. As a first example let

$$K(u) = \int_{S^1} F(x, u)\,\mathrm{d}x,$$

where F defines a real analytic map

$$\{\lambda \in \mathbb{R}\colon\ |\lambda| < R\} \to \mathcal{H}_0^N, \qquad \lambda \mapsto F(\,\cdot\,, \lambda),$$

for some $R > 0$ and $N \geq 1$. Then, with $f = \partial F/\partial\lambda$,

$$\frac{\partial K}{\partial u} = f(x, u) - [f(x, u)]$$

belongs to $\mathcal{H}_0^N$, and the perturbed KdV equation is

$$u_t = -u_{xxx} + 6uu_x + \varepsilon\frac{\mathrm{d}}{\mathrm{d}x} f(x, u).$$

Theorem 1.2 applies after fixing Γ and c for all sufficiently small ε. Of course, F may also depend on ε, if the dependence is, say, continuous.

We remark that perturbations of the KdV equation with a Hamiltonian K as above can be characterized equivalently as local perturbations given by $\frac{\mathrm{d}}{\mathrm{d}x} f(x, u(x))$, where f admits a power series expansion in the second argument,

$$f(x, \lambda) = \sum_{k \geq 0} f_k(x) \lambda^k,$$

convergent in $\mathcal{H}_0^N$. In this case, the perturbed equation is again a partial differential equation, and its Hamiltonian K is given by $K(u) = \int_{S^1} F(x, u)\, \mathrm{d}x$, where F is a primitive of f with respect to λ,

$$F(x, \lambda) = \sum_{k \geq 0} \frac{f_k(x)}{k+1} \lambda^{k+1}.$$

As a second example let

$$K(u) = \int_{S^1} F(x, u_x)\, \mathrm{d}x,$$

where F is as above with $N \geq 3$. More generally, F could also depend on u, but this adds nothing new. Then

$$\frac{\partial K}{\partial u} = -\frac{\mathrm{d}}{\mathrm{d}x} f(x, u_x),$$

with $f = \partial F / \partial \lambda$, belongs to $\mathcal{H}_0^{N-2}$, and the perturbed second KdV equation is

$$u_t = \cdots - \varepsilon \frac{\mathrm{d}^2}{\mathrm{d}x^2} f(x, u_x).$$

To this second example, Theorem 1.3 applies.

Outline of the Proof of Theorems 1.2 and 1.3

A prerequisite for developing a perturbation theory of KAM type is the existence of coordinates with respect to which the variational equations along the unperturbed motions on the invariant tori reduce to constant coefficient form. Often, such coordinates are difficult to construct even locally. Here, they are provided *globally* by Theorem 1.1.

According to Theorem 1.1 the Hamiltonian of the KdV equation on the model space $\hbar_{3/2}$ is of the form

$$H_c = H_c(I_1, I_2, \ldots), \qquad I_n = \frac{1}{2}(x_n^2 + y_n^2).$$

The equations of motion are thus

$$\dot{x}_n = \omega_n(I) y_n, \qquad \dot{y}_n = -\omega_n(I) x_n,$$

with frequencies

$$\omega_n = \omega_{c,n} = \frac{\partial H_c}{\partial I_n}(I), \qquad I = (I_n)_{n\geq 1},$$

that are constant along each orbit. So each orbit is winding around some invariant torus

$$T_I = \left\{ (x, y) \colon x_n^2 + y_n^2 = 2I_n,\ n \geq 1 \right\},$$

where the parameters $I = (I_n)_{n\geq 1}$ are the actions of its initial data.

We are interested in a perturbation theory for families of *finite*-dimensional tori T_I. So we fix an index set $A \subset \mathbb{N}$ of finite cardinality $|A|$, and consider tori with

$$I_n > 0 \quad \Leftrightarrow \quad n \in A.$$

The linearized equations of motion along any such torus have now constant coefficients and are determined by $|A|$ *internal frequencies* $\omega = (\omega_n)_{n\in A}$ and infinitely many *external frequencies* $\Omega = (\omega_n)_{n\notin A}$. Both depend on the $|A|$-dimensional parameter

$$\xi = (I_n^{\mathrm{o}})_{n\in A},$$

since all other components of I vanish in this family.

The KAM theorem for such families of finite dimensional tori requires a number of assumptions, among which the most notorious and unpleasant ones are the so called *nondegeneracy* and *nonresonance conditions*. In this case, they essentially amount to the following. First, the map

$$\xi \mapsto \omega(\xi)$$

from the parameters to the internal frequencies has to be a local homeomorphism, which is Lipschitz in both directions. This is known as *Kolmogorov's condition* in the classical theory. Second, for each $k \in \mathbb{Z}^A$ and $l \in \mathbb{Z}^{\mathbb{N}\smallsetminus A}$ with $1 \leq |l| \leq 2$, the zero set of any of the frequency combinations

$$\langle k, \omega(\xi) \rangle + \langle l, \Omega(\xi) \rangle$$

has to be a set of measure zero. This is sometimes called *Melnikov's condition.*

The verification of these conditions for the KdV Hamiltonian requires some knowledge of its frequencies. One way to obtain this knowledge is to use Riemann surface theory: Krichever proved that the frequency map $\xi \mapsto \omega(\xi)$ is a local diffeomorphism everywhere on the space of A-gap potentials, see [68, 9], and Bobenko & Kuksin showed that the second condition is satisfied in the case $c = 0$ using Schottky uniformization [11]. Here, however, we follow a different and more elementary route to verify these conditions by computing the first coefficients of the Birkhoff normal form of the KdV Hamiltonian, which we explain now.

In classical mechanics the Birkhoff normal form allows to view a Hamiltonian system near an elliptic equilibrium as a small perturbation of an integrable system.

This tool is also applicable in an infinite dimensional setting. Writing

$$u = \sum_{n \neq 0} \gamma_n q_n \mathrm{e}^{2\pi \mathrm{i} n x}$$

with weights $\gamma_n = \sqrt{2\pi\,|n|}$ and complex coefficients $q_{\pm n} = (x_n \mp \mathrm{i} y_n)/\sqrt{2}$, the KdV Hamiltonian becomes

$$H_c = \sum_{n \geq 1} \lambda_n \left| q_n \right|^2 + \sum_{k+l+m=0} \gamma_k \gamma_l \gamma_m q_k q_l q_m$$

on $\hbar_{3/2}$ with

$$\lambda_n = (2\pi n)^3 + 6c \cdot 2\pi n.$$

Thus, at the origin we have an elliptic equilibrium with characteristic frequencies λ_n.

To transform this Hamiltonian into its Birkhoff normal form up to order four two coordinate transformations are required: one to eliminate the cubic term, and one to normalize the resulting fourth order term. Both calculations are elementary. Expressed in real coordinates (x, y) the result is the following.

Theorem 1.4. *There exists a real analytic, symplectic coordinate transformation Φ in a neighbourhood of the origin in $\hbar_{3/2}$, which transforms the KdV Hamiltonian on $\hbar_{3/2}$ into*

$$H_c \circ \Phi = \frac{1}{2} \sum_{n \geq 1} \lambda_n (x_n^2 + y_n^2) - \frac{3}{4} \sum_{n \geq 1} (x_n^2 + y_n^2)^2 + \ldots,$$

where the dots stand for terms of higher order in x and y.

The important fact about the non-resonant Birkhoff normal form is that its coefficients are uniquely determined independently of the normalizing transformation, as long as it is of the form "identity + higher order terms". For this reason, these coefficients are also called *Birkhoff invariants*. Comparing Theorem 1.4 with Theorem 1.1 and viewing Ψ as a *global* transformation into a *complete* Birkhoff normal form we thus conclude that the two resulting Hamiltonians on $\hbar_{3/2}$ must agree up to terms of order four. In other words, the local result provides us with the first terms of the Taylor series expansion of the globally integrable KdV Hamiltonian.

Corollary 1.5. *The canonical transformation Ψ of Theorem 1.1 transforms the KdV Hamiltonian into the Hamiltonian*

$$H_c(I) = \sum_{n \geq 1} \lambda_n I_n - 3 \sum_{n \geq 1} I_n^2 + \ldots,$$

where $I_n = \frac{1}{2}(x_n^2 + y_n^2)$, and the dots stand for higher order terms in (x, y). Thus,

$$\omega_n(I) = \frac{\partial H_c}{\partial I_n}(I) = \lambda_n - 6 I_n + \ldots.$$

Here, λ_n and hence ω_n also depend on c.

By further computing some additional terms of order six in the expansion above, we gain sufficient control over the frequencies ω to verify all nondegeneracy and nonresonance conditions for any c.

Incidentally, the normal form of Theorem 1.4 already suffices to prove the persistence of quasi-periodic solutions of the KdV equation of sufficiently *small amplitude* under small Hamiltonian perturbations. In addition, if the perturbing term $\partial K/\partial u$ is of degree three or more in u, then no small parameter ε is needed to make the perturbing terms small, as it suffices to work in a sufficiently small neighbourhood of the equilibrium solution $u \equiv 0$. We will not expand on this point in this book.

A Remark on the KAM Proof

Previous versions of the KAM theorem for partial differential equations such as [72, 109] were concerned with perturbations that were given by *bounded* nonlinear operators. This was sufficient to handle, among others, nonlinear Schrödinger and wave equations on a bounded interval, see for example [12, 76, 110]. This is *not* sufficient, however, to deal with perturbations of the KdV equation, as here the term

$$\frac{\mathrm{d}}{\mathrm{d}x}\frac{\partial K}{\partial u}$$

is an *unbounded* operator. This entails some subtle difficulties in the proof of the KAM theorem, as we outline now.

Write the perturbed Hamiltonian as

$$H = N + P,$$

where N denotes some integrable normal form and P a general perturbation. The KAM proof employs a rapidly converging iteration scheme of Newton type to handle small divisor problems, and involves an infinite sequence of coordinate transformations. At each step a transformation Φ is constructed as the time-1-map $X_F^t\big|_{t=1}$ of a Hamiltonian vector field X_F that brings the perturbed Hamiltonian $H = N + P$ closer to some new normal form N_+. Its generating Hamiltonian F as well as the correction $\hat{N}$ to the given normal form N are a solution of the linearized equation

$$\{F, N\} + \hat{N} = R,$$

where R is some suitable truncation of the Taylor and Fourier expansion of P. Then Φ takes the truncated Hamiltonian $H' = N + R$ into $H' \circ \Phi = N_+ + R_+$, where $N_+ = N + \hat{N}$ is the new normal form and

$$R_+ = \int_0^1 \{(1-t)\hat{N} + tR, F\} \circ X_F^t \,\mathrm{d}t$$

the new error term arising from R. Accordingly, the full Hamiltonian $H = N + P$ is transformed into $H \circ \Phi = N_+ + R_+ + (P - R) \circ \Phi$. See section 19 for details.

What makes this scheme more complicated than previous ones is the fact that the vector field X_R generated by R represents an *unbounded* operator, whereas the vector field X_F generated by the solution F of the linearized equation has to represent a *bounded* operator to define a *bona fide* coordinate transformation. For most terms in F this presents no problem, because they are obtained from the corresponding terms in R by dividing with a *large divisor*. There is no such smoothing effect, however, for that part of R of the form

$$\frac{1}{2}\sum_{n\notin A} R_n(\theta;\xi)(x_n^2+y_n^2),$$

where $\theta = (\theta_n)_{n\in A}$ are the coordinates on the torus $\mathbb{T}^A$, and ξ the parameters mentioned above. We therefore include these terms in $\hat{N}$ and hence in the new normal form N_+. However, subsequently we have to deal with a generalized, *θ-dependent* normal form

$$N = \sum_{n\in A}\omega_n(\xi)I_n + \frac{1}{2}\sum_{n\notin A}\Omega_n(\theta;\xi)(x_n^2+y_n^2).$$

This, in turn, makes it difficult to obtain solutions of the linearized equation with useful estimates.

In [74] Kuksin obtained such estimates and thus rendered the iterative construction convergent. It requires a delicate discussion of a linear small divisor equation with *large variable* coefficients, which we reproduce in section 23.

Existing Literature

Theorem 1.1 was first given in [5] and [6]. A quite different approach to this result, and the one detailed here, was first presented in [60] and extended to the nonlinear Schrödinger equation in [51]. At the heart of the argument are orthogonality relations which first have been established in the case of the nonlinear Schrödinger equation by McKean & Vaninsky [91, 92].

A version of Theorem 1.2 in the case $c = 0$ is due to Kuksin [71, 74]. In the second paper, he proves a KAM theorem of the type discussed above which is needed to deal with perturbations given by unbounded operators, and combines it with earlier results [9, 11] concerning nonresonance properties of the KdV frequencies and the construction of local coordinates so that the linearized equations of motions along a given torus of finite gap potentials reduce to constant coefficients [71].

The proof of Theorem 1.2 presented in this book is different from the approach in [71, 74]. Instead of the local coordinates constructed in [71] we use the global, real analytic angle-action coordinates given by Theorem 1.1 to obtain quasi-periodic solutions of arbitrary size for sufficiently small and sufficiently regular perturbations of the KdV equation. To verify the relevant nonresonance conditions we follow the line of arguments used in [76] where small quasi-periodic solutions for nonlinear Schrödinger equations were obtained, and explicitly compute the Birkhoff normal form of the KdV Hamiltonian up to order 4 and a few terms of order 6.

We stress again that our results are concerned exclusively with the existence of *quasi-periodic* solutions. Nothing is known about the persistence of almost-periodic solutions. The KAM theory of [108] for such solutions is not applicable here, since the nonlinearities effect a strong, long range coupling among all "modes" in the KdV equation.

There are, however, existence results for certain simplified problems. Bourgain [16, 17] considered the Schrödinger equation

$$\mathrm{i}u_t = u_{xx} - V(x)u - |u|^2 u$$

on $[0, \pi]$ with Dirichlet boundary conditions, depending on some analytic potential V. Given an almost-periodic solution of the linear equation with *very* rapidly decreasing amplitudes and nonresonant frequencies, he showed that the potential V may be modified so that this solution persists for the nonlinear equation. The potential serves as an infinite dimensional parameter, which has to be chosen properly for each initial choice of amplitudes. This result is obtained by iterating the Lyapunov-Schmidt reduction scheme introduced by Craig & Wayne [28].

A similar result was obtained independently in [111] by iterating the KAM theorem about the existence of quasi-periodic solutions. As a result, one obtains for – in a suitable sense – almost all potentials V a set of almost-periodic solutions, which – again in a suitable sense – has density one at the origin. See [111] for more details.

II

Classical Background

In this book we consider the periodic KdV equation as an *infinite dimensional* integrable Hamiltonian system, and subject it to small Hamiltonian perturbations. To this end, we extend many concepts, ideas and notions from the classical *finite dimensional* theory, such as angle-action coordinates, Birkhoff normal forms, and in particular KAM theory.

To set the stage we give a concise review of these notions in the classical setting. To keep things short we omit lengthy proofs and give references instead.

2 Hamiltonian Formalism

The abstract Hamiltonian formalism can be described either in terms of symplectic manifolds, or in terms of Poisson manifolds. The former is more familiar in mathematics, while the latter, used often in physics, is more general. We will give descriptions of both set ups and show that they are equivalent in the case of nondegenerate Poisson structures.

In the following, M denotes a smooth manifold of finite dimension without boundary, where 'smooth' means 'infinitely often differentiable'. We assume M to be connected, but not necessarily compact.

Symplectic Manifolds

Definition 2.1. *A* symplectic form *on a smooth manifold M is a closed, nondegenerate 2-form υ on M. The pair (M, υ) is called a* symplectic manifold.

Here, 'closed' means that $\mathrm{d}\upsilon = 0$. 'Nondegenerate' means that υ is nondegenerate on every tangent space of M, making it a symplectic vector space.

Necessarily, the dimension of a symplectic manifold M is even, since otherwise υ had a nontrivial kernel. Moreover, M is orientable, since the n-fold wedge product $\upsilon \wedge \cdots \wedge \upsilon$ defines a volume form on M.

Being nondegenerate, the symplectic form υ induces an isomorphism

$$\begin{aligned} S\colon\ TM &\to T^*M \\ X &\mapsto \upsilon \circ X \end{aligned}$$

between the tangent and cotangent bundle of M at each point, referred to as the *symplectic structure*, where $\upsilon \circ X = \upsilon(X, \cdot)$ denotes the *inner product* or *contraction* of a form υ and a vector field X. Let

$$J = S^{-1}\colon\ T^*M \to TM$$

be the inverse of this map. Every smooth function $H\colon M \to \mathbb{R}$ then defines a vector field

$$X_H = J\mathrm{d}H$$

on M, which is the unique vector field satisfying

$$\upsilon \circ X_H = \mathrm{d}H.$$

X_H is called the *Hamiltonian vector field* associated with the *Hamiltonian* H, and M its underlying *phase space*. In turn, this vector field defines a flow Φ^t on M called the *Hamiltonian flow* of the Hamiltonian H. We usually denote this flow by X_H^t to indicate its connection with the Hamiltonian H.

It is an immediate consequence of the definitions and the skew symmetry of υ that the Hamiltonian H is constant along the flow lines of its Hamiltonian vector field X_H:

$$\frac{\mathrm{d}}{\mathrm{d}t} H \circ X_H^t = \mathrm{d}H(X_H) = \upsilon(X_H, X_H) = 0.$$

In classical mechanics this is known as the law of conservation of energy.

Given a symplectic form, one defines the *Poisson bracket* of two smooth functions G and H as

$$\{G, H\} = \upsilon(X_G, X_H).$$

It is a skew form on the linear space $C^\infty(M)$ of all smooth functions on M. In view of the definition of Hamiltonian vector fields, we have

$$\{G, H\} = \upsilon(X_G, X_H) = \mathrm{d}G(X_H).$$

Consequently, the flow X_H^t has the property that

$$\dot{F} = \{F, H\}$$

for *any* smooth function F on M, where

$$\dot{F} = \frac{\mathrm{d}}{\mathrm{d}t} F \circ X_H^t \Big|_{t=0} = \mathrm{d}F(X_H)$$

denotes the derivation of F with respect to the vector field X_H.

The Poisson bracket satisfies *Leibniz rule*,

$$\{FG, H\} = F\{G, H\} + G\{F, H\},$$

and the *Jacobi identity*,

$$\{F, \{G, H\}\} + \{G, \{H, F\}\} + \{H, \{F, G\}\} = 0.$$

Being a local property the Jacobi identity is most easily verified in the Darboux coordinates introduced below. Alternatively, one may first show that for *any* nondegenerate 2-form υ one has

$$\{F, \{G, H\}\} + \{G, \{H, F\}\} + \{H, \{F, G\}\} = \mathrm{d}\upsilon(X_F, X_G, X_H), \tag{2.1}$$

see for example [86, p. 85]. Note that to define Hamiltonian vector fields the closedness of the 2-form υ is not required. Rather, the above identity shows that the latter is *equivalent* to the Jacobi identity for its Poisson bracket.

Poisson Manifolds

Instead of describing the evolution of a system directly by a Hamiltonian vector field on the phase space, one may also describe it indirectly by the evolution of so called *observables*, such as smooth functions on the phase space. The starting point here is a Poisson bracket. Let $\mathbb{F} = C^\infty(M)$.

Definition 2.2. *A* Poisson bracket *on a smooth manifold M is a skew-symmetric bilinear map*

$$\{\cdot, \cdot\}\colon\ \mathbb{F} \times \mathbb{F} \to \mathbb{F},$$

which satisfies Leibniz rule and the Jacobi identity. A smooth manifold with a Poisson bracket is called a Poisson manifold.

A flow Φ^t on a Poisson manifold is called a *Poisson system*, if there exists a function H on M, called the *Hamiltonian* of the system, such that

$$\dot{F} = \frac{\mathrm{d}}{\mathrm{d}t} F \circ \Phi^t \bigg|_{t=0} = \{F, H\}$$

for any $F \in \mathbb{F}$. By skew symmetry of the Poisson bracket, $\{F, F\} = 0$ for all F. Hence, again

$$\dot{H} = 0$$

for a Poisson system with Hamiltonian H. The latter, however, is not necessarily uniquely determined by the above characterization.

This description of a Poisson system is implicit. But it can be made explicit in terms of its structure map as follows. By linearity and the Leibniz rule, the map $F \mapsto \{F, H\}$ of $\mathbb{F}$ into itself is a derivation. Therefore, in a finite dimensional setting, there exists a unique vector field X_H, the *Hamiltonian vector field* associated with H, such that

$$\{G,H\} = X_H G = \langle \mathrm{d}G, X_H \rangle,$$

where $\langle \cdot\,,\cdot \rangle$ denotes the dual pairing between T^*M and TM. By skew-symmetry of the Poisson bracket, $\langle \mathrm{d}G, X_H \rangle = -\langle \mathrm{d}H, X_G \rangle$. Moreover, the association between $\mathrm{d}H$ and X_H is linear. Therefore, there exists a unique map

$$K\colon\ T^*M \to TM,$$

called the *Poisson structure*, mapping each fiber T_p^*M linearly into T_pM, such that

$$X_H = K\mathrm{d}H.$$

We then obtain

$$\{F,H\} = \langle \mathrm{d}F, X_H \rangle = \langle \mathrm{d}F, K\mathrm{d}H \rangle.$$

K is skew-symmetric, since the Poisson bracket is.

In an infinite-dimensional setting, a derivation is not necessarily given by a vector field. In this case, the starting point is the structure map itself.

Definition 2.3. *A* Poisson structure *is a map*

$$K\colon\ T^*M \to TM,$$

mapping each fiber T_p^*M linearly to T_pM, such that the associated bracket

$$\{F,G\} = \langle \mathrm{d}F, K\mathrm{d}G \rangle$$

is skew-symmetric and satisfies the Jacobi identity.

Note that this bracket automatically satisfies Leibniz rule, hence is a Poisson bracket.

A Poisson structure is called *nondegenerate*, if it has a trivial kernel. In this case we have an inverse

$$K^{-1}\colon\ TM \to T^*M,$$

and we can define a bilinear form υ on vector fields by

$$\upsilon(X,Y) = \langle K^{-1}X, Y \rangle.$$

This form is skew-symmetric and nondegenerate, since K^{-1} is. A Hamiltonian vector field with respect to υ is then again given by $X_H = K\mathrm{d}H$, since $\upsilon \circ X_H = \mathrm{d}H$. Moreover,

$$\upsilon(X_G, X_H) = \langle \mathrm{d}G, K\mathrm{d}H \rangle = \{G,H\},$$

so υ is also closed in view of (2.1) and Jacobi's identity, hence is a symplectic form.

Thus, a nondegenerate Poisson structure K gives rise to a symplectic structure with $S = K^{-1}$. Conversely, a symplectic structure S defines a nondegenerate Poisson structure with $K = S^{-1}$. Hence, these two notions are equivalent, and in the following we will not distinguish between them.

Lie Brackets and Integrals

The *Lie bracket* of two vector fields X and Y, considered as derivations, is defined as

$$[X,Y] = YX - XY.$$

Alternatively, one has

$$[X,Y] = L_Y X = -L_X Y = -[Y,X],$$

where L_X denotes the Lie-derivative with respect to the vector field X. This bracket is clearly bilinear and skew-symmetric. Moreover, the Lie bracket of two Hamiltonian vector fields is again Hamiltonian:

Proposition 2.4.

$$[X_G, X_H] = X_{\{G,H\}}$$

for any two Hamiltonians G and H on a symplectic manifold.

Proof. With $\{G,H\} = \mathrm{d}G(X_H) = X_H G$ we have

$$\begin{aligned} \upsilon \circ X_{\{G,H\}} &= \mathrm{d}\{G,H\} \\ &= \mathrm{d}L_{X_H} G \\ &= L_{X_H} \mathrm{d}G \\ &= L_{X_H}(\upsilon \circ X_G) \\ &= (L_{X_H}\upsilon) \circ X_G + \upsilon \circ (L_{X_H} X_G) \\ &= \upsilon \circ [X_G, X_H], \end{aligned}$$

since $L_{X_H}\upsilon = 0$ by the Hamiltonian character of X_H. □

A smooth non-constant function G is called an *integral* of a Hamiltonian system with Hamiltonian H, if

$$\{G,H\} = 0.$$

Since $\{G,H\} = X_H G$, this means that G is constant along the flow lines of X_H, which justifies the terminology. By skew symmetry of the bracket, if G is an integral for X_H, then H is an integral for X_G, and one says that the two Hamiltonians G and H are *in involution*.

There may be nontrivial functions $C \in \mathbb{F}$, called *Casimir functions*, with

$$\{C, \cdot\} \equiv 0.$$

They are integrals for any Poisson system and may exist if the Poisson structure is degenerate.

In the nondegenerate case the preceding proposition implies that two Hamiltonians G and H are in involution, if and only if

$$[X_G, X_H] = 0.$$

Since this is equivalent to the fact that the flows of X_G and X_H commute, one also says that the two vector fields X_G and X_H *commute*.

Canonical Transformations

To preserve the Hamiltonian nature of vector fields a diffeomorphism of a symplectic or Poisson manifold has to preserve the underlying structure.

Definition 2.5. *A diffeomorphism Φ of a symplectic manifold is called* symplectic *or a* symplectomorphism, *if it preserves the symplectic form: $\Phi^*\upsilon = \upsilon$.*

A diffeomorphism Φ of a Poisson manifold is called canonical, *if it preserves the Poisson bracket: $\{F,G\} \circ \Phi = \{F \circ \Phi, G \circ \Phi\}$ for any F and G.*

We consider the two cases in detail.

A symplectomorphism transforms Hamiltonian vector fields according to the rule

$$\Phi^* X_H = X_{H\circ\Phi}.$$

Using that $\Phi^*\mathrm{d}H = \mathrm{d}(H \circ \Phi)$ this follows from the calculation

$$\begin{aligned} \upsilon \circ X_{H\circ\Phi} &= \mathrm{d}(H \circ \Phi) \\ &= \Phi^*\mathrm{d}H \\ &= \Phi^*(\upsilon \circ X_H) \\ &= \Phi^*\upsilon \circ \Phi^* X_H = \upsilon \circ \Phi^* X_H. \end{aligned} \tag{2.2}$$

Moreover, Φ then also preserves the associated Poisson bracket, since

$$\begin{aligned} \{G,H\} \circ \Phi &= \Phi^*\upsilon(X_G, X_H) \\ &= \upsilon(\Phi^* X_G, \Phi^* X_H) \\ &= \upsilon(X_{G\circ\Phi}, H_{H\circ\Phi}) = \{G \circ \Phi, H \circ \Phi\}. \end{aligned}$$

So Φ is also canonical.

On the other hand, on a Poisson manifold a canonical transformation Φ also has to preserve the Poisson structure K. This follows by inspecting its definition. If K is nondegenerate and thus defines a symplectic structure as above, then also

$$\begin{aligned} \Phi^* X_H &= (\Phi^* K)(\Phi^*\mathrm{d}H) \\ &= K\mathrm{d}(H \circ \Phi) = X_{H\circ\Phi}, \end{aligned}$$

so Φ also transforms Hamiltonian vector fields properly. Traversing the calculation in (2.2) backwards, this implies that

$$\Phi^*\upsilon \circ \Phi^* X_H = \upsilon \circ \Phi^* X_H$$

for all H. But this in turn implies that $\Phi^*\upsilon = \upsilon$, so Φ is a symplectomorphism with respect to the induced symplectic structure.

The concepts of a canonical transformation and a symplectomorphism generalize in an obvious way to transformations between two different Poisson or symplectic manifolds.

The Standard Example

The standard example of a symplectic manifold is $\mathbb{R}^{2n} = \mathbb{R}^n \times \mathbb{R}^n$ with coordinates $(q, p) = (q_1, \dots, q_n, p_1, \dots, p_n)$, often referred to as positions and moments. The symplectic form is

$$\upsilon_0 = \sum_{i=1}^{n} \mathrm{d}q_i \wedge \mathrm{d}p_i,$$

which in terms of the standard scalar product $\langle \cdot, \cdot \rangle_0$ on Euclidean space is given as

$$\upsilon_0 = \langle \cdot, J_0 \cdot \rangle_0, \qquad J_0 = \begin{pmatrix} 0 & I \\ -I & 0 \end{pmatrix},$$

where I denotes the n-dimensional identity matrix. Note that

$$J_0^{-1} = J_0^T = -J_0, \qquad \langle \cdot, J_0 \cdot \rangle_0 = \langle J_0^{-1} \cdot, \cdot \rangle_0,$$

so the symplectic structure is given by J_0^{-1}, interpreted as a map $T\mathbb{R}^{2n} \to T^*\mathbb{R}^{2n}$.

Expressing $\mathrm{d}H = \langle \nabla H, \cdot \rangle_0$ in terms of the $\langle \cdot, \cdot \rangle_0$-gradient of H, the identity $\upsilon \circ X_H = \mathrm{d}H$ amounts to

$$\begin{aligned} \langle \nabla H, \cdot \rangle_0 &= \mathrm{d}H \\ &= \upsilon \circ X_H \\ &= \langle X_H, J_0 \cdot \rangle_0 \\ &= \langle J_0^{-1} X_H, \cdot \rangle_0. \end{aligned}$$

We obtain

$$X_H = J_0 \nabla H$$

and the equations of motion

$$\dot{q}_i = \frac{\partial H}{\partial p_i}, \qquad \dot{p}_i = -\frac{\partial H}{\partial q_i}.$$

Moreover, for the induced Poisson bracket we find

$$\begin{aligned} \{F, G\} &= \langle J_0 \nabla F, J_0^2 \nabla G \rangle_0 \\ &= \langle \nabla F, J_0 \nabla G \rangle_0 \\ &= \langle F_q, G_p \rangle_0 - \langle F_p, G_q \rangle_0, \end{aligned}$$

and a transformation Φ is symplectic, if $\langle D\Phi \cdot , J_0 D\Phi \cdot \rangle_0 = \langle \cdot, J_0 \cdot \rangle_0$, or

$$D\Phi^T J_0 D\Phi = J_0.$$

So everything comes out as it should.

Similarly, the standard example of a Poisson manifold is the same manifold $\mathbb{R}^{2n} = \mathbb{R}^n \times \mathbb{R}^n$ with Poisson bracket

$$\{F, G\}_0 = \langle F_q, G_p \rangle_0 - \langle F_p, G_q \rangle_0.$$

Given a Hamiltonian H, the coordinate functions evolve by

$$\dot{q}_i = \{q_i, H\}_0 = H_{p_i}, \qquad \dot{p}_i = \{p_i, H\}_0 = -H_{q_i},$$

as before. Moreover, $\{F, G\}_0 = \langle \mathrm{d}F, J_0 \nabla G \rangle$, so the Poisson structure is J_0, interpreted as a map from $T^*\mathbb{R}^{2n}$ to $T\mathbb{R}^{2n}$. This map is invertible, and the associated symplectic form is

$$\upsilon(X, Y) = \langle J_0^{-1} X, Y \rangle = \langle X, J_0 Y \rangle,$$

as it ought to be.

A variant of the standard example is obtained when the position coordinates are identified modulo 2π and are thus *angular* coordinates. The phase space is then $\mathbb{T}^n \times \mathbb{R}^n$, where

$$\mathbb{T}^n = \mathbb{R}^n / 2\pi \mathbb{Z}^n$$

denotes the n-dimensional torus. The coordinates are usually denoted by

$$(\theta, I) = (\theta_1, \ldots, \theta_n, I_1, \ldots, I_n)$$

and are called *angle-action coordinates*. The standard symplectic two-form is then

$$\upsilon_0 = \sum_{i=1}^{n} \mathrm{d}\theta_i \wedge \mathrm{d}I_i.$$

Darboux Coordinates

In the linear case all symplectic vector spaces of the same dimension are symplectically isomorphic [55, p. 6]. This, of course, is no longer true for nonlinear symplectic manifolds. But Darboux's theorem states that this is still true *locally* around every point of a symplectic manifold.

Theorem 2.6 (Darboux). *Locally a symplectic manifold (M, υ) of dimension $2n$ is symplectomorphic to an open subset of $(\mathbb{R}^{2n}, \upsilon_0)$.*

That is, given any point p in M, there is a neighbourhood W of p in M and a diffeomorphism $\Phi\colon V \to W$ of an open set V in $\mathbb{R}^{2n}$ onto W such that

$$\Phi^* \upsilon = \upsilon_0.$$

The coordinates provided by Φ are called *Darboux coordinates*. For a proof see for example [55, p. 10–11].

3 Liouville Integrable Systems

Integrable systems are particular Hamiltonian systems that can be solved for any initial data by quadratures – whence the name. For this the system has to admit sufficiently many conserved quantities in involution. It turns out that for a system of n degrees of freedom, n independent integrals in involution suffice.

From now on, all manifolds, mappings and functions are assumed to be real analytic, unless otherwise stated.

Lagrangian Foliations

A family of m functions $F_1, \dots, F_m$ on M is called *independent*, if their 1-forms $\mathrm{d}F_1, \dots, \mathrm{d}F_m$ are linearly independent at every point in M.

Definition 3.1. *A Hamiltonian system on a symplectic manifold M of dimension $2n$ is called* integrable (in the sense of Liouville), *if its Hamiltonian H admits n independent integrals $F_1, \dots, F_n$ in involution. That is,*

(i) $\{H, F_i\} = 0$ *for* $1 \le i \le n$,

(ii) $\{F_i, F_j\} = 0$ *for* $1 \le i, j \le n$, *and*

(iii) $\mathrm{d}F_1 \wedge \cdots \wedge \mathrm{d}F_n \neq 0$

everywhere on M.

Example A. In standard angle-action coordinates (θ, I) on $\mathbb{T}^n \times \mathbb{R}^n$ any Hamiltonian of the form $H = H(I)$ is integrable with integrals

$$F_i = I_i, \qquad 1 \le i \le n.$$

They are everywhere in involution and independent, and the Hamiltonian is in fact a function of these integrals. This example will be discussed below.

Example B. In standard cartesian coordinates (q, p) on $\mathbb{R}^n \times \mathbb{R}^n$ any Hamiltonian of the form

$$H = H(q_1^2 + p_1^2, \dots, q_n^2 + p_n^2)$$

is integrable with integrals

$$F_i = q_i^2 + p_i^2, \qquad 1 \le i \le n.$$

More precisely, these functions are integrals in involution everywhere on $\mathbb{R}^n \times \mathbb{R}^n$, but are independent only on the dense open subset, where none of them vanishes. This example will be discussed in the next section.

To give a geometric description of an integrable system consider first an arbitrary number of smooth independent functions $F_1, \dots, F_m$ on M. Let

$$F = (F_1, \dots, F_m) \colon \quad M \to \mathbb{R}^m.$$

This map is a submersion, and every value is a regular value. Every nonempty leaf

$$M^c = F^{-1}(c) = \{\, p \in M \colon F(p) = c\,\}$$

is therefore a smooth submanifold of M of codimension m, and the whole manifold M is foliated into these leaves.

To give a symplectic description of the tangent space of M^c, define the skew-orthogonal complement of a set V of vectors to be the linear space $V^{\angle}$ of all vectors X such that $\upsilon \circ X$ vanishes on V. Then, by the definition of Hamiltonian vector fields,

$$T_pM^c = \bigcap_{1\le i\le m} \ker \mathrm{d}F_i = \bigcap_{1\le i\le m} X_{F_i}^{\angle}(p) = V_F^{\angle}(p),$$

where $V_F = \operatorname{span}(X_{F_1}, \dots, X_{F_m})$. If the functions F_i are also in involution, then the following result applies.

Lemma 3.2. *Suppose the functions $F = (F_1, \dots, F_m)$ define a foliation of M with leaves $M^c = F^{-1}(c)$. Then the following statements are equivalent.*

(i) *The functions $F_1, \dots, F_m$ are in involution: $\{F_i, F_j\} = 0$ for $1 \le i, j \le m$.*

(ii) *The Hamiltonian vector fields X_{F_i} are everywhere tangent to the leaves of F. That is, $X_{F_i}(p) \in T_pM^c$ for $1 \le i \le m$ and $p \in M^c$.*

(iii) *The leaves M^c are co-isotropic submanifolds of M. That is, $T_p^{\angle}M^c \subseteq T_pM^c$ at every point p in M^c.*

Proof. All these statements are different interpretations of the condition that

$$\{F_i, F_j\} = \mathrm{d}F_i(X_{F_j}) = \upsilon(X_{F_i}, X_{F_j})$$

vanishes for $1 \le i, j \le m$. □

From $T_p^{\angle}M^c \subseteq T_pM^c$ and the dimension formula for orthogonal complements in general we conclude that $m \le n$. So on a symplectic manifold M of dimension $2n$ there are at most n independent functions in involution. Moreover, if their number is indeed n, then

$$T_p^{\angle}M^c = T_pM^c$$

everywhere, so each leaf is a so called *Lagrangian submanifold* of M. The latter is, by definition, a submanifold of maximal possible dimension such that the restriction of the symplectic form to it vanishes.

Corollary 3.3. *If $F_1, \dots, F_n$ are n independent functions in involution on M, then the map $F = (F_1, \dots, F_n)$ defines a foliation of M into Lagrangian submanifolds $M^c = F^{-1}(c)$.*

The converse is also true. If $F = (F_1, \dots, F_n)$ defines a foliation of M into Lagrangian submanifolds, then the F_i are independent and in involution. This follows immediately from the preceding lemma.

Now suppose the Hamiltonian H admits $F_1, \ldots, F_n$ as independent integrals. From $\{H, F_i\} = 0$ we conclude that

$$0 = \{F_i, H\} = \mathrm{d}F_i(X_H),$$

so its Hamiltonian vector field X_H is tangent to the leaves M^c. It follows that these are *invariant manifolds* with respect to its flow, and we obtain the following geometric picture of an integrable Hamiltonian system.

Corollary 3.4. *A Hamiltonian system is integrable in the sense of Liouville if and only if it admits a foliation of its phase space into invariant Lagrangian submanifolds.*

It is clear from the preceding discussion that there is nothing intrinsic about a particular integrable Hamiltonian admitting $F_1, \ldots, F_n$ as integrals. Any other Hamiltonian in involution with them is integrable as well. This leads one to consider the Poisson algebra associated with those integrals.

Definition 3.5. *The* Poisson algebra *associated with an integrable Hamiltonian system with independent integrals $F = (F_1, \ldots, F_n)$ is the space*

$$\mathcal{A}(F) = \{G \in \mathbb{F} \colon \{G, F_i\} = 0 \text{ for } 1 \le i \le n\}$$

of all functions in $\mathbb{F} = C^\omega(M)$ in involution with $F_1, \ldots, F_n$.

This is an algebra with multiplication given by the Poisson bracket, since by the Jacobi identity,

$$\begin{aligned} 0 &= \{\{G, H\}, F_i\} + \{\{H, F_i\}, G\} + \{\{F_i, G\}, H\} \\ &= \{\{G, H\}, F_i\} \end{aligned}$$

for any G and H in $\mathcal{A}(F)$ and any F_i. In fact, we have

$$\{G, H\} = 0, \qquad G, H \in \mathcal{A}(F),$$

since X_G and X_H are tangent to any leaf of the associated foliation of M, and the latter are Lagrangian manifolds. Thus, $\{G, H\} = \upsilon(X_G, X_H) = 0$.

We also note that any smooth function of functions in $\mathcal{A}(F)$ is again in $\mathcal{A}(F)$. For if $G_1, \ldots, G_m$ are constant along the flow lines of any X_{F_i}, then so is any compound function $f(G_1, \ldots, G_m)$, whence it is in involution with $F_1, \ldots, F_n$.

The Liouville-Arnold-Jost Theorem

Suppose the Hamiltonian H is integrable in the sense of Liouville with integrals $F_1, \ldots, F_n$ in involution. Liouville showed that locally around every point one can introduce standard symplectic coordinates (q, p) in such a way that

$$H = H(p).$$

Thus, the coordinates $p_1, \dots, p_n$ become integrals. Under an additional topological assumption there is a global version of Liouville's result due to Arnold [4] in the case of the standard symplectic space. Jost [58] generalized it to arbitrary symplectic manifolds and also removed an unnecessary assumption from Arnold's result. The following is therefore known as the *Liouville-Arnold-Jost Theorem*, while Arnold himself refers to it as *Liouville's Theorem*.

First we consider n commuting integrals by themselves.

Theorem 3.6 (Liouville-Arnold-Jost). *Let (M, υ) be a symplectic manifold of dimension $2n$, and let $F = (F_1, \dots, F_n)$ be n independent functions in involution on M. Suppose one of the leaves of F, say*

$$M^0 = F^{-1}(0),$$

is compact and connected. Then

(i) *M^0 is an n-dimensional embedded torus, and*

(ii) *there exist an open neighbourhood U of M^0 in M, an open neighbourhood D of 0 in $\mathbb{R}^n$, and a diffeomorphism*

$$\Psi: \ \mathbb{T}^n \times D \to U$$

introducing angle-action coordinates with

$$\Psi^* \upsilon = \upsilon_0, \qquad \Psi^* M^0 = \mathbb{T}^n \times \{0\},$$

such that the functions $F_i \circ \Psi$ are independent of the angular coordinates.

There is nothing intrinsic about a particular set of integrals $F = (F_1, \dots, F_n)$, and we may always replace them by other integrals in the Poisson algebra $\mathcal{A}(F)$. Using this freedom we have the following addendum to the Liouville-Arnold-Jost theorem.

Addendum to Theorem 3.6. *Let B be the range of F on U. Restricting the neighbourhoods D and U, if necessary, there also exists a diffeomorphism*

$$\Xi: \ B \to D,$$

such that the functions $\Xi \circ F_i \circ \Psi$ are the coordinate functions on D.

Now assume the Hamiltonian H is integrable with integrals $F_1, \dots, F_n$ satisfying the assumptions of the Liouville-Arnold-Jost Theorem. Replacing them by other integrals if necessary and denoting the coordinates by (θ, I), we can assume that

$$\tilde{F}_i = F_i \circ \Psi = I_i.$$

The transformed Hamiltonian $\tilde{H} = H \circ \Psi$ then satisfies

$$0 = \{\tilde{H}, \tilde{F}_i\} = \frac{\partial \tilde{H}}{\partial \theta_i}, \qquad 1 \le i \le n.$$

Hence we find that $\tilde{H} = \tilde{H}(I)$ is independent of the angular coordinates as in example A above. Thus we have the following corollary, which is also often referred to as the Liouville-Arnold-Jost Theorem.

Corollary 3.7 (Liouville-Arnold-Jost). *Suppose the Hamiltonian H is integrable in the sense of Liouville. If one of its leaves is compact and connected, then a neighbourhood U of it is completely foliated into invariant n-dimensional Lagrangian tori, and one can introduce angle-action coordinates by a transformation*

$$\Psi\colon\ \mathbb{T}^n \times D \to U, \qquad \Psi^* \upsilon = \upsilon_0,$$

such that the transformed Hamiltonian $H \circ \Psi$ is a function of the actions alone.

Thus example A is *typical* for an integrable system with compact leaves, at least locally around any given leaf. For the question of *global* existence of angle-action coordinates see for example [37].

It is evident from the preceding discussion that one set of angle-action coordinates works not only for a particular integrable Hamiltonian, but indeed for all Hamiltonians in the associated algebra $\mathcal{A}$. Thus, *one* coordinate transformation introduces angle-action coordinates simultaneously for *all* Hamiltonians in $\mathcal{A}$.

Moreover, in these coordinates, $\mathcal{A}$ consists of all functions in involution with $I_1, \ldots, I_n$, so we have

$$\mathcal{A} = \{G \in \mathbb{F}\colon G = G(I)\}.$$

In an arbitrary coordinate system, $\mathcal{A}$ consists of all analytic functions that are functions of the integrals $F_1, \ldots, F_n$.

Kronecker Tori

Consider now an integrable Hamiltonian $H = H(I)$ in angle-action coordinates. The equations of motion are

$$\dot{\theta}_i = \omega_i(I), \qquad \dot{I}_i = 0,$$

where

$$\omega_i(I) = \frac{\partial H}{\partial I_i}(I), \qquad 1 \le i \le n.$$

They are easily integrated, whence the name *integrable system*. Their general solution is

$$\theta(t) = \theta^{\mathrm{o}} + \omega(I^{\mathrm{o}})t, \qquad I(t) = I^{\mathrm{o}}.$$

Every solution curve is a straight line, which, due to the identification of the angular coordinates θ modulo 2π, is winding around the underlying invariant torus

$$T_{I^{\mathrm{o}}} = \mathbb{T}^n \times \{I^{\mathrm{o}}\}$$

with constant angular velocities, or *frequencies*

$$\omega(I^{\mathrm{o}}) = (\omega_1(I^{\mathrm{o}}), \ldots, \omega_n(I^{\mathrm{o}})).$$

They completely determine the dynamics on this torus which consists of parallel translations.

Such tori play an eminent role in dynamics and have a variety of special names. They are called *Kronecker tori*, or *parallel* or *rotational tori*. One also speaks of *tori with parallel*, *linear*, or *rotational flow*, among others. The associated frequencies are often simply called the *frequencies of the invariant torus*.

In other coordinates a Kronecker torus usually does not look that simple. We therefore make the following definition, which applies to vector fields in general.

Definition 3.8. *Let X be a smooth vector field on a manifold M of arbitrary dimension. An invariant n-torus T of X is called a* Kronecker torus, *or* torus with linear flow, *if there exists a diffeomorphism*

$$\Phi\colon\ \mathbb{T}^n \to T,$$

such that $\Phi^* X$ is a constant n-vector ω on $\mathbb{T}^n$, the frequency vector *of the Kronecker torus. The pair (T, ω) is called a* Kronecker torus with frequencies ω.

Put differently, $\Phi\colon \mathbb{T}^n \to T$ is an embedding into the manifold M, which is one-to-one and onto and introduces angular coordinates θ in such a way that the equations of motion on T reduce to $\dot{\theta} = \omega$. Consequently, through Φ the flow of X on T is conjugate to the family of parallel translations by ωt on $\mathbb{T}^n$:

$$\begin{array}{ccc} \mathbb{T}^n & \xrightarrow{\ \Phi^t_\omega\ } & \mathbb{T}^n \\ {\scriptstyle\Phi}\big\downarrow & & \big\downarrow{\scriptstyle\Phi} \\ T & \xrightarrow{\ X^t\ } & T \end{array}$$

with $\Phi^t_\omega(\theta) = \theta + \omega t$.

We point out that the frequencies ω of a Kronecker torus are not uniquely determined. Applying a torus automorphism $A\colon \mathbb{T}^n \to \mathbb{T}^n$ defined by a matrix $A \in \mathrm{SL}(n, \mathbb{Z})$, a so called *unimodular matrix*, we can always replace ω by $A\omega$. What is uniquely determined, however, is the *frequency module*

$$M(\omega) = \left\{ \langle k, \omega \rangle : k \in \mathbb{Z}^n \right\}$$

consisting of all integer combinations of the frequencies $\omega = (\omega_1, \ldots, \omega_n)$ with integer coefficients.

From a geometrical point of view an integrable Hamiltonian system around a compact connected leaf is thus completely foliated into an n-parameter family of invariant tori with linear flow, the parameters being provided by suitable integrals of the Hamiltonian system. Moreover, all these tori are Lagrangian submanifolds.

From an analytical point of view all solution curves on an invariant Kronecker torus T with frequencies ω are represented as

$$\Phi(\theta^{\mathrm{o}} + \omega t), \qquad \theta^{\mathrm{o}} \in \mathbb{T}^n.$$

Each coordinate function, and more generally every observable along such a solution curve, is therefore what is called a *quasi-periodic* function of t.

Definition 3.9. *A continuous function* $q\colon \mathbb{R} \to \mathbb{R}$ *is called* quasi-periodic with frequencies $\omega = (\omega_1, \dots, \omega_n)$, *if there exists a continuous function*

$$Q\colon\ \mathbb{T}^n \to \mathbb{R},$$

called the hull *of* q, *such that* $q(t) = Q(\omega t)$ *for all* $t \in \mathbb{R}$.

In other words, there exists a continuous function Q on $\mathbb{R}^n$ with period 2π in each of its arguments $\theta = (\theta_1, \dots, \theta_n)$ such that

$$q(t) = Q(\omega_1 t, \dots, \omega_n t)$$

for all t. If $n = 1$, then q is a periodic function of period $2\pi/\omega_1$. So the notion of a quasi-periodic function generalizes that of a periodic function.

If Q is sufficiently smooth, say of class C^{2n+1}, then it is represented by its multi-dimensional Fourier series as

$$Q(\theta) = \sum_{k\in\mathbb{Z}^n} \hat{Q}_k \mathrm{e}^{\mathrm{i}\langle k,\theta\rangle},$$

with its Fourier coefficients $\hat{Q}_k$ given by

$$\hat{Q}_k = \frac{1}{(2\pi)^n} \int_{\mathbb{T}^n} Q(\theta)\mathrm{e}^{-\mathrm{i}\langle k,\theta\rangle}\mathrm{d}\theta.$$

We then obtain

$$q(t) = \sum_{k\in\mathbb{Z}^n} \hat{q}_k \mathrm{e}^{\mathrm{i}\langle k,\omega\rangle t}, \qquad \hat{q}_k = \hat{Q}_k.$$

Thus a quasi-periodic function has a *frequency spectrum*, which is precisely the frequency module $M(\omega)$ of ω.

Hence, from an analytical point of view the solutions of an integrable Hamiltonian system near a compact connected leaf are all quasi-periodic functions of t, which can be represented in the form

$$\Phi(\theta^{\mathrm{o}} + \omega(I^{\mathrm{o}})t, I^{\mathrm{o}}) = \sum_{k\in\mathbb{Z}^n} \hat{\Phi}_k(I^{\mathrm{o}})\mathrm{e}^{\mathrm{i}\langle k,\theta^{\mathrm{o}}\rangle}\, \mathrm{e}^{\mathrm{i}\langle k,\omega(I^{\mathrm{o}})\rangle t},$$

where $\theta^{\mathrm{o}} \in \mathbb{T}^n$, I^{o} varies over some domain in $\mathbb{R}^n$, and Φ is a smooth map into the symplectic manifold M with period 2π in its first n arguments.

Resonant and Nonresonant Tori

The flow on a Kronecker torus is rather simple. Yet its topological and metrical properties differ sharply depending on arithmetical properties of its frequencies ω. There are essentially two cases.

Case 1. The frequencies ω are *nonresonant*, or *rationally independent*:

$$\langle k, \omega \rangle \neq 0 \qquad \text{for all } 0 \neq k \in \mathbb{Z}^n.$$

Then, on this torus, each orbit is dense, the flow is ergodic, and the torus itself is minimal.

Case 2. The frequencies are *resonant*, or *rationally dependent*: there exist integer relations

$$\langle k, \omega \rangle = 0 \qquad \text{for some } 0 \neq k \in \mathbb{Z}^n.$$

The prototype is $\omega = (\omega_1, \ldots, \omega_m, 0, \ldots, 0)$ with $n - m \geq 1$ trailing zeroes and nonresonant $(\omega_1, \ldots, \omega_m)$. In this case the torus decomposes into an $n-m$-parameter family of identical invariant m-tori. Each orbit is dense on such a lower dimensional torus, but not in the entire Kronecker torus.

A special case arises when there exist $n-1$ independent resonant relations. Then each frequency $\omega_1, \ldots, \omega_n$ is an integer multiple of a fixed non-zero frequency ω_*, and the whole torus is filled by *periodic orbits* with one and the same period $2\pi/\omega_*$.

In an integrable system the frequencies on the tori may or may not vary with the torus, depending on the nature of the *frequency map*

$$I \mapsto \omega(I).$$

If it is *nondegenerate* in the sense that

$$\det \frac{\partial \omega}{\partial I} = \det \frac{\partial^2 H}{\partial I^2} \neq 0,$$

then this map is a local diffeomorphism, and "the frequencies ω effectively depend on the amplitudes I". Nonresonant and resonant tori of all types then each form dense subsets in phase space. Indeed, the resonant ones sit among the nonresonant ones like the rational numbers among the irrational numbers.

4 Birkhoff Integrable Systems

Another type of integrable system arises in the study of equilibria of Hamiltonian systems, the so called Birkhoff integrable systems. On one hand, this type is more special than the Liouville type, as it suffices to look at a neighbourhood of a single point. On the other hand, it is also more general, as this neighbourhood is foliated into invariant tori of any dimension between 0 and n.

Birkhoff Normal Forms

Consider an isolated equilibrium of a Hamiltonian system on some symplectic manifold, that is, an isolated singular point of the Hamiltonian vector field. Choosing

Darboux coordinates we may transfer this equilibrium to the origin in the standard symplectic space $\mathbb{R}^n \times \mathbb{R}^n$ with symplectic form $\upsilon_0 = \langle \cdot , J_0 \cdot \rangle$ and coordinates $u = (q, p)$. Dropping an irrelevant additive constant the Hamiltonian then starts with quadratic terms,

$$H = \frac{1}{2} \langle Au, u \rangle + \dots,$$

where A is the symmetric $2n \times 2n$-Hessian of H at 0. The equations of motion are

$$\dot{u} = J_0 Au + \dots,$$

where here and in the following, the dots stand for terms of higher order in u.

We focus on the case of an *elliptic equilibrium*. That is, the spectrum of the linearized system $\dot{u} = J_0 Au$ is purely imaginary:

$$\operatorname{spec}(J_0 A) = \{\pm \mathrm{i}\lambda_1, \dots, \pm \mathrm{i}\lambda_n\},$$

with real numbers $\lambda_1, \dots, \lambda_n$. Assuming this spectrum to be simple, there exists a linear symplectic change of coordinates that brings the quadratic part of the Hamiltonian into the normal form

$$\langle Au, u \rangle = \sum_{i=1}^{n} \lambda_i (q_i^2 + p_i^2),$$

where for simplicity we denote the new coordinates by the same symbols. The associated equations of motion are

$$\dot{q}_i = \lambda_i p_i, \qquad \dot{p}_i = -\lambda_i q_i.$$

This is a system of n uncoupled harmonic oscillators, each moving in its own invariant plane with frequencies $\lambda_1, \dots, \lambda_n$, respectively. The total motion of all oscillators combined is quasi-periodic in time.

Incidentally, the same normal form can be obtained, if the Hamiltonian is definite around the origin, without assuming the spectrum to be simple [55, section 1.7].

Having put the quadratic terms into normal form, this normalization process can be pushed to higher order, if additional nonresonance conditions are satisfied. The result is known as the Birkhoff normal form of a Hamiltonian.

Definition 4.1. *The frequencies $\lambda_1, \dots, \lambda_n$ are* nonresonant up to order m, *if*

$$\sum_{i=1}^{n} k_i \lambda_i \neq 0 \quad \textit{whenever} \quad 1 \leq \sum_{i=1}^{n} |k_i| \leq m,$$

where $k_1, \dots, k_n$ are arbitrary integers and $m \geq 1$. They are nonresonant, *if they are nonresonant of any finite order, or equivalently, if $k_1\lambda_1 + \dots + k_n\lambda_n$ vanishes only when all the integer coefficients vanish.*

Definition 4.2. *A Hamiltonian H is in* Birkhoff normal form up to order m, *if it is of the form*

$$H = N_2 + N_4 + \cdots + N_m + H_{m+1} + \ldots,$$

where the N_k, $2 \leq k \leq m$, are homogeneous polynomials of order k, which are actually functions of $q_1^2 + p_1^2, \ldots, q_n^2 + p_n^2$, and where $H_{m+1} + \ldots$ stands for arbitrary terms of order strictly greater than m. If this holds for any m, the Hamiltonian is simply said to be in Birkhoff normal form.

Note that if m is odd, then N_m is zero, and the last nontrivial term in the normal form is at most of order $m - 1$.

Theorem 4.3 (Birkhoff normal form of order m). *Let $H = N_2 + \ldots$ be a real analytic Hamiltonian around the origin in standard symplectic coordinates with quadratic part*

$$N_2 = \frac{1}{2} \sum_{i=1}^{n} \lambda_i (q_i^2 + p_i^2).$$

If the frequencies $\lambda_1, \ldots, \lambda_n$ are nonresonant up to order $m \geq 3$, then there exists a real analytic symplectic transformation $\Phi = \mathrm{id} + \ldots$ such that

$$H \circ \Phi = N_2 + N_4 + \ldots N_m + H_{m+1} + \ldots$$

is in Birkhoff normal form up to order m. Moreover, the normal form terms are uniquely defined as long as the normalizing transformation is of the form id $+ \ldots$.

The theorem is proven by applying a succession of symplectic transformations, which from $H = N_2 + H_3 + H_4 + \ldots$ eliminate H_3, normalize H_4, eliminate H_5, and so on, until finally H_m is normalized. See the proof of Theorem 14.2 for the first two steps of this procedure, and for example [26, 98] for the general result. It is clear from this construction, that the normalizing transformation is by no means uniquely determined. In contrast, the normal form *is*, as long as the quadratic terms are kept fixed.

If the frequencies $\lambda_1, \ldots, \lambda_n$ are nonresonant to any order, then this normalization process can be carried to any order as well. The resulting symplectic transformation, however, is in general no longer convergent in any neighbourhood of the origin and can only be given a meaning as a formal power series. Indeed, Siegel showed that for this transformation to be convergent, infinitely many analytic conditions have to be satisfied [123].

Theorem 4.4 (Formal Birkhoff normal form). *Let $H = N_2 + \ldots$ be a real analytic Hamiltonian as in the preceding theorem. If the frequencies $\lambda_1, \ldots, \lambda_n$ are nonresonant, then there exists a symplectic transformation $\Phi = \mathrm{id} + \ldots$, represented by a formal power series, such that*

$$H \circ \Phi = N_2 + N_4 + \ldots$$

is in Birkhoff normal form as a formal power series.

The Rüssmann-Vey-Ito Theorem

If some transformation into Birkhoff normal form were convergent, then the resulting Hamiltonian would be integrable in a neighbourhood of the origin, the integrals in involution being $q_1^2 + p_1^2, \ldots, q_n^2 + p_n^2$ or any functions of these, as in example B in the previous section. These integrals are *functionally independent* in the sense, that their 1-forms are linearly independent on a dense open subset.

It turns out that a certain converse is also true. If a Hamiltonian with a *nonresonant* elliptic equilibrium admits n functionally independent integrals in involution, then the formal transformation into Birkhoff normal form is convergent, hence the Hamiltonian itself is integrable. Such a result was first proven by Rüssmann [115] for systems of two degrees of freedom. It was later extended by Vey [132] to higher degrees of freedom. Finally Ito [56] removed a certain nondegeneracy assumption from Vey's theorem. See also [140] for further results.

Again, as in the case of the Liouville-Arnold-Jost Theorem, this is not so much a statement about a particular Hamiltonian, but rather a statement about a Poisson algebra of Hamiltonians. To formulate this result, we make the following definition.

Definition 4.5. *A Poisson algebra $\mathcal{A}$ of Hamiltonians on a symplectic manifold M is said to be* nonresonant *at a point m in M, if it contains a Hamiltonian with a nonresonant elliptic equilibrium at m.*

It turns out that then m is an equilibrium for *any* Hamiltonian in the algebra.

Theorem 4.6 (Rüssmann-Vey-Ito [56, 115, 132]). *Let $F = (F_1, \ldots, F_n)$ be n functionally independent functions in involution in a neighbourhood of a point m on a symplectic manifold M of dimension $2n$. If their associated Poisson algebra $\mathcal{A}(F)$ is nonresonant at m, then one can introduce symplectic coordinates (q, p) around m so that $\mathcal{A}(F)$ consists of all functions in $\mathbb{F} = C^\omega(M)$, which are actually functions of $q_1^2 + p_1^2, \ldots, q_n^2 + p_n^2$.*

Thus, *one* symplectic transformation puts *all* Hamiltonians simultaneously into Birkhoff normal form, which are in involution with $F_1, \ldots, F_n$. Among those, many even have resonant frequencies at the equilibrium.

Incidentally, the assumption that the functions $F_1, \ldots, F_n$ are in involution can be dropped. This property follows if all other conditions are satisfied [56].

For a particular Hamiltonian we have the following corollary. But first we make a definition.

Definition 4.7. *A $2n$-dimensional Hamiltonian is called* Birkhoff integrable *near an equilibrium, if it admits n functionally independent integrals in involution in some neighbourhood of the equilibrium.*

Corollary 4.8. *Suppose the Hamiltonian H is Birkhoff integrable near an elliptic equilibrium. If the frequencies at the equilibrium are nonresonant, then one can introduce symplectic coordinate (q, p) around it so that the Hamiltonian is a function of $q_1^2 + p_1^2, \ldots, q_n^2 + p_n^2$ alone.*

We refer to coordinates of this type as *Birkhoff coordinates.*

Orbit Structure

Consider now an integrable Hamiltonian in convergent Birkhoff normal form,

$$H = H(I_1, \dots, I_n)$$

with $2I_i = q_i^2 + p_i^2$ for $1 \le i \le n$. The equations of motions are

$$\dot{q}_i = \omega_i(I)p_i, \qquad \dot{p}_i = -\omega_i(I)q_i,$$

where

$$\omega_i(I) = \frac{\partial H}{\partial I_i}(I_1, \dots, I_n)$$

are constant along each orbit, as the I_i are integrals.

Again, each orbit is the superposition of the motions of n oscillators, each moving in its own invariant plane with fixed frequencies $\omega_1, \dots, \omega_n$, which now depend on $I_1, \dots, I_n$ in a usually nonlinear fashion. This orbit is winding around the underlying invariant torus

$$\mathcal{T}_{I^o} = \left\{ (q, p) \colon q_i^2 + p_i^2 = 2I_i^o \text{ for } 1 \le i \le n \right\}.$$

Hence, the entire motion is again quasi-periodic. But in contrast to a Liouville integrable Hamiltonian, the dimension of these tori is not fixed, but varies. Indeed,

$$\dim \mathcal{T}_{I^o} = \operatorname{card} \left\{ I_i^o \colon I_i^o > 0 \right\}$$

is the number of positive amplitudes, or oscillators actually in motion.

For example, on the dense open subset where all oscillators are excited the tori have maximal dimension, and we may introduce angle-action coordinates (θ, I) by the symplectic transformation

$$q_i = \sqrt{2I_i} \cos\theta_i, \qquad p_i = \sqrt{2I_i} \sin\theta_i$$

for $1 \le i \le n$ to obtain a Liouville integrable Hamiltonian as in the previous section.

If, however, only a smaller number of oscillators is excited, say

$$I_1^o, \dots, I_m^o > 0, \qquad I_{m+1}^o, \dots, I_n^o = 0,$$

with $1 \le m < n$, then the motion is confined to an m-dimensional torus, and we may introduce angle-action coordinates only for the first m modes. We obtain an integrable Hamiltonian of the form

$$\begin{aligned} H &= H(I_1, \dots, I_m, q_{m+1}^2 + p_{m+1}^2, \dots, q_n^2 + p_n^2) \\ &= \sum_{i=1}^{m} \lambda_i I_i + \frac{1}{2} \sum_{i=m+1}^{n} \lambda_i (q_i^2 + p_i^2) + \dots \ . \end{aligned}$$

The equations of motion are

$$\dot{\theta}_i = \omega_i(I), \qquad \dot{I}_i = 0$$

for $1 \leq i \leq m$, and

$$\dot{q}_i = \omega_i(I) p_i, \qquad \dot{p}_i = -\omega_i(I) q_i$$

for $m+1 \leq i \leq n$, with frequencies $\omega_i(I)$ defined as before.

This system features a family of m-dimensional invariant tori

$$\mathbb{T}^m \times \{I^o\} \times \{0\} \times \{0\},$$

depending on m parameters $I^o = (I_1^o, \ldots, I_m^o, 0, \ldots, 0)$. On each torus the motion is described by m *internal frequencies* $\omega_1(I^o), \ldots, \omega_m(I^o)$. Their normal space is described by the remaining cartesian coordinates, whose origin is an elliptic equilibrium with characteristic frequencies $\omega_{m+1}(I^o), \ldots, \omega_n(I^o)$, the *external frequencies* of the torus. These so called *lower dimensional elliptic invariant tori* will play a central role in extending the results of the classical KAM theory described in the next section to an infinite dimensional system such as the perturbed KdV equation.

5 KAM Theory

Integrable systems are the exception, not the rule. But many interesting Hamiltonian systems may be viewed as small perturbations of an integrable system. Examples are the planetary system we live in, the motion of a free particle on a slightly deformed surface of revolution, or a Hamiltonian in Birkhoff normal form up to a finite order. Thus the question arises: What happens to a foliation of invariant tori with their quasi-periodic motions under small perturbations of the Hamiltonian?

The Classical KAM Theorem

Consider a Hamiltonian in angle-action coordinates (θ, I) of the form

$$H = H_0(I) + H_\varepsilon(\theta, I),$$

where H_0 is the unperturbed Liouville integrable Hamiltonian, and H_ε is a general perturbation. For simplicity we assume the latter to be of the form

$$H_\varepsilon = \varepsilon H_1(\theta, I),$$

so that ε measures the size of the perturbation. We recall that the unperturbed system is said to be *nondegenerate*, if the frequency map

$$I \mapsto \omega(I) = \frac{\partial H_0}{\partial I}(I)$$

is a local diffeomorphism everywhere. This is sometimes called *Kolmogorov's condition*. Under this assumption, the frequencies vary with the actions in a locally one-to-one manner.

The first result goes back to Poincaré and is of a negative nature. He observed that the *resonant* tori are in general *destroyed* by an arbitrarily small perturbation. In particular, out of a torus with an $n-1$-parameter family of periodic orbits, usually only *finitely* many periodic orbits survive a perturbation, while the others disintegrate and give way to chaotic behavior. So in a nondegenerate system a *dense* set of tori is usually destroyed. This, in particular, implies that a generic Hamiltonian system is *not integrable* [85].

A dense set of tori being destroyed there seems to be little hope for other tori to survive. Indeed, until the fifties it was a common belief that arbitrarily small perturbations can turn an integrable system into an ergodic one on each energy surface. In the twenties there even appeared an – erroneous – proof of this "ergodic hypothesis" by Fermi.

But in 1954 Kolmogorov [65] observed that the converse is true – the *majority* of tori survives. He stated the persistence of those Kronecker systems, whose frequencies ω are not only nonresonant, but are *strongly nonresonant* or *diophantine* in the sense that there exist constants $\alpha > 0$ and $\tau > 0$ such that

$$|\langle k,\omega\rangle| \geq \frac{\alpha}{|k|^\tau} \qquad \text{for all } 0 \neq k \in \mathbb{Z}^n.$$

Such a condition is called a *diophantine* or *small divisor condition*, as the expressions $\langle k,\omega\rangle$ enter into the denominators of formal series expansion of quasi-periodic solutions of the perturbed system also known as *Lindstedt series*.

The existence of such frequencies is easy to verify. Fix τ, and let Δ_α denote the set of all $\omega \in \mathbb{R}^n$ satisfying these infinitely many conditions with given $\alpha > 0$. If $\tau > n-1$, then for any bounded subset $\Omega \subset \mathbb{R}^n$ one has the straightforward Lebesgue measure estimate

$$\operatorname{meas}(\Omega \setminus \Delta_\alpha) = O(\alpha).$$

Hence, *almost every* ω in $\mathbb{R}^n$ belongs to *some* set Δ_α with $\alpha > 0$.

But although almost all frequencies are strongly nonresonant, it is not true that almost all tori survive a given perturbation, no matter how small ε. The reason is that the parameter α in the small divisor condition limits the size of the perturbation through the condition

$$\varepsilon \ll \alpha^2.$$

Conversely, under a given small perturbation of size ε, only those Kronecker tori with frequencies ω in Δ_α with

$$\alpha \gg \sqrt{\varepsilon},$$

do survive. Thus, we can not allow α to vary, but have to *fix* it in advance.

To state the KAM theorem, we therefore single out from a bounded domain Ω in $\mathbb{R}^n$ the subsets

$$\Omega_\alpha \subset \Omega, \qquad \alpha > 0,$$

whose frequencies belong to Δ_α and also have at least distance α to the boundary of Ω. These, like Δ_α, are *Cantor sets*: they are closed, perfect and *nowhere dense*, hence of first Baire category. But they also have *large Lebesgue measure*:

$$\operatorname{meas}(\Omega \setminus \Omega_\alpha) = O(\alpha),$$

provided the boundary of Ω is piecewise smooth.

The main theorem of Kolmogorov, Arnold and Moser can now be stated as follows.

Theorem 5.1 (The Classical KAM Theorem [2, 65, 96]). *Suppose the Hamiltonian*

$$H = H_0 + H_\varepsilon$$

is real analytic on the closure of $\mathbb{T}^n \times D$, where D is a bounded domain in $\mathbb{R}^n$. If the integrable Hamiltonian H_0 is nondegenerate and its frequency map a diffeomorphism $D \to \Omega$, then there exists a constant $\delta > 0$ such that for

$$|\varepsilon| < \delta\alpha^2$$

all Kronecker tori $(\mathbb{T}^n, \omega)$ of the unperturbed system with $\omega \in \Omega_\alpha$ persist as Lagrangian tori, being only slightly deformed. Moreover, they depend in a Lipschitz continuous way on ω and fill the phase $\mathbb{T}^n \times D$ up to a set of measure $O(\alpha)$.

Note that given a perturbation of size ε one may choose $\alpha \sim \sqrt{\varepsilon}$, so the deformed tori fill the phase space at least up to a set of measure $O(\sqrt{\varepsilon})$.

It is an immediate and important consequence of the KAM theorem that small perturbations of nondegenerate Hamiltonians are *not ergodic*, as the Kronecker tori form an invariant set, which is neither of full nor of zero measure. Thus the ergodic hypothesis of the twenties was wrong.

Since its conception the KAM theorem has been generalized and extended in numerous ways, and all its assumptions have been relaxed. We mention just a few of these developments, for more extensive discussions and references we point for example to [13, 20, 79, 81] and the references therein.

First, neither the perturbation nor the integrable Hamiltonian need to be real analytic. It suffices that they are differentiable of class C^l with

$$l > 2\tau + 2 > 2n$$

to prove the persistence of individual tori [99, 105, 120]. For their Lipschitz dependence some more regularity is required [106]. For optimal regularity results, in particular for the closely related problem of the existence of invariant curves for area preserving twist maps, we refer to the work of Herman [53, 54]. Unfortunately, most of his deep results appeared only in the form of manuscripts.

The nondegeneracy condition may also be relaxed. Originally, Kolmogorov's condition was used to completely control the frequencies, so that their diophantine estimates can be preserved under perturbation. But in fact, it suffices that the intersection of the range of the frequency map with any hyperplane has measure zero. Then, after perturbation, one can still find sufficiently many diophantine frequencies, albeit they are now not known *a priori*.

For example, if it happens that

$$\frac{\partial H_0}{\partial I} = (\omega_1(I_1), \dots, \omega_n(I_1))$$

is a function of I_1 alone and thus completely degenerate in the above sense, it suffices to require that

$$\det \left(\frac{\partial^j \omega_i}{\partial I_1^j} \right)_{1 \le i,j \le n} \neq 0,$$

see Xiu, You & Qiu [137]. For further results see for example the papers of Cheng & Sun, Rüssmann and Sevryuk [22, 118, 119, 122].

Finally, the Hamiltonian nature of the equations of motion is almost indispensable for a large family of tori to persist. Analogous results are true for reversible systems [20, 100, 106]. But in any event the system has to be *conservative*. Any kind of dissipation immediately destroys the Cantor family of tori, although isolated ones may persist.

Lower Dimensional Tori

The classical KAM theorem is concerned with the persistence of Kronecker tori with strongly nonresonant frequencies in a nondegenerate system. It does not apply to resonant tori, which are foliated into lower dimensional tori and usually break up, giving rise to chaotic motions.

Those lower dimensional tori do not vanish entirely, however. Consider a resonant Kronecker torus, decomposing into an $n - m$-parameter family of identical m-dimensional minimal tori with $1 \le m \le n - 1$. In suitable coordinates its frequencies are $\omega = (\omega_1, \dots, \omega_m, 0, \dots, 0)$, where

$$\hat{\omega} = (\omega_1, \dots, \omega_m)$$

is nonresonant.

For $m = 1$ in particular, such a torus is foliated into identical closed orbits. Bernstein & Katok [8] showed that in a convex system at least n of them survive any sufficiently small perturbation. For $m = n - 1 > 1$, the analogous result for nondegenerate systems is due to Cheng [21]. He showed that at least two tori survive, one elliptic and one hyperbolic generically, provided $\hat{\omega}$ is strongly nonresonant.

For the intermediate cases with $1 < m < n-1$, only partial result are known such as [23, 129]. These require additional assumptions for the unperturbed system or its perturbation. The long standing conjecture is that at least $n - m + 1$, and generically 2^{n-m}, invariant m-tori always survive in a nondegenerate system, when $\hat{\omega}$ is strongly nonresonant. That is, their number should be equal to the number of critical points of smooth functions on the torus $\mathbb{T}^{n-m}$.

But in a Birkhoff integrable system, one also encounters another type of lower dimensional tori: the *elliptic* tori described in the previous section. Families of such tori of dimension m form smooth $2m$-dimensional submanifolds, and one may ask as well about the persistence of such families under small perturbations.

A special case arises for $m = 1$, where we encounter families of *periodic* orbits filling two-dimensional discs containing the equilibrium. Already Lyapunov showed that such discs persist, being only slightly deformed, if at the equilibrium the frequency of the periodic orbit is not in resonance with all other frequencies. More precisely, if $\lambda_1, \dots, \lambda_n$ are these frequencies, and the disc of the unperturbed periodic orbits is contained in the plane associated with the frequency λ_j, then this disc persists slightly deformed if

$$\frac{\lambda_i}{\lambda_j} \notin \mathbb{Z}, \qquad i \neq j.$$

This condition is also known as *Lyapunov's condition*. Here, no small divisors occur, and standard analytic methods suffice [124].

For $2 \leq m \leq n-1$, however, small divisors do occur. The first results for lower dimensional elliptic tori were announced by Melnikov [93] in the sixties of the last century, but no proofs were given. Meanwhile, Moser [97] proved the existence of quasi-periodic solutions for a large class of parameter-dependent ordinary differential equations that included the case of a Hamiltonian elliptic invariant torus of dimension $m = n-1$. The case $m < n-1$, however, was not accessible by his method, since it required *all* frequencies to be kept fixed. Finally, Eliasson [38] established the general result by allowing the frequencies to vary with the perturbation.

To motivate the formulation of this result consider a Birkhoff integrable Hamiltonian in standard coordinates,

$$H = H_0(I_1, \dots, I_n), \qquad I_i = \frac{1}{2}(q_i^2 + p_i^2).$$

Fixing $1 \leq m \leq n-1$, we are interested in the persistence of the family of m-dimensional tori

$$\mathcal{T}_{I^{\circ}} = \left\{ (q, p) \colon q_i^2 + p_i^2 = 2I_i^{\circ} \text{ for } 1 \leq i \leq n \right\}$$

with, say, $I_1^{\circ}, \dots, I_m^{\circ} > 0$ and $I_{m+1}^{\circ}, \dots, I_n^{\circ} = 0$.

Using an approach first used by Moser [97] we introduce angle-action coordinates around *each individual* torus $\mathcal{T}_{I^{\circ}}$ by setting

$$q_i = \sqrt{2(\xi_i + I_i)} \cos\theta_i, \qquad p_i = \sqrt{2(\xi_i + I_i)} \sin\theta_i$$

for $1 \leq i \leq m$ depending on the amplitudes

$$\xi = (\xi_1, \dots, \xi_m) = (I_1^{\circ}, \dots, I_m^{\circ}),$$

while keeping the other cartesian coordinates. This way, we introduce its position as an m-dimensional *parameter* ξ. Expanding H_0 up to terms of first order in the I_i and second order in the q_i, p_i, we obtain an integrable Hamiltonian

$$N_0 = \sum_{i=1}^{m} \omega_i(\xi) I_i + \frac{1}{2} \sum_{i=m+1}^{n} \omega_i(\xi)(q_i^2 + p_i^2)$$

with

$$\omega_i(\xi) = \frac{\partial H_0}{\partial I_i}(I_1^{\mathrm{o}}, \ldots, I_m^{\mathrm{o}}, 0, \ldots, 0).$$

It turns out that the discarded higher order terms may be regarded as just another perturbation. Thus, the new starting point of the perturbation theory is the family of integrable Hamiltonians N_0 above on the phase space

$$\mathbb{T}^m \times \mathbb{R}^m \times \mathbb{R}^{n-m} \times \mathbb{R}^{n-m}$$

with coordinates (θ, I, q, p), depending on the m-dimensional parameter ξ. Its equations of motion admit for each ξ the invariant m-torus

$$T_0 = \mathbb{T}^m \times \{0\} \times \{0\} \times \{0\}$$

with internal and external frequencies $\omega = (\omega_1, \ldots, \omega_m)$ and $\Omega = (\omega_{m+1}, \ldots, \omega_n)$, respectively, all depending on ξ.

We are interested in the persistence of this torus under small perturbations of the Hamiltonian N_0 for a large set of parameters. Thus, instead of proving the existence of a large family of invariant tori in *one* Hamiltonian system, we establish the existence of *one* invariant torus for a large family of Hamiltonian systems. These two approaches are essentially equivalent, but the latter has a number of advantages. Among others, the unperturbed system is very simple, and the control of the frequencies is separated from the actions.

As usual, we need some nondegeneracy to obtain a KAM theorem.

Definition 5.2. *The parameter-dependent family of Hamiltonians N_0 given above is* nondegenerate, *if the map*

$$\xi \mapsto \omega(\xi)$$

is a local diffeomorphism everywhere on its domain of definition, and if

$$\xi \mapsto \langle k, \omega(\xi) \rangle + \langle l, \Omega(\xi) \rangle \not\equiv 0$$

for all $(k, l) \in \mathbb{Z}^m \times \mathbb{Z}^{n-m}$ with $1 \leq |l| \leq 2$.

The first condition is tantamount to the usual Kolmogorov condition and makes sure that there is complete control over the internal frequencies ω. The second condition, also known as *Melnikov's condition*, makes sure that the additional divisors $\langle k, \omega(\xi) \rangle + \langle l, \Omega(\xi) \rangle$ arising in the perturbation theory are not locked in complete resonance.

Consider now a perturbation $N_0 + P_\varepsilon$ of N_0 that is real analytic in the space coordinates as well as the parameters and is of the order of ε, say,

$$P_\varepsilon = \varepsilon P_1(\theta, I, q, p, \xi).$$

Theorem 5.3. *Suppose the Hamiltonian*

$$H = N_0 + P_\varepsilon$$

is real analytic in a fixed neighbourhood of $T_0 \times \Pi$, where Π is a closed bounded m-dimensional parameter domain of positive Lebesgue measure. If N_0 is nondegenerate, then for all sufficiently small ε there exists a Cantor subset Π_ε of parameters, such that for each parameter value in Π_ε the perturbed system admits an elliptic invariant Kronecker torus close to T_0. Moreover, $\operatorname{meas}(\Pi \setminus \Pi_\varepsilon) \to 0$ *as* $\varepsilon \to 0$.

A much more general and detailed formulation of this theorem is given and proven in chapter V, so we will not go into this here.

We point out that in general we can not make explicit the rate with which the measure of $\Pi \setminus \Pi_\varepsilon$ tends to zero. The reason is that the external frequencies are not completely under control due to the smaller numbers of parameters – see section 22 for these matters. But if the perturbation problem does arise from considering a Birkhoff normal form of finite order, then one indeed has

$$\operatorname{meas}(\Pi \setminus \Pi_\varepsilon) = O(\sqrt{\varepsilon}),$$

see [107].

Recent significant extension of this theorem are due to Rüssmann [119] and You [138]. In particular, the latter relaxed Melnikov's condition to the requirement that

$$\langle k, \omega(\xi) \rangle - \Omega_i(\xi) \not\equiv 0$$

for all $k \in \mathbb{Z}^m$ and $m + 1 \leq i \leq n$. This is the analogue for quasi-periodic solutions to Lyapunov's condition for periodic solutions. This way, one can also allow for *multiple* external frequencies, which is not possible with Melnikov's condition.

Infinite Dimensional Systems

Since its conception there has been a great interest in extending the classical KAM theory to infinite dimensional systems in order to apply it to certain nonlinear partial differential equations in Hamiltonian form and models in mathematical physics, among others. The objective is to find quasi-periodic, or more generally, *almost-periodic solutions*. The latter can be thought of as "quasi-periodic solutions with an infinite dimensional frequency module". More formally, they can be characterized as the uniform limits of trigonometric polynomials – see for example [63].

In the seventies some special results were obtained by Nikolenko, Ware, Zehnder and others extending to infinite dimensions Siegel's famous result about the linearization of complex analytic vector fields in the neighbourhood of an elliptic equilibrium – see [102, 123, 134, 139] and in particular [103] and the references therein.

Nikolenko applied these results to evolution equations in Banach spaces including nonlinear heat and Schrödinger equations. However, these results are restricted to perturbations of linear systems by higher order terms in a sufficiently small neighbourhood of the equilibrium, with the frequencies of the linear system satisfying infinitely many small divisor conditions. Moreover, some results such as [102] do not apply to the *real* analytic case.

In general the difficulties arise from the fact that for any n-vector of frequencies $\omega = (\omega_1, \ldots, \omega_n)$,

$$\min_{0<|k|_\infty \le K} |\langle k, \omega \rangle| \le \frac{|\omega|}{K^{n-1}}$$

by Dirichlet's theorem. Hence, as the number n of frequencies tends to infinity, the small divisor bounds tend to zero rapidly even when the frequencies ω_n are allowed to grow at some power of n. As a result, the bound on admissible perturbations tends to zero exponentially in n at best, and a simple minded extension of the classical KAM theorem to infinite dimensions does not apply to any perturbation different from zero.

The situation is different, if the frequencies ω_n tend to infinity *very rapidly*, say, hyperexponentially with n. In this case the classical diophantine estimates still hold, and the classical KAM proof works also in this case, when norms are properly chosen. Such a result was obtained by Chierchia & Perfetti [24] for Hamiltonians of the form "kinetic energy + potential energy" in the so called isochronous case, where $\omega_n = I_n$, establishing the existence of almost-periodic solutions on invariant tori of maximal dimension. But the strong growth condition on the frequencies obviously excludes applications to realistic problems.

In mathematical physics one is interested in lattices of harmonic oscillators which are subject to anharmonic coupling forces of finite or short range, or of hierarchical structure. Their frequencies are usually assumed to be independent, identically distributed random variables with bounded or unbounded range. Such infinite dimensional systems are well approximable by finite dimensional ones, and one can show that for sufficiently rapidly decaying amplitude distributions the majority of frequency distributions gives rise to almost-periodic motions of the entire ensemble of oscillators.

These developments were initiated by the works of Vittot & Bellissard [133] and Fröhlich, Spencer & Wayne [45], and a unified approach and improved results were given in [108]. The idea is to exploit the strongly localized structure of the perturbations affected by the couplings, which decay very rapidly with the distance of the oscillators involved. This allows to employ small divisor estimates of the form

$$|\langle k, \omega \rangle| \ge \frac{\alpha}{\Delta(|k|)\Delta([k])},$$

where Δ is a *Rüssmann approximation function* [19, 116, 117] such as

$$\Delta(t) = \exp\left(\frac{t}{\log^{1+\beta} t}\right), \qquad \beta > 0,$$

and $[k]$ represents the "weight" of k such as the cardinality or diameter of its support, or something similar. See [108] for details and examples.

The upshot is that again the classical KAM theorem extends to this situation establishing the existence of infinite dimensional, albeit strongly localized invariant tori with almost-periodic motions. Moreover, these tori fill some nontrivial part of the phase space.

Neither of these results, however, applies to nonlinear partial differential equations of Hamiltonian type, such as nonlinear Schrödinger or wave equations, or perturbed KdV equations. On one hand, the perturbations are not sufficiently localized to allow a rapid approximation by finite dimensional systems. On the other hand, the frequencies grow only polynomially at most, which is far too slow for the other approach.

Essentially, up to now there is no general KAM theorem to handle the effects of small divisors including arbitrary combinations of infinitely many frequencies in systems arising from pde's.

But in such systems there are also families of finite dimensional elliptic invariant tori with quasi-periodic motions forming real analytic submanifolds which fill the underlying phase space densely. A KAM theorem for such tori along the lines of Theorem 5.3 involves a substantially smaller set of small divisor conditions. Besides the usual *Kolmogorov conditions*

$$|\langle k,\omega\rangle| \geq \frac{\alpha}{|k|^\tau}$$

for $k \neq 0$ for the finitely many internal frequencies $\omega = (\omega_1, \ldots, \omega_n)$, there are the *Lyapunov conditions*

$$|\langle k,\omega\rangle - \Omega_i| \geq \frac{\alpha}{1+|k|^\tau}$$

and the *Melnikov conditions*

$$\left|\langle k,\omega\rangle - \Omega_i \pm \Omega_j\right| \geq \frac{\alpha}{1+|k|^\tau}$$

involving the infinitely many external frequencies $\Omega = (\Omega_1, \Omega_2, \ldots)$. Thus, at most *two* of them enter these conditions at a time, making them essentially finite dimensional and manageable.

The first results in this direction are due to Kuksin and Wayne [70, 135]. In subsequent developments Kuksin applied them for example to nonlinear wave and Schrödinger equations on intervals with Dirichlet boundary conditions and perturbed KdV equations. We point to his books [72, 75] for a comprehensive presentation and further references.

In this book we will develop this theory in full detail following the exposition in [109]. Therefore we will not go into further details here.

It turns out, however, that in order to meet all three conditions above the external frequencies should grow asymptotically like

$$\Omega_i \sim i^d, \qquad d \geq 1.$$

A linear growth rate $\Omega_i \sim i$ seems to be the limiting case, and a slower growth rate with $d < 1$ does not suffice, since resonances are no longer sufficiently separated. This restricts the theory essentially to nonlinear pde on *one*-dimensional domains.

Another restriction arises from Melnikov's condition, which in particular requires the external frequencies to be separated from each other:

$$\left|\Omega_i - \Omega_j\right| \geq \alpha, \qquad i \neq j.$$

As a consequence one can not deal with periodic boundary conditions, where the frequencies are asymptotically double. This restriction, however, was recently alleviated by Chierchia & You [25], who observed that it suffices to require that the Ω_i asymptotically form finite clusters of the same size and structure.

Melnikov's condition, either in the form given above or in the form formulated by Chierchia & You, is required to put the variational equations of motion along a lower dimensional invariant torus into constant coefficient form. This allows the classical KAM procedure of successive transformations of the *entire* system of equations to proceed. It does not, however, enter directly into the construction of the invariant torus with quasi-periodic motions itself. For example, it does not show up in Lyapunov's theorem about the existence of families of periodic solutions near an elliptic equilibrium.

Indeed, in the nineties Craig & Wayne [28] extended Lyapunov's classical result and constructed Cantor discs of periodic solutions for a nonlinear wave equation with periodic boundary conditions. They used a Lyapunov-Schmidt reduction scheme for the external components together with small divisor estimates in the spirit of KAM theory. This way they could replace Melnikov's condition by a somewhat implicit, but more general condition about the distribution of all frequencies.

Their approach was considerably extended by Bourgain, who not only obtained quasi-periodic solutions for Schrödinger equations in this way, but also periodic and quasi-periodic solutions for some two-dimensional Schrödinger equations to which the standard KAM theory does not apply. See [14, 15, 16] and in particular [17]. In contrast to the latter, however, this approach does not establish the linear stability of the solutions so constructed.

So far, all these extensions and generalizations are restricted to perturbations which are given by *bounded* nonlinear operators, as explained in section 16 on page 137. They do *not* apply to perturbations affected by *unbounded* operators, as is the case for perturbed KdV equations. In this book we therefore choose to develop the underlying KAM theory along the classical lines.

All these results concern the existence of quasi-periodic solutions filling finite dimensional invariant tori in an infinite dimensional phase space. Almost nothing is known, however, about the existence of almost-periodic solutions for nonlinear pde. It seems that the nonlinearities affect a strong, long range coupling which is beyond the control of the current techniques.

There are, however, some existence results for a simplified problem. Bourgain [16] considered the Schrödinger equation

$$\mathrm{i}u_t = u_{tt} - V(x)u - |u|^2 u$$

on $[0, \pi]$ with Dirichlet boundary conditions, depending on an analytic potential V. Given an almost-periodic solution of the linear equation with small and *very* rapidly decaying amplitudes and nonresonant frequencies he showed that the potential can be modified so that this solution persists in the nonlinear equation. A similar result was obtained in [111].

In both cases the construction of *quasi*-periodic solutions is *iterated* infinitely often, increasing at each step the number of frequencies involved. The potential V plays the role of infinitely many external parameters which can be adjusted along this process. This renders the example somewhat academic.

III

Birkhoff Coordinates

6 Background and Results

In this chapter we consider the KdV equation

$$q_t = -q_{xxx} + 6qq_x$$

on the space $L^2(S^1)$ of 1-periodic functions on the real line. Our aim is to show that the subspace

$$L_0^2(S^1) = \{q \in L^2(S^1) : [q] = 0\}$$

of functions with vanishing mean value $[q] = \int_{S^1} q(x)\, dx$ admits global Birkhoff coordinates which linearize the KdV equation. More precisely, our aim is to show that there are canonical coordinates $(x_n, y_n)_{n \geq 1}$ on $L_0^2(S^1)$ in which the KdV equation for smooth initial data takes the form

$$\begin{aligned} \dot{x}_n &= \omega_n y_n, \\ \dot{y}_n &= -\omega_n x_n, \end{aligned}$$

with frequencies $(\omega_n)_{n \geq 1}$, which depend only on the *actions*

$$I_n = \tfrac{1}{2}(x_n^2 + y_n^2).$$

In particular, the latter are preserved quantities. Indeed, the same coordinates are Birkhoff coordinates for *any* Hamiltonian in the KdV hierarchy.

Integrals for KdV

To state the result we review some basic facts about the KdV equation and fix some notations. For any integer $N \geq 0$ we introduce the Sobolev space

$$\mathcal{H}^N = \{q \in L^2(S^1; \mathbb{R}) : \|q\|_N < \infty\}$$

of real valued functions q on $S^1 = \mathbb{R}/\mathbb{Z}$ with finite norm $\|q\|_N$, where

$$\|q\|_N^2 = |\hat{q}(0)|^2 + \sum_{k\in\mathbb{Z}} |k|^{2N} \, |\hat{q}(k)|^2$$

is defined in terms of the discrete Fourier transform $\hat{q}$ of q, $q(x) = \sum_{k\in\mathbb{Z}} \hat{q}(k)\mathrm{e}^{2\pi i k x}$. In particular, we have $\mathcal{H}^0 = L^2(S^1)$ with norm $\|\cdot\| = \|\cdot\|_0$. We endow $\mathcal{H}^N$ with the Poisson structure proposed by Gardner,

$$\{F,G\} = \int_{S^1} \frac{\partial F}{\partial q(x)} \frac{\mathrm{d}}{\mathrm{d}x} \frac{\partial G}{\partial q(x)} \, \mathrm{d}x, \tag{6.1}$$

where F and G are differentiable functions on $\mathcal{H}^N$ with L^2-gradients in $\mathcal{H}^1$. The KdV equation with periodic boundary conditions can then be expressed as a Hamiltonian system,

$$\frac{\partial q}{\partial t} = \frac{\mathrm{d}}{\mathrm{d}x} \frac{\partial H}{\partial q},$$

with the KdV Hamiltonian

$$H(q) = \int_{S^1} \left(\tfrac{1}{2} q_x^2 + q^3\right) \mathrm{d}x.$$

The mean value function $M = [\,\cdot\,]$ is a Casimir function for the Poisson structure $\mathrm{d}/\mathrm{d}x$, that is,

$$\frac{\mathrm{d}}{\mathrm{d}x} \frac{\partial M}{\partial q} = 0$$

everywhere on $L^2(S^1)$. So M commutes with every Hamiltonian on $L^2(S^1)$. Therefore, $L_0^2(S^1)$ is an invariant subspace for *every* flow defined in terms of this Poisson structure, so in particular for the KdV equation.

Besides the mean value, the evolution of the KdV equation admits infinitely many independent conserved quantities, or *integrals*. An elegant way to construct such a family is by way of the so called Lax pair formalism. Lax [77] observed that under sufficient regularity assumptions, q is a solution of the KdV equation if and only if

$$\frac{\mathrm{d}}{\mathrm{d}t} L = [B, L],$$

where

$$L = -\frac{\mathrm{d}^2}{\mathrm{d}x^2} + q$$

denotes the Schrödinger operator with potential q, considered on [0, 2] with periodic boundary conditions, and $[B, L] = BL - LB$ denotes the commutator of L with the anti-symmetric operator

$$B = -4\frac{\mathrm{d}^3}{\mathrm{d}x^3} + 3q\frac{\mathrm{d}}{\mathrm{d}x} + 3\frac{\mathrm{d}}{\mathrm{d}x} q.$$

It follows by an elementary calculation that the flow of

$$\frac{\mathrm{d}}{\mathrm{d}t}U = BU, \qquad U(0) = I,$$

defines a family of *unitary* operators $U(t)$ such that

$$U^*(t)L(t)U(t) = L(0),$$

where $L(t)$ denotes the Schrödinger operator with potential $q(t, \cdot)$. Consequently, the spectrum of $L(t)$ is independent of t, and the flow of the KdV equation defines an isospectral deformation on the space of sufficiently smooth functions q in L^2 when considered as potentials of the Schrödinger operator L. Therefore, the corresponding spectral data of q provide conserved quantities of the KdV equation.

Let us recall the basic properties of the Schrödinger operator L with potential q on $[0, 2]$ with periodic boundary conditions – see also appendix B. It is well known [80, 82, 84] that its *periodic spectrum*, denoted $\operatorname{spec}(q)$, is pure point and consists of an unbounded sequence of *periodic eigenvalues*

$$\lambda_0(q) < \lambda_1(q) \le \lambda_2(q) < \lambda_3(q) \le \lambda_4(q) < \dots .$$

Equality or inequality may occur in every place with a '$\le$'-sign. The asymptotic behavior of the periodic eigenvalues for $q \in L^2$ is

$$\lambda_{2n-1}(q), \lambda_{2n}(q) = n^2\pi^2 + [q] + \ell^2(n),$$

where $\ell^2(n)$ stands for the n-th term of a square summable sequence.

By Floquet theory, the periodic spectrum also determines the spectrum of L considered on the real line $\mathbb{R}$ with L^2-integrability condition. The latter is the union

$$[\lambda_0, \lambda_1] \cup [\lambda_2, \lambda_3] \cup [\lambda_4, \lambda_5] \cup \dots$$

of the *spectral bands* $[\lambda_{2n}, \lambda_{2n+1}]$. Between two adjacent bands one has the *spectral gaps* $(\lambda_{2n-1}, \lambda_{2n})$. By the above inequalities, these gaps are pairwise disjoint. But each of them may degenerate to an empty set, when $\lambda_{2n} = \lambda_{2n-1}$, in which case one speaks of a *collapsed gap*. Its length,

$$\gamma_n = \lambda_{2n} - \lambda_{2n-1}, \qquad n \ge 1,$$

is called the *n-th gap length* of the spectrum. By the asymptotic behavior of the eigenvalues, $\gamma_n = \ell^2(n)$.

In view of the Lax pair formalism the periodic eigenvalues $\lambda_n(q)$ and hence the gap lengths $\gamma_n(q)$ are conserved quantities under the evolution of the KdV equation. However, λ_{2n}, λ_{2n-1} and γ_n are not smooth functions of q whenever the corresponding gap collapses. In contrast, the *squared gap lengths*

$$\gamma_n^2 = (\lambda_{2n} - \lambda_{2n-1})^2, \qquad n \ge 1,$$

are real analytic integrals of the KdV equation. Moreover, Garnett & Trubowitz [48] showed that together with the average $[q]$, they uniquely determine the periodic spectrum of q.

The space $L_0^2(S^1)$ thus decomposes into the *isospectral sets*

$$\mathrm{Iso}(q) = \{\, p \in L_0^2(S^1) \colon\ \mathrm{spec}(p) = \mathrm{spec}(q) \,\},$$

which are invariant under the KdV flow and may also be characterized by the gap lengths,

$$\mathrm{Iso}(q) = \{\, p \in L_0^2(S^1) \colon\ \text{gap length}(p) = \text{gap length}(q) \,\}.$$

As shown by McKean & Trubowitz [89], these are homeomorphic to compact connected *tori*, whose dimension equals the number of positive gap lengths and is infinite generically – see also Proposition B.14. Moreover, the asymptotic behavior of the gap lengths characterizes the regularity of a potential in exactly the same way as its Fourier coefficients do [84]. Therefore, for any q in $L_0^2(S^1)$,

$$q \in \mathcal{H}_0^N \quad \text{iff} \quad \mathrm{Iso}(q) \subset \mathcal{H}_0^N$$

for each $N \geq 1$, where

$$\mathcal{H}_0^N = \mathcal{H}^N \cap L_0^2.$$

Hence also the phase space $\mathcal{H}_0^N$ decomposes into a collection of tori of varying dimension which are *invariant* under the KdV flow.

In classical mechanics the existence of a foliation of the phase space into Lagrangian invariant tori is tantamount, at least locally, to the existence of angle-action coordinates. This is the content of the Liouville-Arnold-Jost theorem as described in section 3. In the infinite dimensional setting of the KdV equation, however, such coordinates can not exist due to the ubiquity of finite gap potentials.

Recall that a potential $q \in L_0^2$ is called a *finite gap potential*, if only a finite number of its spectral gaps are open, while all others are collapsed. It is a fact that finite gap potentials are real analytic and dense in $\mathcal{H}_0^N$ for each $N \geq 0$ – see [49] or [84, Theorem 3.4.3]. As a consequence, the dimension of the foliation of $\mathcal{H}_0^N$ into invariant tori is nowhere constant. In particular, arbitrarily close to an infinite dimensional invariant torus $\mathrm{Iso}(p)$, there are finite dimensional tori of arbitrary dimension. Consequently, there are no angle-action coordinates in any neighbourhood of $\mathrm{Iso}(p)$.

Nevertheless, we are going to show that the spectral data of a potential can be used to construct actions I_n and angles θ_n on dense open subsets of L_0^2 in such a way that the associated *Birkhoff coordinates*

$$x_n = \sqrt{2I_n}\cos\theta_n, \qquad y_n = \sqrt{2I_n}\sin\theta_n,$$

extend real analytically to *all* of L_0^2 and give rise to a global symplectic coordinate system. The actions can be viewed as rescalings of the squared gap lengths, and we will prove all of the above statements about squared gap lengths for these actions instead. For example,

$$\operatorname{Iso}(q) = \left\{ p \in L_0^2(S^1) \colon \text{actions}(p) = \text{actions}(q) \right\},$$

that is, isospectral sets may be characterized by these actions.

Actions and Angles

We now try to motivate the definitions of actions and angles given in the next two sections. To this end we introduce some more concepts and notations. Some more details are given in appendix B.

Let $y_1(x, \lambda, q)$ and $y_2(x, \lambda, q)$ be the standard *fundamental solution* of

$$-y'' + qy = \lambda y,$$

defined by the initial conditions

$$\begin{aligned} y_1(0, \lambda, q) &= 1, \qquad & y_2(0, \lambda, q) &= 0, \\ y_1'(0, \lambda, q) &= 0, \qquad & y_2'(0, \lambda, q) &= 1. \end{aligned}$$

Let $\Delta(\lambda, q) = y_1(1, \lambda, q) + y_2'(1, \lambda, q)$ be its associated discriminant. The periodic spectrum of q is precisely the zero set of the entire function $\Delta^2(\lambda, q) - 4$, and we have the product representation

$$\Delta^2(\lambda) - 4 = 4(\lambda_0 - \lambda) \prod_{n \geq 1} \frac{(\lambda_{2n} - \lambda)(\lambda_{2n-1} - \lambda)}{n^4 \pi^4},$$

see for example [89]. Hence, this function is also a spectral invariant.

The square root of $\Delta^2(\lambda) - 4$ is defined on the Riemann surface

$$\Sigma(q) = \left\{ (\lambda, z) \colon z^2 = \Delta^2(\lambda, q) - 4 \right\} \subset \mathbb{C}^2,$$

which is a hyperelliptic surface whose genus is precisely the number of open gaps of q. It may be viewed as two copies of the complex plane slit open along $(-\infty, \lambda_0)$ and each open gap $(\lambda_{2n-1}, \lambda_{2n})$ and then glued together crosswise along the slits. This Riemann surface is another spectral invariant associated with q.

To define angles we also need to consider the spectrum of the differential operator $L = -\mathrm{d}^2/\mathrm{d}x^2 + q$ with Dirichlet boundary conditions on the interval $[0, 1]$. This spectrum consists of an unbounded sequence of *Dirichlet eigenvalues*

$$\mu_1(q) < \mu_2(q) < \mu_3(q) < \ldots,$$

which satisfy

$$\lambda_{2n-1}(q) \leq \mu_n(q) \leq \lambda_{2n}(q)$$

for all $n \geq 1$. Thus, the n-th Dirichlet eigenvalue μ_n is always contained in the n-th interval $[\lambda_{2n-1}, \lambda_{2n}]$.

By the Wronskian identity,

$$\Delta^2(\mu_n) - 4 = (y_1(1,\mu_n) - y_2'(1,\mu_n))^2.$$

With any Dirichlet eigenvalue μ_n we can therefore uniquely and analytically associate a sign by defining

$$\sqrt[*]{\Delta^2(\mu_n) - 4} = y_1(1,\mu_n) - y_2'(1,\mu_n). \tag{6.2}$$

This in turn defines the point

$$\mu_n^* = \left(\mu_n, \sqrt[*]{\Delta^2(\mu_n) - 4}\right) \tag{6.3}$$

on the Riemann surface $\Sigma(q)$, which is sometimes referred to as a Dirichlet divisor.

McKean and Trubowitz [89] observed that these Dirichlet divisors provide coordinates on isospectral sets. Each set $\operatorname{Iso}(q)$ is homeomorphic to the product of two-fold coverings of the intervals $[\lambda_{2n-1}(q), \lambda_{2n}(q)]$, $n \geq 1$, with endpoints identified, and each point on $\operatorname{Iso}(q)$ is uniquely described by the positions of its Dirichlet eigenvalues μ_n within these intervals and the signs of $\sqrt[*]{\Delta^2(\mu_n) - 4}$. See Proposition B.14 for the details.

The Dirichlet eigenvalues $(\mu_n)_{n\geq 1}$ can be complemented to a symplectic coordinate system on L_0^2 by introducing the quantities

$$\kappa_n(q) = 2\log\,(-1)^n y_2'(1,\mu_n(q),q), \qquad n \geq 1,$$

[43, 130]. Then

$$q \mapsto (\hat{\mu}_n(q), \kappa_n(q))_{n\geq 1},$$

where $\hat{\mu}_n = \mu_n - n^2\pi^2$, defines a real analytic diffeomorphism of L_0^2 into a suitable Hilbert space of sequences, and one computes that

$$\begin{aligned} \{\mu_n, \mu_m\} &= 0, \\ \{\kappa_n, \mu_m\} &= \delta_{nm}, \\ \{\kappa_n, \kappa_m\} &= 0, \end{aligned}$$

for all $n, m \geq 1$. See [112, Theorem 2.8] for a detailed proof. Hence, the new variables are canonical, and the induced symplectic structure is given by $\mathrm{d}\alpha$, where

$$\alpha = \sum_{m\geq 1} \kappa_m \,\mathrm{d}\mu_m.$$

We may now define actions in analogy to Arnold's formula in the finite dimensional case [3, section 50]. Set

$$I_n = \frac{1}{2\pi}\int_{c_n} \alpha = \frac{1}{2\pi}\sum_{m\geq 1}\int_{c_n} \kappa_m \,\mathrm{d}\mu_m,$$

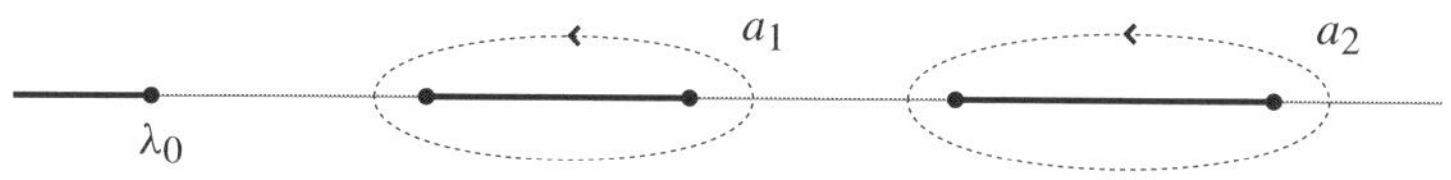

Figure 1 *a*-cycles

where c_n is a cycle on the invariant torus $\mathrm{Iso}(q)$ corresponding to μ_n. As $\mathrm{d}\mu_m = 0$ for $m \neq n$ along c_n, actually

$$I_n = \frac{1}{2\pi} \int_{c_n} \kappa_n \, \mathrm{d}\mu_n = -\frac{1}{2\pi} \int_{c_n} \mu_n \, \mathrm{d}\kappa_n.$$

Along c_n, μ_n moves from λ_{2n-1} to λ_{2n} and back to λ_{2n-1}, with $\sqrt[*]{\Delta^2(\mu_n) - 4}$ changing its sign at the turning point. With

$$y_2'(1, \mu_n) = \frac{1}{2} \left(\Delta(\mu_n) - \sqrt[*]{\Delta^2(\mu_n) - 4} \right),$$

one calculates

$$\begin{aligned} \mathrm{d}\kappa_n &= 2 \frac{\mathrm{d} y_2'(1, \mu_n)}{y_2'(1, \mu_n)} \\ &= 2\dot{\Delta}(\mu_n) \frac{1 - \Delta(\mu_n) / \sqrt[*]{\Delta^2(\mu_n) - 4}}{\Delta(\mu_n) - \sqrt[*]{\Delta^2(\mu_n) - 4}} \, \mathrm{d}\mu_n \\ &= -\frac{2\dot{\Delta}(\mu_n)}{\sqrt[*]{\Delta^2(\mu_n) - 4}} \, \mathrm{d}\mu_n. \end{aligned} \tag{6.4}$$

So altogether we may write

$$I_n = \frac{2}{\pi} \int_{\lambda_{2n-1}}^{\lambda_{2n}} \mu \frac{\dot{\Delta}(\mu)}{\sqrt{\Delta^2(\mu) - 4}} \, \mathrm{d}\mu,$$

with the sign of the root to be chosen properly.

Now Flaschka & McLaughlin [43] observed that the last integral can be rewritten as a contour integral on the Riemann surface $\Sigma(q)$,

$$I_n = \frac{1}{\pi} \int_{a_n} \lambda \frac{\dot{\Delta}(\lambda)}{\sqrt{\Delta^2(\lambda) - 4}} \, \mathrm{d}\lambda,$$

where a_n denotes a contour around the lift of $[\lambda_{2n-1}, \lambda_{2n}]$ on the canonical sheet of $\Sigma(q)$ as indicated in figure 1. This way, the actions I_n are determined entirely by the spectral data of q. See also [44, 131].

Now let us pretend that we can complement the actions I_n by canonically conjugate angles θ_n. Then the canonical one-form α can also be written as

$$\alpha = \sum_{m\geq 1} I_m \,\mathrm{d}\theta_m + \mathrm{d}S$$

with some exact one-form $\mathrm{d}S$. Let us also assume that the restriction of $\mathrm{d}S$ to any isospectral torus $\mathrm{Iso}(q)$ vanishes, that is, S is a function of the actions alone. Then, formally,

$$\mathrm{d}\theta_n \Big|_{\mathrm{Iso}(q)} = \frac{\partial}{\partial I_n} \alpha \Big|_{\mathrm{Iso}(q)} \stackrel{\mathrm{def}}{=} \alpha_n .$$

Integrating along any path on $\mathrm{Iso}(q)$ from some fixed starting point q_0 we thus get

$$\theta_n = \int_{q_0}^{q} \mathrm{d}\theta_n = \int_{q_0}^{q} \alpha_n .$$

This integral is independent of the path chosen since $\mathrm{d}\alpha = 0$ on $\mathrm{Iso}(q)$.

To make this more explicit we look again at the (μ, κ)-coordinate system. Fix as q_0 the unique potential on $\mathrm{Iso}(q)$ with

$$\mu_k(q_0) = \lambda_{2k-1}(q_0), \qquad k \geq 1.$$

Then we can move from q_0 to q by moving μ_k from λ_{2k-1} to $\mu_k(q)$ successively for $k = 1, 2, \ldots$. For each move, the proper sheet of the two-folded covering of $[\lambda_{2k-1}, \lambda_{2k}]$ is determined by the sign of $\sqrt[*]{\Delta^2(\mu_k) - 4}$, that is, by the point μ_k^*. This way we obtain, somewhat formally,

$$\theta_n = \sum_{k\geq 1} \int_{\lambda_{2k-1}}^{\mu_k} \alpha_n . \tag{6.5}$$

We still have to get hold of the integrals in this identity. Expressing α again in terms of the (μ, κ)-coordinates we can proceed as above for the actions. Formally,

$$\frac{1}{2\pi} \int_{\lambda_{2k-1}}^{\mu_k} \alpha_n = \frac{1}{2\pi} \int_{\lambda_{2k-1}}^{\mu_k} \frac{\partial \kappa_k}{\partial I_n} \,\mathrm{d}\mu_k .$$

Similarly to (6.4) one computes

$$\frac{\partial \kappa_k}{\partial I_n} = -\frac{2 \partial \Delta / \partial I_n |_{\mu_k}}{\sqrt[*]{\Delta^2(\mu_k) - 4}}$$

and again notes that the last integral can be rewritten as a path integral on the Riemann surface $\Sigma(q)$. That is,

$$\frac{1}{2\pi} \int_{\lambda_{2k-1}}^{\mu_k} \frac{\partial \kappa_k}{\partial I_n} \,\mathrm{d}\mu_k = \frac{1}{2\pi} \int_{\lambda_{2k-1}}^{\mu_k^*} \beta_n ,$$

with the on $\Sigma(q)$ holomorphic one-form

$$\beta_n = -\frac{2\partial\Delta/\partial I_n}{\sqrt{\Delta^2(\lambda)-4}}\,\mathrm{d}\lambda.$$

To identify β_n note that by the definition of I_m,

$$\frac{\partial I_m}{\partial I_n} = \frac{1}{2\pi}\int_{c_m} \alpha_n = \delta_{mn},$$

and hence

$$\frac{1}{2\pi}\int_{a_m} \beta_n = \delta_{mn}$$

for all $m, n \geq 1$. But these properties uniquely characterize a holomorphic one-form on $\Sigma(q)$, and it is actually fairly straightforward to construct them in the first place. More precisely, using the implicit function theorem we will construct entire functions $\psi_1, \psi_2, \ldots$ such that

$$\frac{1}{2\pi}\int_{a_m} \frac{\psi_n(\lambda)}{\sqrt{\Delta^2(\lambda)-4}}\,\mathrm{d}\lambda = \delta_{mn}$$

for all $m, n \geq 1$ – see appendix D. Then we can rewrite each integral $\int_{\lambda_{2k-1}}^{\mu_k} \alpha_n$ in (6.5) as a Riemann surface integral $\int_{\lambda_{2k-1}}^{\mu_k^*} \beta_n$, and we *define* the n-th angle as

$$\theta_n = \sum_{k\geq 1} \int_{\lambda_{2k-1}}^{\mu_k^*} \frac{\psi_n(\lambda)}{\sqrt{\Delta^2(\lambda)-4}}\,\mathrm{d}\lambda \mod 2\pi$$

for each open gap $(\lambda_{2n-1}, \lambda_{2n})$. By a slight abuse of terminology, we may refer to the map $q \mapsto \theta(q) = (\theta_n(q))$ as the *Abel map* [101].

It turned out that these angles linearize the KdV equation [32, 33, 35, 36, 57, 88, 89, 90]. That is, if $t \mapsto q^t$ is a solution curve of the KdV equation with $q^{\mathrm{o}} = q$, then q^t evolves on $\mathrm{Iso}(q^{\mathrm{o}})$ in such a way that for every n corresponding to an open gap,

$$\theta_n(q^t) = \theta_n(q^{\mathrm{o}}) + \omega_n t$$

with a certain frequency ω_n. We will reprove this fact by verifying that the I_n and θ_n are indeed angle-action coordinates for the KdV equation. This will also justify the assumption on $\mathrm{d}S$ made above.

Birkhoff Coordinates

The action variables I_n are real analytic on all of L_0^2, while each angle θ_n is real analytic modulo 2π on the dense open domain $L_0^2 \setminus D_n$, with $D_n = \{q : \gamma_n(q) = 0\}$. We show that the associated Birkhoff coordinates

$$x_n = \sqrt{2I_n}\cos\theta_n, \qquad y_n = \sqrt{2I_n}\sin\theta_n$$

are real analytic on all of L_0^2 and give rise to a global coordinate system, in which the KdV Hamiltonian is a function of the actions alone. Indeed, the same holds simultaneously for *any* Hamiltonian in the KdV hierarchy on suitable subspaces of L_0^2. This may be considered as an instance of an infinite dimensional version of the Rüssmann-Vey-Ito theorem of section 4 concerning the existence of transformations into integrable Birkhoff normal forms [56, 115, 132].

To state this result, we introduce the model space of real sequences

$$\hbar_r = \ell_r^2 \times \ell_r^2$$

with elements $(\boldsymbol{x}, \boldsymbol{y})$, where

$$\ell_r^2 = \left\{ \boldsymbol{x} \in \ell^2(\mathbb{N}; \mathbb{R}) \colon \ \|\boldsymbol{x}\|_r^2 = \sum_{n \geq 1} n^{2r} \, |x_n|^2 < \infty \right\}.$$

We endow this space with the standard Poisson structure, for which $\{x_n, y_m\} = \delta_{nm}$, while all other brackets vanish.

Theorem 6.1. *There exists a diffeomorphism $\Omega \colon L_0^2 \to \hbar_{1/2}$ with the following properties.*

(i) *Ω is one-to-one, onto, bi-analytic, and preserves the Poisson bracket.*

(ii) *For each $N \geq 0$, the restriction of Ω to $\mathcal{H}_0^N$, denoted by the same symbol, is a map*

$$\Omega \colon \ \mathcal{H}_0^N \to \hbar_{N+1/2},$$

which is one-to-one, onto, and bi-analytic as well.

(iii) *The coordinates $(\boldsymbol{x}, \boldsymbol{y})$ in $\hbar_{3/2}$ are global Birkhoff coordinates for the KdV equation. That is, the transformed KdV Hamiltonian $H \circ \Omega^{-1}$ depends only on $x_n^2 + y_n^2$, $n \geq 1$, with $(\boldsymbol{x}, \boldsymbol{y})$ being canonical coordinates in $\hbar_{3/2}$.*

(iv) *The same holds for any other Hamiltonian in the KdV hierarchy, considered on a subspace $\mathcal{H}_0^N$ with appropriate N.*

Further properties of the map Ω are formulated in Theorems 9.8, 11.9 and 11.10 below. They are all collected in a single theorem, Theorem 12.6 on page 108, at the end of this chapter.

Given that $(\boldsymbol{x}, \boldsymbol{y})$ are Birkhoff coordinates for each Hamiltonian H in the KdV hierarchy on the appropriate subspace of $\hbar_{1/2}$, the corresponding equations of motions are simply

$$\dot{x}_n = \omega_n(\boldsymbol{I}) y_n, \qquad \dot{y}_n = -\omega_n(\boldsymbol{I}) x_n,$$

where

$$\omega_n(\boldsymbol{I}) = \frac{\partial H'}{\partial I_n}(\boldsymbol{I}), \qquad \boldsymbol{I} = (I_n)_{n \geq 1} = \tfrac{1}{2} \, (x_n^2 + y_n^2)_{n \geq 1},$$

are the frequencies associated with the transformed Hamiltonian $H' = H \circ \Omega^{-1}$. Thus, the KdV equations are rewritten as infinite-dimensional systems of integrable ordinary differential equations.

Furthermore, using Riemann bilinear relations we obtain explicit formulas and asymptotic expansions of the frequencies of the first three Hamiltonians in the KdV hierarchy, which we need for the KAM results in chapter IV.

Outline of Proof

The proof of the theorem splits into four parts. First we define actions I_n and angles θ_n as indicated above. We show that each I_n is real analytic on L_0^2, while each θ_n is real analytic only on the dense open domain $L_0^2 \setminus D_n$, when taken modulo 2π.

Next, we define the associated cartesian coordinates x_n and y_n as usual. Although defined originally only on $L_0^2 \setminus D_n$, we show that they extend real analytically to a complex neighbourhood W of L_0^2. Surely, the angle θ_n blows up when γ_n collapses, but this blow up is compensated by the rate at which I_n vanishes in the process. In particular, for real q the resulting limit will vanish.

For complex potentials, however, this limit will typically not vanish, since the associated Schrödinger operator is no longer self-adjoint. As a consequence, it may happen that $\lambda_{2n} = \lambda_{2n-1}$, but $\mu_n \neq \lambda_{2n}$. In such a case, x_n and y_n will not vanish.

Then we show that the thus defined map $\Omega\colon q \mapsto (\boldsymbol{x}, \boldsymbol{y})$ is a diffeomorphism between L_0^2 and $\hbar_{1/2}$. The main problem here is to verify that $d_q\Omega$ is a linear isomorphism at *every* point q. This is done with the help of orthogonality relations among the coordinates, which are nothing but their Poisson brackets. For the nonlinear Schrödinger equation the corresponding orthogonality relations have first been established by McKean & Vaninsky [91, 92]. It turned out that many of their ideas can also be used in the case of the KdV equation.

We also verify that each Hamiltonian in the KdV hierarchy becomes a function of the actions alone, using their characterization in terms of the asymptotic expansion of the discriminant Δ as $\lambda \to -\infty$ and hence as spectral invariants.

Finally, we verify that Ω preserves the Poisson bracket. To this end it is more convenient to look at the associated symplectic structures. This way, we only need to establish the regularity of the gradient of θ_n at special points, not everywhere. Thus, we equivalently show that Ω is a symplectomorphism. This will complete the proof of the main results of Theorem 6.1.

Some Notations and Notions

In the sequel we will need to consider various root functions, and it will be important to fix the proper branch in each case. The principal branch of the square root on $\mathbb{C} \setminus (-\infty, 0]$ is denoted by $\sqrt[+]{\lambda}$, so

$$\sqrt[+]{\lambda} > 0 \quad \text{for} \quad \lambda > 0.$$

If the radicand λ is obviously real and nonnegative, however, we simply write $\sqrt{\lambda}$.

On $\mathbb{C} \setminus [-1, 1]$ we define a "standard" branch $\sqrt[s]{1-\lambda^2}$ by

$$\sqrt[s]{1-\lambda^2} = \mathrm{i}\lambda \sqrt[+]{1-\lambda^{-2}} \quad \text{for} \quad |\lambda| > 1. \tag{6.6}$$

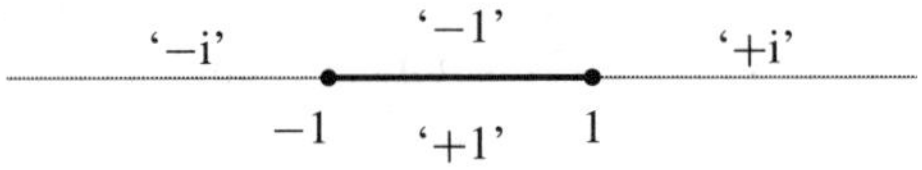

Figure 2 Signs of $\sqrt[s]{1-\lambda^2}$

Thus,

$$\sqrt[s]{1-\lambda^2} > 0 \quad \text{for} \quad \mathrm{i}\lambda > 0,$$

as illustrated in figure 2. We then extend this definition continuously to more general quadratic radicands $(b-\lambda)(\lambda-a)$ with $a \neq b$. To avoid ambiguities we require that $a \prec b$ in the *lexicographic ordering*, where

$$a \prec b \quad \Leftrightarrow \quad \begin{cases} \operatorname{Re} a < \operatorname{Re} b \\ \quad \text{or} \\ \operatorname{Re} a = \operatorname{Re} b \ \text{ and } \ \operatorname{Im} a \leq \operatorname{Im} b. \end{cases} \tag{6.7}$$

Setting $\gamma = b - a \neq 0$ and $\tau = (b+a)/2$ we then define

$$\begin{aligned} \sqrt[s]{(b-\lambda)(\lambda-a)} &= \frac{\gamma}{2} \sqrt[s]{1-w^2} \\ &= \mathrm{i}(\lambda-\tau) \sqrt[+]{1-w^{-2}}, \qquad w = \frac{\lambda-\tau}{\gamma/2}, \end{aligned} \tag{6.8}$$

for λ not on the segment from a to b. The last expression also makes sense when $a = b$ and gives

$$\sqrt[s]{(a-\lambda)(\lambda-a)} = \mathrm{i}(\lambda-a). \tag{6.9}$$

Typically, we have $a = \lambda_{2n-1}$ and $b = \lambda_{2n}$.

We define a "canonical" branch $\sqrt[c]{\Delta^2(\lambda)-4}$ on $\mathbb{C} \smallsetminus \bigcup_{n\geq 0}(\lambda_{2n-1}, \lambda_{2n})$, the complex plane slit open along the interval $(-\infty, \lambda_0)$ and each open gap $(\lambda_{2n-1}, \lambda_{2n})$, by requiring that for real q,

$$\mathrm{i} \sqrt[c]{\Delta^2(\lambda)-4} > 0 \quad \text{for} \quad \lambda \in (\lambda_0, \lambda_1). \tag{6.10}$$

For complex q this root is defined by continuous extension. See figure 3 for an illustration. Finally, we denote by $\sqrt{\Delta^2(\lambda)-4}$ the root on the hyperelliptic surface $\Sigma(q)$.

The L^2-*gradient* of a differentiable function $F\colon L^2 \to \mathbb{C}$ is denoted by

$$\partial F = \frac{\partial F}{\partial q},$$

and is that function on L^2 satisfying

$$dF(v) = \int_0^1 \partial F \cdot v \, \mathrm{d}x.$$

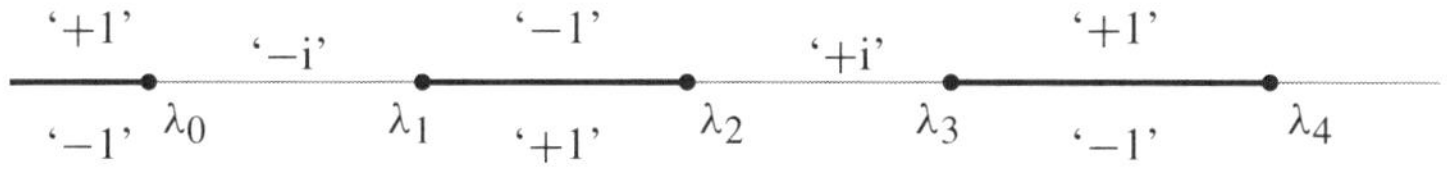

Figure 3 Signs of $\sqrt[c]{\Delta^2(\lambda) - 4}$ for real q

The functions we consider, such as the actions and angles, naturally extend from L_0^2 to L^2, so their gradients in this sense are also well defined. However, those functions are invariant under translations of q. Consequently,

$$[\partial F] = \int_0^1 \partial F \,\mathrm{d}x = \langle \partial F, 1 \rangle = \frac{\mathrm{d}}{\mathrm{d}c} F(q+c)\bigg|_0 = 0, \tag{6.11}$$

that is, their gradients belong to L_0^2. So their gradients as functions on L_0^2 and on L^2 are actually the same, and we need not distinguish between them.

For standard notations concerning spectra and eigenfunctions of Schrödinger operators we refer to appendix B. Finally, all neighbourhoods are assumed to be connected without saying so explicitly.

7 Actions

In this section we define action variables I_n by the formula of Flaschka & McLaughlin [43] derived in the previous section and prove some regularity and asymptotic properties. In particular, we are interested in analyticity properties. This requires that we not only consider real, but also complex potentials q in some small complex neighbourhood W of L_0^2 within $L_{0,\mathbb{C}}^2 = \{q \in L_{\mathbb{C}}^2 \colon [q] = 0\}$.

For $q \in L_{0,\mathbb{C}}^2$, denote by $y_1(x,\lambda,q)$ and $y_2(x,\lambda,q)$ the usual fundamental solution of the Schrödinger equation

$$-y'' + qy = \lambda y,$$

and let

$$\Delta(\lambda, q) = y_1(1,\lambda,q) + y_2'(1,\lambda,q)$$

be its associated discriminant. The spectrum of the operator $-\mathrm{d}^2/\mathrm{d}x^2 + q$ with periodic boundary conditions on the interval $[0,2]$ is the zero set of the entire function $\Delta^2(\lambda,q) - 4$. It is a sequence of complex numbers, which we order lexicographically as in (6.7) so that

$$\lambda_0 \prec \lambda_1 \prec \cdots \prec \lambda_{2n-1} \prec \lambda_{2n} \prec \ldots .$$

For real q, the periodic eigenvalues λ_n are real, and this ordering amounts to the usual ordering of the periodic eigenvalues, in which case they are continuous functions of q. For non-real q, however, λ_n is *not* a continuous function of q, since eigenvalues may change their position in the ordering discontinuously as indicated in figure 4.

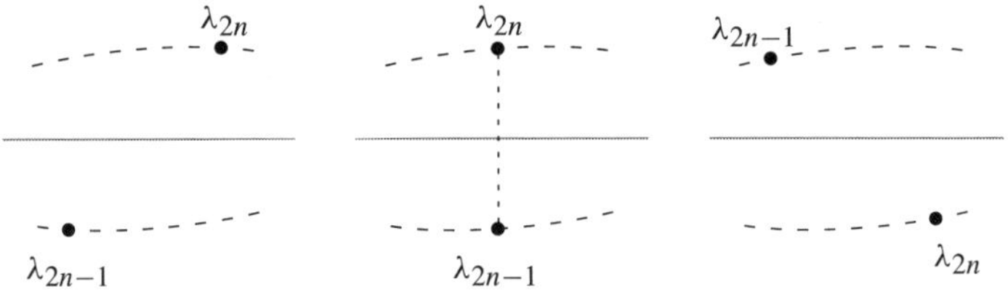

Figure 4 Labeling of periodic eigenvalues as q varies

Restricting ourselves to a sufficiently small complex neighbourhood W of L_0^2, however, we can always ensure that the closed intervals

$$G_n = \{(1-t)\lambda_{2n-1} + t\lambda_{2n} : 0 \le t \le 1\}, \qquad n \ge 1,$$

as well as

$$G_0 = \{(1-\mathrm{e}^{-t}) + \mathrm{e}^{t}\lambda_0 : \ -\infty < t \le 0\}$$

are disjoint from each other. Due to the asymptotic behavior of the periodic eigenvalues, they admit mutually disjoint neighbourhoods $U_n \subset \mathbb{C}$, called *isolating neighbourhoods*. Moreover, inside each U_n we may choose a circuit Γ_n around G_n with counterclockwise orientation as illustrated in figure 5. Both U_n and Γ_n may be chosen to be locally independent of q.

Following Flaschka & McLaughlin [43] we now define actions for the KdV equation by

$$I_n = \frac{1}{\pi}\int_{a_n} \lambda \frac{\dot{\Delta}(\lambda)}{\sqrt{\Delta^2(\lambda)-4}}\,\mathrm{d}\lambda, \qquad n \ge 1,$$

where a_n is the lift of the circuit Γ_n to that sheet of $\Sigma(q)$ on which $\sqrt{\Delta^2(\lambda)-4}$ is given by the c-root. Hence, this definition is equivalent to

$$I_n = \frac{1}{\pi}\int_{\Gamma_n} \lambda \frac{\dot{\Delta}(\lambda)}{\sqrt[c]{\Delta^2(\lambda)-4}}\,\mathrm{d}\lambda, \qquad n \ge 1.$$

By Cauchy's theorem, I_n does not depend on the choice of Γ_n as long as it stays inside U_n.

In the following we may need to shrink the complex neighbourhood W of L_0^2 within $L_{0,\mathbb{C}}^2$ several times, but we will denote it by the same symbol throughout.

Theorem 7.1. *There exists a complex neighbourhood W of L_0^2 such that for each $n \ge 1$, the function I_n is analytic on W with L^2-gradient*

$$\partial I_n = -\frac{1}{\pi}\int_{\Gamma_n} \frac{\partial \Delta(\lambda)}{\sqrt[c]{\Delta^2(\lambda)-4}}\,\mathrm{d}\lambda. \tag{7.1}$$

This L^2-gradient has mean value zero. Moreover, on the real subspace L_0^2, each function I_n is real, nonnegative, and vanishes if and only if γ_n vanishes.

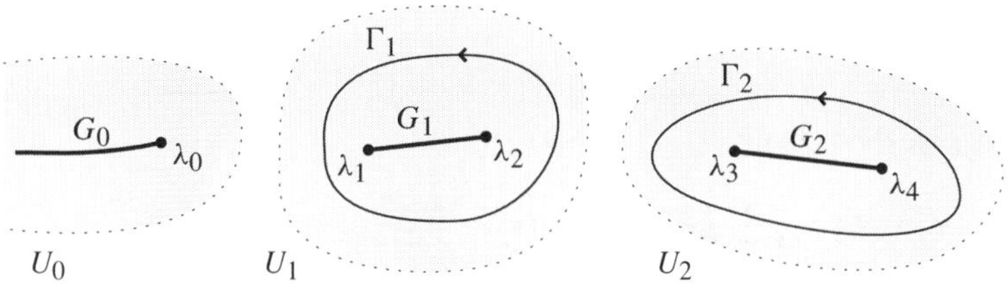

Figure 5 Isolating neighbourhoods

Proof. Choose the complex neighbourhood W of L_0^2 as above. Locally on W, the contours of integration Γ_n can be chosen to be independent of q. As Δ is an analytic function of λ and q, and $\sqrt[c]{\Delta^2(\lambda)-4}$ is analytic in a neighbourhood of Γ_n, the function I_n is clearly analytic in q.

Since I_n is invariant under translations of q, the gradient of I_n has mean value zero in view of (6.11). To obtain its representation we observe that on the interval $[\lambda_{2n-1}, \lambda_{2n}]$, we have $(-1)^n\Delta(\lambda) \geq 2$ and hence

$$(-1)^n\Delta(\lambda) \pm \sqrt{\Delta^2(\lambda)-4} > 0$$

for real q. Therefore, on a sufficiently small complex neighbourhood W_n of L_0^2 and a circuit Γ_n sufficiently close to $[\lambda_{2n-1}, \lambda_{2n}]$, the principle branch of the logarithm

$$\phi(\lambda) = \log{(-1)^n}\left(\Delta(\lambda) - \sqrt[c]{\Delta^2(\lambda)-4}\right)$$

is well defined along Γ_n. Since

$$\dot{\phi}(\lambda) = -\dot{\Delta}(\lambda)/\sqrt[c]{\Delta^2(\lambda)-4},$$

partial integration gives

$$I_n = \frac{1}{\pi}\int_{\Gamma_n} \log{(-1)^n}\left(\Delta(\lambda) - \sqrt[c]{\Delta^2(\lambda)-4}\right) \mathrm{d}\lambda. \tag{7.2}$$

Again keeping Γ_n fixed and taking the gradient with respect to q we obtain the above formula for ∂I_n on W_n. As both sides of (7.1) are analytic on W, this identity holds everywhere on W.

To prove the remaining assertions we observe that in view of the existence of the primitive $\phi(\lambda)$,

$$\int_{\Gamma_n} \frac{\dot{\Delta}(\lambda)}{\sqrt[c]{\Delta^2(\lambda)-4}}\,\mathrm{d}\lambda = 0.$$

With $\dot\lambda_n$ denoting the root of $\dot\Delta$ near λ_{2n-1} and λ_{2n}, we can therefore also write

$$I_n = \frac{1}{\pi}\int_{\Gamma_n} (\lambda - \dot\lambda_n)\frac{\dot\Delta(\lambda)}{\sqrt[c]{\Delta^2(\lambda)-4}}\,\mathrm{d}\lambda.$$

For real q we then obtain

$$I_n = \frac{2}{\pi}\int_{\lambda_{2n-1}}^{\lambda_{2n}} (-1)^{n-1}(\lambda - \dot\lambda_n)\frac{\dot\Delta(\lambda)}{\sqrt[+]{\Delta^2(\lambda)-4}}\,\mathrm{d}\lambda$$

by shrinking the contour of integration to the real interval and taking into account the definition of the c-root. Since $\operatorname{sign}(\lambda - \dot\lambda_n)\dot\Delta(\lambda) = (-1)^{n-1}$ on $[\lambda_{2n-1}, \lambda_{2n}]$, the integrand is non-negative, and the claimed results follow. □

Next we show that each action is not only a continuous, but even a *compact function* on L_0^2 in the following sense: if q converges weakly to p, then $I_n(q)$ converges to $I_n(p)$ for each n.

Lemma 7.2. *Each function I_n, $n \geq 1$, is compact on L_0^2.*

Proof. The periodic eigenvalues λ_m are compact functions on L_0^2 – the proof is the same as for the Dirichlet eigenvalues in [112, p. 32]. The same holds for the discriminant Δ and its λ-derivative $\dot\Delta$, considered as maps from q to functions on the complex plane with the topology of uniform convergence on bounded subsets of $\mathbb{C}$ [112, p. 18]. If q converges weakly to p in L_0^2, then eventually we may choose the contour Γ_n to be independent of q, and conclude that

$$\begin{aligned}
\lim_{q\rightharpoonup p} I_n(q) &= \lim_{q\rightharpoonup p}\frac{1}{\pi}\int_{\Gamma_n} \lambda\frac{\dot\Delta(\lambda,q)}{\sqrt[c]{\Delta^2(\lambda,q)-4}}\,\mathrm{d}\lambda \\
&= \frac{1}{\pi}\int_{\Gamma_n} \lambda\lim_{q\rightharpoonup p}\frac{\dot\Delta(\lambda,q)}{\sqrt[c]{\Delta^2(\lambda,q)-4}}\,\mathrm{d}\lambda \\
&= \frac{1}{\pi}\int_{\Gamma_n} \lambda\frac{\dot\Delta(\lambda,p)}{\sqrt[c]{\Delta^2(\lambda,p)-4}}\,\mathrm{d}\lambda \\
&= I_n(p).
\end{aligned}$$

as required. □

Let

$$D_n = \{q \in W : \gamma_n(q) = 0\}$$

be the subvariety of potentials with collapsed n-th gap in the complex neighbourhood W of L_0^2 of Theorem 7.1. This subvariety is analytic, as it may also be written as

$$D_n = \{q \in W : \Delta(\tau_n(q)) = 2(-1)^n,\ \dot\Delta(\tau_n(q)) = 0\}.$$

As I_n and γ_n^2 are analytic on W, their quotient is analytic on $W \setminus D_n$. We show that this quotient extends analytically to all of W to a nonvanishing function.

Theorem 7.3. *There exists a complex neighbourhood W of L_0^2 such that the quotient I_n/γ_n^2 extends analytically to W for all $n \geq 1$ and satisfies*

$$8\pi n \frac{I_n}{\gamma_n^2} = 1 + O\left(\frac{\log n}{n}\right)$$

locally uniformly on W. Moreover,

$$\xi_n = \sqrt[+]{\frac{8I_n}{\gamma_n^2}} = \frac{1}{\sqrt{\pi n}}\left(1 + O\left(\frac{\log n}{n}\right)\right)$$

is well defined as a real analytic, nonvanishing function on W. In particular, at $q = 0$, we have $\xi_n = 1/\sqrt{\pi n}$ for all $n \geq 1$.

Remark. The theorem also implies that the squared gap lengths are real analytic functions on L_0^2.

Proof. Let W be the complex neighbourhood of L_0^2 of the preceding theorem. We show that I_n/γ_n^2 extends continuously from $W \setminus D_n$ to all of W and is weakly analytic when restricted to D_n. By Theorem A.6 we may then conclude that I_n/γ_n^2 is analytic on all of W.

Recall the product expansions

$$\Delta^2(\lambda) - 4 = 4(\lambda_0 - \lambda) \prod_{n \geq 1} \frac{(\lambda_{2n} - \lambda)(\lambda_{2n-1} - \lambda)}{n^4\pi^4},$$

$$\dot{\Delta}(\lambda) = -\prod_{n \geq 1} \frac{\dot{\lambda}_n - \lambda}{n^2\pi^2},$$

where $\dot{\lambda}_n$ are the roots of $\dot{\Delta}$. Their asymptotics are

$$\dot{\lambda}_n = \tau_n + O(\gamma_n^2 \log n/n)$$

by Proposition B.13, where $\tau_n = (\lambda_{2n} + \lambda_{2n-1})/2$. Along the circuit Γ_n we then can write

$$\frac{\dot{\Delta}(\lambda)}{\sqrt[c]{\Delta^2(\lambda) - 4}} = \frac{1}{2\pi n} \frac{\lambda - \dot{\lambda}_n}{\sqrt[s]{(\lambda_{2n} - \lambda)(\lambda - \lambda_{2n-1})}} \chi_n(\lambda)$$

with

$$\chi_n(\lambda) = (-1)^{n-1} \frac{\pi n}{\sqrt[+]{\lambda - \lambda_0}} \prod_{m \neq n} \frac{\dot{\lambda}_m - \lambda}{\sqrt[+]{(\lambda_{2m} - \lambda)(\lambda_{2m-1} - \lambda)}},$$

where the principle branches in the last expression are well defined for λ near G_n. To check the correctness of the sign in these expansions note that χ_n is nonnegative for real q and refer to the definitions (6.8) and (6.10) of the s-root and the c-root.

With the formula for I_n in the preceding proof we then get, for q in $W \setminus D_n$,

$$\begin{aligned} I_n &= \frac{1}{\pi}\int_{\Gamma_n} (\lambda - \dot{\lambda}_n)\frac{\dot{\Delta}(\lambda)}{\sqrt[c]{\Delta^2(\lambda)-4}}\,\mathrm{d}\lambda \\ &= \frac{1}{2\pi^2 n}\int_{\Gamma_n} \frac{(\lambda-\dot{\lambda}_n)^2}{\sqrt[s]{(\lambda_{2n}-\lambda)(\lambda-\lambda_{2n-1})}}\,\chi_n(\lambda)\,\mathrm{d}\lambda \\ &= \frac{\gamma_n^2}{8\pi^2 n}\int_{\Gamma_n'} (\zeta-\delta_n)^2\chi_n(\tau_n+\zeta\gamma_n/2)\,\frac{\mathrm{d}\zeta}{\sqrt[s]{1-\zeta^2}} \end{aligned}$$

upon the substitution $\lambda = \tau_n + \zeta\gamma_n/2$ with $\delta_n = 2(\dot{\lambda}_n - \tau_n)/\gamma_n$, where Γ_n' is some circuit in $\mathbb{C}$ around $[-1, 1]$. Thus,

$$8\pi n\,\frac{I_n}{\gamma_n^2} = \frac{1}{\pi}\int_{\Gamma_n'} (\zeta-\delta_n)^2\chi_n(\tau_n+\zeta\gamma_n/2)\,\frac{\mathrm{d}\zeta}{\sqrt[s]{1-\zeta^2}}$$

on $W \setminus D_n$.

The right hand side of the last identity is continuous on all of W including D_n, since when γ_n tends to 0, also δ_n tends to zero by Lemma B.13, so it tends to

$$\frac{1}{\pi}\int_{\Gamma_n'} \chi_n(\tau_n)\frac{\zeta^2\,\mathrm{d}\zeta}{\sqrt[s]{1-\zeta^2}} = \chi_n(\tau_n)\,\frac{2}{\pi}\int_{-1}^{1}\frac{t^2\,\mathrm{d}t}{\sqrt{1-t^2}} = \chi_n(\tau_n).$$

But χ_n and τ_n are analytic on D_n and even on W. Thus, we may apply Theorem A.6 and conclude that I_n/γ_n^2 extends analytically to all of W.

By Lemma L.2, $\chi_n(\lambda) = 1 + O(\log n/n)$ for λ near G_n locally uniformly on W. Together with the last two displayed identities and in view of the asymptotics of δ_n we thus conclude that

$$8\pi n\,\frac{I_n}{\gamma_n^2} = 1 + O\left(\frac{\log n}{n}\right)$$

locally uniformly on W.

For $q = 0$ we have $\gamma_n = 0$, $\delta_n = 0$, and $\chi_n(\tau_n) = 1$ for all $n \geq 1$. Hence, for the analytic extension of I_n/γ_n^2 we actually have

$$8\pi n\frac{I_n}{\gamma_n^2} = 1, \qquad n \geq 1,$$

at $q = 0$.

Finally, on the real subspace L_0^2, we have locally uniformly as $n \to \infty$

$$\begin{aligned} 0 < 8\pi n\frac{I_n}{\gamma_n^2} &= \frac{2}{\pi}\int_{-1}^{1} (t-\delta_n)^2\chi_n(\tau_n+t\gamma_n/2)\,\frac{\mathrm{d}t}{\sqrt{1-t^2}} \\ &\to \lim_{n\to\infty}\chi_n(\tau_n) = 1. \end{aligned}$$

Therefore, by choosing the complex neighbourhood W sufficiently small we can assure that the real part of nI_n/γ_n^2 is positive and locally uniformly bounded away from zero for all $n \geq 1$. Consequently,

$$\xi_n = \sqrt[+]{\frac{8I_n}{\gamma_n^2}}$$

is well defined, real analytic, and positive for real q in W for all $n \geq 1$. □

8 Angles

Next we define angular coordinates θ_n for potentials q in W. More precisely, the n-th angle θ_n is defined for those q whose n-th gap is not collapsed. The starting point is the following theorem, which we will prove in appendix D and is due to [6] in this form. For an earlier version see [90].

Theorem 8.1. *There exists a complex neighbourhood W of L_0^2 such that for each q in W there exist entire functions ψ_n, $n \geq 1$, satisfying*

$$\frac{1}{2\pi}\int_{\Gamma_m} \frac{\psi_n(\lambda)}{\sqrt[c]{\Delta^2(\lambda)-4}}\,\mathrm{d}\lambda = \delta_{mn} \tag{8.1}$$

for all $m \geq 1$. These functions depend analytically on λ and q and admit a product representation

$$\psi_n(\lambda) = \frac{2}{\pi n}\prod_{m\neq n}\frac{\sigma_m^n - \lambda}{m^2\pi^2},$$

whose complex roots σ_m^n depend real analytically on q and satisfy

$$\sigma_m^n = \tau_m + O\left(\frac{\gamma_m^2}{m}\right)$$

locally uniformly on W and uniformly in n, where $\tau_m = (\lambda_{2m} + \lambda_{2m-1})/2$.

We now turn to the definition of the angles formally derived in section 6. Recall that $D_n = \{q \in W : \gamma_n(q) = 0\}$. For $q \in W \setminus D_n$ define

$$\theta_n(q) = \eta_n(q) + \sum_{k\neq n}\beta_k^n(q),$$

where

$$\eta_n(q) = \int_{\lambda_{2n-1}}^{\mu_n^*}\frac{\psi_n(\lambda)}{\sqrt{\Delta^2(\lambda)-4}}\,\mathrm{d}\lambda \mod 2\pi$$

and

$$\beta_k^n(q) = \int_{\lambda_{2k-1}}^{\mu_k^*} \frac{\psi_n(\lambda)}{\sqrt{\Delta^2(\lambda) - 4}}\,\mathrm{d}\lambda,$$

and where μ_n^* is given by (6.3). As we will show, the paths of integration are arbitrary on the Riemann surface $\Sigma(q)$ as long as their projections onto $\mathbb{C}$ stay inside an isolating neighbourhood of the corresponding interval G_n as described above on page 64. We call such paths *admissible*.

Note that the function η_n is considered as a function on $W \setminus D_n$ taking values in the cylinder $\mathbb{C}/2\pi\mathbb{Z}$ rather than $\mathbb{C}$, whereas the β_k^n are considered as functions taking values in $\mathbb{C}$.

We begin by showing that these functions are well defined in the sense that they are independent of the path of integration. Indeed, the β_k^n are well defined on all of W.

Lemma 8.2. (i) *For any $k \neq n$, the function β_k^n is well defined on all of W and vanishes at $q = 0$.*

(ii) *For $n \geq 1$, the function η_n is well defined on $W \setminus D_n$.*

Proof. Consider β_k^n for $k \neq n$. By the product expansions for $\Delta^2(\lambda) - 4$ and $\psi_n(\lambda)$, we have, for λ near G_k,

$$\frac{\psi_n(\lambda)}{\sqrt{\Delta^2(\lambda) - 4}} = \frac{\sigma_k^n - \lambda}{\sqrt{(\lambda_{2k} - \lambda)(\lambda - \lambda_{2k-1})}}\,\zeta_k^n(\lambda) \tag{8.2}$$

with

$$\zeta_k^n(\lambda) = \frac{\pi n}{(\sigma_n^n - \lambda)\,\sqrt[+]{\lambda - \lambda_0}} \prod_{m \neq k} \frac{\sigma_m^n - \lambda}{\sqrt[+]{(\lambda_{2m} - \lambda)(\lambda_{2m-1} - \lambda)}},$$

where we set $\sigma_n^n = \tau_n$, and where $\sqrt{\Delta^2(\lambda) - 4}$ and $\sqrt{(\lambda_{2k} - \lambda)(\lambda - \lambda_{2k-1})}$ are understood as functions on a neighbourhood around the lift of $[\lambda_{2k-1}, \lambda_{2k}]$ onto the Riemann surface $\Sigma(q)$. If $\gamma_k \neq 0$, then the factor

$$\frac{\sigma_k^n - \lambda}{\sqrt{(\lambda_{2k} - \lambda)(\lambda - \lambda_{2k-1})}}$$

is integrable along any admissible path. If $\gamma_k = 0$, then $\lambda_{2k} = \sigma_k^n = \lambda_{2k-1}$, and this factor equals $\pm\mathrm{i}$. So the integrand of β_k^n is an analytic function on each sheet of $\Sigma(q)$ around $[\lambda_{2k-1}, \lambda_{2k}]$. Consequently, the function β_k^n is well defined in both cases.

The integral is independent of any admissible path of integration, since by (8.1),

$$\int_{\lambda_{2k-1}}^{\lambda_{2k}} \frac{\psi_n(\lambda)}{\sqrt{\Delta^2(\lambda) - 4}}\,\mathrm{d}\lambda = 0, \qquad k \neq n. \tag{8.3}$$

It vanishes when $\mu_k = \lambda_{2k-1}$, so in particular when $q = 0$. This proves part (i).

As to η_n the integral exists along an admissible path as long as $\lambda_{2n-1} \neq \lambda_{2n}$. It is well defined modulo 2π by (8.1) for $k = n$. This proves part (ii). □

Next we prove analyticity.

Lemma 8.3. (i) *The functions* β_k^n, $k \neq n$, *are real analytic on* W.
(ii) *The functions* η_n, $n \geq 1$, *are real analytic on* $W \setminus D_n$, *if taken modulo* π.

Remark. The values of η_n have to be taken modulo π due to the discontinuities of the periodic eigenvalues as functions of q when q is not real, which in turn is due to their lexicographic ordering. Along a curve in $W \setminus D_n$, the eigenvalues λ_{2n} and λ_{2n-1} may jump discontinuously as indicated in figure 4, causing η_n to jump by $\pm\pi$. This, of course, does not happen in the real space $L_0^2 \setminus D_n$. There, η_n *is* real analytic when taken modulo 2π.

Proof. In W consider the two subsets

$$D_k = \{q \in W : \gamma_k(q) = 0\},$$
$$E_k = \{q \in W : \mu_k(q) \in \{\lambda_{2k}(q), \lambda_{2k-1}(q)\}\}.$$

Like the former, the latter is an analytic subvariety of W, since it may also be written in the form

$$E_k = \left\{q \in W : \Delta(\mu_k(q)) = 2(-1)^k\right\}.$$

Our plan is to prove that β_k^n is analytic on $W \setminus (D_k \cup E_k)$ and•continuous on W, and that its restrictions to D_k and E_k are each weakly analytic there. In view of Theorem A.6 it then follows that β_k^n is analytic on W.

Outside of D_k, λ_{2k} and λ_{2k-1} are simple eigenvalues, and locally there exist analytic functions λ_k^+ and λ_k^- such that the sets $\{\lambda_k^+, \lambda_k^-\}$ and $\{\lambda_{2k}, \lambda_{2k-1}\}$ are equal. In view of (8.3) and the substitution $\lambda = \lambda_k^- + z$ we then can write

$$\beta_k^n = \int_{\lambda_k^-}^{\mu_k^*} \frac{\psi_n(\lambda)}{\sqrt{\Delta^2(\lambda) - 4}}\, \mathrm{d}\lambda = \int_0^{\mu_k^* - \lambda_k^-} \frac{\psi_n(\lambda_k^- + z)}{\sqrt{z}\sqrt{D(z)}}\, \mathrm{d}z,$$

where

$$D(z) = \frac{\Delta^2(\lambda_k^- + z) - 4}{z}$$

is analytic near $z = 0$ with $D(0) \neq 0$. If we integrate along an admissible path not going through λ_k^+, then $D(z)$ does not vanish and $\psi_n(\lambda_k^- + z)/\sqrt{D(z)}$ is smooth and locally analytic on $W \setminus (D_k \cup E_k)$. As μ_k^* and λ_k^- are analytic on this set as well, we conclude that the latter integral representing β_k^n is analytic on $W \setminus (D_k \cup E_k)$.

Next we show that β_k^n is weakly analytic when restricted to either D_k or E_k. In view of (8.3), $\beta_k^n \big| E_k = 0$, so on E_k this is obvious. On D_k,

$$\lambda_{2k} = \lambda_{2k-1} = \tau_k = \sigma_k^n,$$

and with (8.2) we can write

$$\beta_k^n = \int_{\tau_k}^{\mu_k^*} \frac{\psi_n(\lambda)}{\sqrt{\Delta^2(\lambda) - 4}}\, \mathrm{d}\lambda = \pm\mathrm{i} \int_{\tau_k}^{\mu_k} \zeta_k^n(\lambda)\, \mathrm{d}\lambda.$$

As μ_k is analytic, this integral is an analytic function on D_k. Altogether we conclude that these restrictions of β_k^n are weakly analytic on D_k and on E_k.

It remains to prove that β_k^n is continuous on all of W. Clearly, β_k^n is continuous on $W \setminus (D_k \cup E_k)$. One shows that it is continuous on $E_k \setminus D_k$ and $D_k \setminus E_k$. The continuity on $D_k \cap E_k$ follows from (8.2) and the estimate

$$\sigma_k^n - \tau_k = O(\gamma_k^2/k).$$

By Theorem A.6, β_k^n is then analytic on W. Obviously, it is real valued on L_0^2.

The proof for η_n is analogous and even simpler, since we only need to consider the domain $W \setminus D_n$. In view of (8.1) we have

$$\int_{\lambda_{2n-1}}^{\lambda_{2n}} \frac{\psi_n(\lambda)}{\sqrt{\Delta^2(\lambda)-4}}\,\mathrm{d}\lambda = \pm\pi \tag{8.4}$$

for the straight line integral, so as above we can write

$$\eta_n = \int_{\lambda_n^-}^{\mu_n^*} \frac{\psi_n(\lambda)}{\sqrt{\Delta^2(\lambda)-4}}\,\mathrm{d}\lambda \quad \mathrm{mod}\ \pi.$$

We conclude that modulo π, the function η_n is analytic outside of $D_n \cup E_n$, and continuous outside D_n. Since η_n mod π vanishes on E_n, it is weakly analytic on E_n, and we are done. □

Lemma 8.4.

$$\beta_k^n = O\left(\frac{|\gamma_k| + |\mu_k - \tau_k|}{k\,|n-k|}\right)$$

uniformly in k and n with $k \neq n$ on bounded subsets of W.

Proof. By (8.3),

$$\beta_k^n = \int_{\lambda_{2k-1}}^{\mu_k^*} \frac{\psi_n(\lambda)}{\sqrt{\Delta^2(\lambda)-4}}\,\mathrm{d}\lambda = \int_{\lambda_{2k}}^{\mu_k^*} \frac{\psi_n(\lambda)}{\sqrt{\Delta^2(\lambda)-4}}\,\mathrm{d}\lambda.$$

The following argument is not affected if one interchanges the roles of λ_{2k-1} and λ_{2k}. Therefore we may assume in the following that $|\mu_k - \lambda_{2k-1}| \le |\mu_k - \lambda_{2k}|$.

For λ near G_k we have

$$\frac{\psi_n(\lambda)}{\sqrt{\Delta^2(\lambda)-4}} = \frac{\sigma_k^n - \lambda}{\sqrt{(\lambda_{2k}-\lambda)(\lambda-\lambda_{2k-1})}}\,\zeta_k^n(\lambda)$$

by equation (8.2). The infinite product in the representation of the function ζ_k^n is uniformly bounded in view of the asymptotic behavior of the σ_m^n and Lemma L.2, so we immediately obtain, uniformly in k and n with $k \neq n$,

$$|\zeta_k^n(\lambda)| = O\left(\frac{n}{k\,|n^2-k^2|}\right) = O\left(\frac{1}{k\,|n-k|}\right)$$

for λ near G_k. Moreover, if we integrate along a straight line ℓ from λ_{2k-1} to μ_k on the sheet of $\Sigma(q)$ determined by μ_k^*, then we have

$$\sqrt{\frac{\sigma_k^n - \lambda}{\lambda_{2k} - \lambda}} = O(1),$$

since $|\mu_k - \lambda_{2k-1}| \le |\mu_k - \lambda_{2k}|$ and $\sigma_k^n = \tau_k + O(\gamma_k^2)$. Thus, it remains to show that

$$\int_{\lambda_{2k-1}}^{\mu_k^*} \sqrt{\frac{\lambda - \sigma_k^n}{\lambda - \lambda_{2k-1}}}\, \mathrm{d}\lambda = O(|\gamma_k| + |\mu_k - \tau_k|),$$

when integrating along the straight line ℓ.

But this follows with the simple substitution $\lambda = \lambda_{2k-1} + t(\mu_k - \lambda_{2k-1})$. Setting $\varepsilon = |\sigma_k^n - \lambda_{2k-1}|$ and $\delta = |\mu_k - \lambda_{2k-1}|$ we obtain the bound

$$\int_0^1 \frac{\sqrt{\varepsilon + \delta}}{\sqrt{\delta}\sqrt{t}}\, \delta\, \mathrm{d}t = 2\sqrt{\varepsilon + \delta}\sqrt{\delta} \le \varepsilon + 2\delta.$$

As $\varepsilon = O(|\gamma_k|)$ and $\delta = O(|\gamma_k| + |\mu_k - \tau_k|)$, the claim follows. □

It is immediate by the preceding estimates that the sum $\sum_{k \ne n} \beta_k^n$ converges locally uniformly on W to a real analytic function on W, which is locally uniformly of order $1/n$. Altogether we obtain the following result.

Theorem 8.5. *The function*

$$\beta_n = \sum_{k \ne n} \beta_k^n$$

is real analytic on W and locally uniformly of order $O(1/n)$. The angle function

$$\theta_n = \eta_n + \sum_{k \ne n} \beta_k^n$$

is real analytic on the real space $L_0^2 \setminus D_n$ and extends to a real analytic function on $W \setminus D_n$ when taken modulo π. The gradients of β_k^n, η_n and θ_n all have mean value zero.

Proof. It remains to prove the statement about the mean value. Inspecting the formula for β_k^n one verifies that β_k^n, η_n and hence θ_n are invariant under translations of the potential when considered on $L^2 \setminus D_n$. Hence, their gradients have mean value zero by (6.11). □

9 Cartesian Coordinates

In the preceding two sections we defined actions and angles,

$$I_n = \frac{1}{8}\xi_n^2\gamma_n^2, \qquad \theta_n = \eta_n + \sum_{k\neq n} \beta_k^n,$$

for potentials q in L_0^2 and $L_0^2 \smallsetminus D_n$, respectively. We showed that these are real analytic on a complex neighbourhood W of L_0^2 and on $W \smallsetminus D_n$, respectively. We did not show yet, however, that they are *coordinates*, nor did we show that they are canonical variables. This will be done in sections 11 and 12.

Here we introduce and study the associated rectangular variables x_n and y_n. As usual, on the real space $L_0^2 \smallsetminus D_n$ they are defined as

$$x_n = \sqrt{2I_n}\cos\theta_n, \qquad y_n = \sqrt{2I_n}\sin\theta_n,$$

where the choice of sin and cos is made so that $\mathrm{d}x_n \wedge \mathrm{d}y_n = \mathrm{d}I_n \wedge \mathrm{d}\theta_n$. This definition extends to the complex domain $W \smallsetminus D_n$ by setting

$$x_n = \frac{\xi_n\gamma_n}{2}\cos\theta_n, \qquad y_n = \frac{\xi_n\gamma_n}{2}\sin\theta_n.$$

The main result of this section is that these functions are in fact well defined and *real analytic* on all of W. This holds even in those points, where the angles θ_n are not well defined.

To attack this problem, recall that the functions ξ_n and $\beta_n = \theta_n - \eta_n$ have already been shown to be real analytic on W. Therefore we focus our attention on the complex functions

$$z_n^{\pm} = \gamma_n \mathrm{e}^{\pm\mathrm{i}\eta_n},$$

which so far are defined on $W \smallsetminus D_n$. We first show that $z_n^{\pm}$ are analytic on $W \smallsetminus D_n$. This is not entirely obvious, since γ_n has discontinuities.

Lemma 9.1. *The functions $z_n^{\pm} = \gamma_n \mathrm{e}^{\pm\mathrm{i}\eta_n}$ are analytic on $W \smallsetminus D_n$.*

Proof. Locally around every point in $W \smallsetminus D_n$ there exist analytic functions λ_n^+ and λ_n^- such that as sets,

$$\{\lambda_n^-(q), \lambda_n^+(q)\} = \{\lambda_{2n-1}(q), \lambda_{2n}(q)\}.$$

Let

$$\gamma_n^- = \lambda_n^+ - \lambda_n^-, \qquad \eta_n^- = \int_{\lambda_n^-}^{\mu_n^*} \frac{\psi_n(\lambda)}{\sqrt{\Delta^2(\lambda)-4}}\,\mathrm{d}\lambda.$$

Depending on whether $\lambda_n^- = \lambda_{2n-1}$ or $\lambda_n^- = \lambda_{2n}$, respectively, we then have

$$\gamma_n = \gamma_n^-, \qquad \eta_n = \eta_n^-,$$

or

$$\gamma_n = -\gamma_n^-, \qquad \eta_n = \int_{\lambda_n^+}^{\mu_n^*} \frac{\psi_n(\lambda)}{\sqrt{\Delta^2(\lambda)-4}}\,\mathrm{d}\lambda = \eta_n^- + \pi \bmod 2\pi$$

in view of (8.4). In either case we obtain

$$\gamma_n \mathrm{e}^{\pm \mathrm{i}\eta_n} = \gamma_n^- \mathrm{e}^{\pm \mathrm{i}\eta_n^-}.$$

The right hand side is analytic, which proves the lemma. □

In analogy with the representation (8.2) we write

$$\frac{\psi_n(\lambda)}{\sqrt{\Delta^2(\lambda)-4}} = \frac{\zeta_n(\lambda)}{\sqrt{(\lambda_{2n}-\lambda)(\lambda-\lambda_{2n-1})}} \tag{9.1}$$

for λ near G_n, where

$$\zeta_n(\lambda) = (-1)^{n-1} \frac{\pi n}{\sqrt[+]{\lambda-\lambda_0}} \prod_{m\neq n} \frac{\sigma_m^n - \lambda}{\sqrt[+]{(\lambda_{2m}-\lambda)(\lambda_{2m-1}-\lambda)}},$$

and the two roots in (9.1) are understood as functions on a neighbourhood of the lift of $[\lambda_{2n-1}, \lambda_{2n}]$ onto $\Sigma(q)$ related to each other by this identity. The principal branches in the last identity are well defined for λ near G_n.

Lemma 9.2. *For $\mu \in G_n$,*

$$\zeta_n(\mu) = 1 + O\left(\frac{|\gamma_n|}{n}\right)$$

locally uniformly on W and uniformly in $n \geq 1$.

Proof. First let q be real with $\gamma_n > 0$. In view of Theorem 8.1 and the definition of the c-root in (6.10) we have

$$(-1)^{n-1} \int_{\lambda_{2n-1}}^{\lambda_{2n}} \frac{\psi_n(\lambda)}{\sqrt[+]{\Delta^2(\lambda)-4}}\,\mathrm{d}\lambda = \pi$$

for the straight line integral from λ_{2n-1} to λ_{2n}. On this line, both $(-1)^{n-1}\psi_n$ and ζ_n are positive. With (9.1) we thus obtain, for every $\mu \in G_n$,

$$\begin{aligned}
\pi &= \int_{\lambda_{2n-1}}^{\lambda_{2n}} \frac{\zeta_n(\lambda)}{\sqrt[+]{(\lambda_{2n}-\lambda)(\lambda-\lambda_{2n-1})}}\,\mathrm{d}\lambda \\
&= \int_{\lambda_{2n-1}}^{\lambda_{2n}} \frac{\zeta_n(\mu)}{\sqrt[+]{(\lambda_{2n}-\lambda)(\lambda-\lambda_{2n-1})}}\,\mathrm{d}\lambda \\
&\qquad + \int_{\lambda_{2n-1}}^{\lambda_{2n}} \frac{\zeta_n(\lambda)-\zeta_n(\mu)}{\sqrt[+]{(\lambda_{2n}-\lambda)(\lambda-\lambda_{2n-1})}}\,\mathrm{d}\lambda \\
&= \pi\zeta_n(\mu) + O\big(\sup_{\lambda\in G_n} |\zeta_n(\lambda)-\zeta_n(\mu)|\big).
\end{aligned}$$

Hence,

$$\zeta_n(\mu) = 1 + O\big(\sup_{\lambda \in G_n} |\zeta_n(\lambda) - \zeta_n(\mu)|\big)$$

for $\mu \in G_n$.

In view of Proposition L.2, ζ_n is uniformly bounded on a neighbourhood of radius $O(n)$ around G_n. This bound holds uniformly in n and locally uniformly on W, no matter if $\gamma_n \neq 0$ or not. Therefore, by Cauchy's estimate,

$$|\zeta_n(\lambda) - \zeta_n(\mu)| \leq \frac{M}{n} |\lambda - \mu| \leq \frac{M}{n} |\gamma_n|$$

for $\lambda, \mu \in G_n$ with a uniform constant M. This proves the claim for real q.

For complex q in W the preceding identities remain true at least up to a sign. Hence, by the continuity of ζ_n in λ and q, the claimed statement must also be valid for complex q . □

We now investigate the limiting behavior of $z_n^\pm$ as the n-th gap collapses. This limit is different from zero when the limit potential is in the open set

$$X_n = \{q \in W : \mu_n(q) \notin G_n(q)\}.$$

This set does not intersect the real space L_0^2, since $\mu_n \in [\lambda_{2n-1}, \lambda_{2n}]$ for real q.

Let

$$\chi_n(q) = \int_{\tau_n}^{\mu_n} \frac{\zeta_n(\lambda) - \zeta_n(\tau_n)}{\lambda - \tau_n} \, \mathrm{d}\lambda.$$

The integral exists due to the analyticity of ζ_n in λ and is analytic in q. To facilitate the statement of the following result we define a sign $\varepsilon_n = \pm 1$ for potentials q in X_n in such a way that

$$\frac{\psi_n(\mu_n)}{\sqrt[*]{\Delta^2(\mu_n) - 4}} = \frac{\varepsilon_n \zeta_n(\mu_n)}{\sqrt[s]{(\lambda_{2n} - \mu_n)(\mu_n - \lambda_{2n-1})}}. \tag{9.2}$$

Note that the sign of the s-root is well defined, since $\mu_n \notin G_n$ for $q \in X_n$.

Lemma 9.3. *As $q \notin D_n$ tends to $p \in D_n \cap X_n$,*

$$\gamma_n \mathrm{e}^{\pm \mathrm{i} \eta_n} \to -2(1 \pm \varepsilon_n)(\mu_n - \tau_n) \mathrm{e}^{\pm \varepsilon_n \chi_n},$$

where ε_n is defined in (9.2).

Proof. As X_n is open and $p \in X_n \cap D_n$, we have $q \in X_n$ for all q sufficiently near p. Also, $q \notin D_n$ by assumption.

For $q \in X_n \setminus D_n$ we can write, modulo 2π,

$$\begin{aligned} \eta_n &= \int_{\lambda_{2n-1}}^{\mu_n^*} \frac{\psi_n(\lambda)}{\sqrt{\Delta^2(\lambda) - 4}} \, \mathrm{d}\lambda \\ &= \varepsilon_n \int_{\lambda_{2n-1}}^{\mu_n} \frac{\zeta_n(\lambda)}{\sqrt[s]{(\lambda_{2n} - \lambda)(\lambda - \lambda_{2n-1})}} \, \mathrm{d}\lambda \end{aligned}$$

$$
\begin{aligned}
&= \varepsilon_n \int_{\lambda_{2n-1}}^{\mu_n} \frac{\zeta_n(\lambda_{2n-1})}{\sqrt[s]{(\lambda_{2n}-\lambda)(\lambda-\lambda_{2n-1})}}\,\mathrm{d}\lambda \\
&\quad + \varepsilon_n \int_{\lambda_{2n-1}}^{\mu_n} \frac{\zeta_n(\lambda)-\zeta_n(\lambda_{2n-1})}{\sqrt[s]{(\lambda_{2n}-\lambda)(\lambda-\lambda_{2n-1})}}\,\mathrm{d}\lambda \\
&= \eta_n' + \eta_n'' \mod 2\pi.
\end{aligned}
$$

The limiting behavior of the second term η_n'' is straightforward. If $q \to p$, then $\gamma_n \to 0$ and so for $\lambda \neq \tau_n$,

$$
\sqrt[s]{(\lambda_{2n}-\lambda)(\lambda-\lambda_{2n-1})} \to \mathrm{i}(\lambda-\tau_n)
$$

by (6.9). Hence, by the definition of χ_n,

$$
\mathrm{i}\eta_n'' \to \varepsilon_n \int_{\tau_n}^{\mu_n} \frac{\zeta_n(\lambda)-\zeta_n(\tau_n)}{\lambda-\tau_n}\,\mathrm{d}\lambda = \varepsilon_n \chi_n(p) \mod 2\pi.
$$

Consequently,

$$
\mathrm{e}^{\mathrm{i}\eta_n''} \to \mathrm{e}^{\varepsilon_n \chi_n(p)} \quad \text{as} \quad q \to p.
$$

Now consider η_n'. By the substitution $\lambda = \tau_n + z\gamma_n/2$ and the definition (6.8),

$$
\int_{\lambda_{2n-1}}^{\mu_n} \frac{\mathrm{d}\lambda}{\sqrt[s]{(\lambda_{2n}-\lambda)(\lambda-\lambda_{2n-1})}} = \int_{-1}^{\rho_n} \frac{\mathrm{d}z}{\sqrt[s]{1-z^2}} = \phi(\rho_n)
$$

with

$$
\rho_n = \frac{\mu_n-\tau_n}{\gamma_n/2}, \qquad \phi(w) = \int_{-1}^{w} \frac{\mathrm{d}z}{\sqrt[s]{1-z^2}}.
$$

It follows that $\mathrm{e}^{\mathrm{i}\phi(w)} = -w + \mathrm{i}\sqrt[s]{1-w^2}$, as both sides are analytic, univalent functions on $\mathbb{C}\setminus[-1,1]$, which have the same limit at -1 and satisfy the same differential equation

$$
\frac{f'(w)}{f(w)} = \frac{\mathrm{i}}{\sqrt[s]{1-w^2}}.
$$

Hence, writing

$$
\begin{aligned}
\exp(\mathrm{i}\eta_n') &= \exp(\mathrm{i}\varepsilon_n\phi(\rho_n)\zeta_n(\lambda_{2n-1})) \\
&= \exp(\mathrm{i}\phi(\rho_n)\varepsilon_n)\exp(\mathrm{i}\varepsilon_n\phi(\rho_n)\hat{\zeta}_n)
\end{aligned}
$$

with $\hat{\zeta}_n = \zeta_n(\lambda_{2n-1}) - 1$ we obtain for $q \in X_n \setminus D_n$

$$
\gamma_n \mathrm{e}^{\mathrm{i}\eta_n'} = \gamma_n \left(-\rho_n + \mathrm{i}\sqrt[s]{1-\rho_n^2}\right)^{\varepsilon_n} \mathrm{e}^{\mathrm{i}\varepsilon_n\phi(\rho_n)\hat{\zeta}_n}.
$$

Now let $q \to p$. Then $\gamma_n \to 0$, while $v_n = \mu_n - \tau_n$ tends to a limit different from zero, and hence $|\rho_n| \to \infty$. For the first factor on the right hand side of the above equation we then obtain by (6.6)

$$\begin{aligned}\gamma_n\left(-\rho_n+\mathrm{i}\sqrt[s]{1-\rho_n^2}\right)^{\varepsilon_n} &= \gamma_n\left(-\rho_n+\mathrm{i}\varepsilon_n\sqrt[s]{1-\rho_n^2}\right)\\ &= \gamma_n\left(-\rho_n-\varepsilon_n\rho_n\sqrt[+]{1-\rho_n^{-2}}\right)\\ &= -2\upsilon_n-2\upsilon_n\varepsilon_n\sqrt[+]{1-\rho_n^{-2}}\\ &\to -2\upsilon_n(1+\varepsilon_n).\end{aligned}$$

As to the second factor we observe that for $|\rho_n|$ large,

$$\begin{aligned}|\phi(\rho_n)| &= \left|\int_{-1}^{\rho_n}\frac{\mathrm{d}z}{\sqrt[s]{1-z^2}}\right|\\ &\le c+\int_1^{|\rho_n|}\frac{\mathrm{d}t}{\sqrt{t-1}\sqrt{t+1}}\\ &= O\left(\sqrt{|\rho_n|}\right).\end{aligned} \tag{9.3}$$

Since $\hat{\zeta}_n = O(|\gamma_n|/n)$ by Lemma 9.2 we thus conclude that $\phi(\rho_n)\hat{\zeta}_n \to 0$ and so

$$\mathrm{e}^{\mathrm{i}\varepsilon_n\phi(\rho_n)\hat{\zeta}_n} \to 1 \quad \text{as} \quad q\to p.$$

Together with the result for $\mathrm{e}^{\mathrm{i}\eta_n''}$ we thus arrive at

$$\gamma_n\mathrm{e}^{\mathrm{i}\eta_n} \to -2\upsilon_n(1+\varepsilon_n)\mathrm{e}^{\varepsilon_n\chi_n(p)}$$

as claimed. The limit of $\gamma_n\mathrm{e}^{-\mathrm{i}\eta_n}$ is a simple variation of the preceding argument. □

In view of the preceding result it is natural to extend the functions $z_n^{\pm}$ to D_n by defining

$$z_n^{\pm} = \begin{cases} -2(1\pm\varepsilon_n)(\mu_n-\tau_n)\mathrm{e}^{\pm\varepsilon_n\chi_n} & \text{on } D_n\cap X_n,\\ 0 & \text{on } D_n\setminus X_n.\end{cases} \tag{9.4}$$

The main result is then the following.

Theorem 9.4. *The functions $z_n^{\pm} = \gamma_n\mathrm{e}^{\pm\mathrm{i}\eta_n}$, as extended above, are analytic on W. Moreover,*

$$z_n^{\pm} = O(|\gamma_n|+|\mu_n-\tau_n|)$$

locally uniformly on W.

Proof. We apply Theorem A.6 to the function $z_n^{\pm}$ on the domain W with the analytic subvariety D_n.

By Lemma 9.1 these functions are analytic on $W\setminus D_n$. By a simple inspection of the formula for $z_n^{\pm}$ they are also continuous in every point of $D_n\cap X_n$ and of $D_n\setminus X_n$. Thus, the functions $z_n^{\pm}$ are continuous on all of W.

It remains to show that they are weakly analytic when restricted to D_n. Let D be a one-dimensional complex disc contained in D_n. If the center of D is in X_n, then the entire disc D is in X_n, if chosen sufficiently small. The analyticity of $z_n^\pm = \gamma_n e^{\pm i\eta_n}$ on D is then evident from formula (9.4), the definition of χ_n, and the local constancy of ε_n on X_n.

If the center of D does not belong to X_n, then consider the analytic function $\mu_n - \tau_n$ on D. This function either vanishes identically on D, in which case $z_n^\pm$ vanishes identically, too. Or it vanishes in only finitely many points. Outside these points, D is in X_n, hence $z_n^\pm$ is analytic there. By continuity and analytic continuation, these functions are analytic on all of D.

We thus have shown that $z_n^\pm$ are analytic on D. The result now follows with Theorem A.6.

To prove the claimed estimate we continue with the notations of the previous proof and recall that on $X_n \setminus D_n$,

$$\gamma_n e^{i\eta_n} = \gamma_n \left(-\rho_n + i\varepsilon_n \sqrt[s]{1-\rho_n^2}\right) e^{i\varepsilon_n \phi(\rho_n)\hat{\zeta}_n} e^{i\eta_n''}$$

with $\rho_n = 2(\mu_n - \tau_n)/\gamma_n$ and

$$\eta_n'' = \varepsilon_n \int_{\lambda_{2n-1}}^{\mu_n} \frac{\zeta_n(\lambda) - \zeta_n(\lambda_{2n-1})}{\sqrt[s]{(\lambda_{2n} - \lambda)(\lambda - \lambda_{2n-1})}} \,\mathrm{d}\lambda \mod 2\pi.$$

In view of the analyticity and local uniform boundedness of ζ_n by Lemma 9.2 we have $e^{i\eta_n''} = O(1)$ locally uniformly. Similarly, $\phi(\rho_n)\hat{\zeta}_n = O(1)$ locally uniformly in view of (9.3). Finally, for $|\rho_n| \le 1$ we have

$$\left|\gamma_n \left(-\rho_n \pm i \sqrt[s]{1-\rho_n^2}\right)\right| \le 2\,|\gamma_n|,$$

while for $|\rho_n| > 1$, using (6.8) and $\gamma_n \rho_n = 2(\mu_n - \tau_n)$,

$$\begin{aligned}\left|\gamma_n \left(-\rho_n \pm i \sqrt[s]{1-\rho_n^2}\right)\right| &= 2\,|\mu_n - \tau_n| \left|1 \pm \sqrt[+]{1-\rho_n^{-2}}\right| \\ &\le 6\,|\mu_n - \tau_n|.\end{aligned}$$

Altogether, on $X_n \setminus D_n$,

$$\gamma_n \left(-\rho_n + i\varepsilon_n \sqrt[s]{1-\rho_n^2}\right) \le 6(|\gamma_n| + |\mu_n - \tau_n|).$$

This establishes the estimate for $z_n = \gamma_n e^{i\eta_n}$ on $X_n \setminus D_n$. By continuity, it extends in a locally uniform fashion to all of W. The argument for $\gamma_n e^{-i\eta_n}$ is completely analogous. □

Before returning to the definition of x_n and y_n, we determine the gradients of $z_n^\pm$ at points in $L_0^2 \cap D_n$, which exist by analyticity. For the definition of the normalized Dirichlet eigenfunctions g_n, see appendix B.

Theorem 9.5. *At a point in $L_0^2 \cap D_n$,*

$$\partial z_n^{\pm} = h_n^2 - g_n^2 \pm 2\mathrm{i} g_n h_n,$$

where g_n denotes the normalized eigenfunction for the Dirichlet eigenvalue μ_n and h_n that periodic eigenfunction orthonormal to g_n which is positive at 0. These gradients are in $\mathcal{H}^2$, have mean value zero and satisfy the asymptotic estimates

$$\partial z_n^{\pm} = 2\cos 2\pi n x \pm 2\mathrm{i}\sin 2\pi n x + O\left(\frac{1}{n}\right).$$

These hold locally uniformly on L_0^2 in the sense that at each point q only those n with $\gamma_n(q) = 0$ are taken into account. At $q = 0$,

$$\partial z_n^{\pm} = 2\cos 2\pi n x \pm 2\mathrm{i}\sin 2\pi n x$$

for all $n \geq 1$.

Proof. Consider z_n^+. To compute its gradient at $p \in L_0^2 \cap D_n$, we approximate p by potentials q in

$$B_n = \left\{ q \in L_0^2 \setminus D_n : \mu_n = \tau_n \text{ and sign } \sqrt[*]{\Delta^2(\mu_n) - 4} = (-1)^{n-1} \right\}.$$

This set is not empty by Theorem B.14. For such q we have

$$\frac{\psi_n(\mu_n)}{\sqrt[*]{\Delta^2(\mu_n) - 4}} = \frac{\zeta_n(\mu_n)}{\sqrt[+]{(\lambda_{2n} - \mu_n)(\mu_n - \lambda_{2n-1})}},$$

as sign $\psi_n(\mu_n) = (-1)^{n-1}$ and sign $\zeta_n(\mu_n) = 1$.

Going through the calculations in the proof of Lemma 9.3 with the latter expression in place of (9.2) we then have

$$\eta_n = \int_{\lambda_{2n-1}}^{\mu_n} \frac{\zeta_n(\lambda)}{\sqrt[+]{(\lambda_{2n} - \lambda)(\lambda - \lambda_{2n-1})}} \, \mathrm{d}\lambda \mod 2\pi.$$

Further on, all arguments remain valid for $\lambda \in G_n$, if we set $\varepsilon_n = 1$ and replace all s-roots by positive roots. So as before we can write $z_n^+ = \gamma_n \mathrm{e}^{\mathrm{i}\eta_n} = \gamma_n \mathrm{e}^{\mathrm{i}\eta_n'} \cdot \mathrm{e}^{\mathrm{i}\eta_n''}$, where both factors on the right hand side are real analytic on W. It follows that

$$\partial z_n^+ = \lim_{B_n \ni q \to p} \partial\left(\gamma_n \mathrm{e}^{\mathrm{i}\eta_n'}\right),$$

since $\gamma_n \mathrm{e}^{\mathrm{i}\eta_n'} = 0$ for $p \in L_0^2 \cap D_n$ by the preceding theorem and $\mathrm{e}^{\mathrm{i}\eta_n''} = 1$. Moreover, with $v_n = \mu_n - \tau_n = \rho_n\gamma_n/2$ and $\hat{\zeta}_n = \zeta_n(\lambda_{2n-1}) - 1$,

$$\begin{aligned}
\gamma_n \mathrm{e}^{\mathrm{i}\eta_n'} &= \gamma_n \left(-\rho_n + \mathrm{i}\sqrt[+]{1 - \rho_n^2}\right)^{\zeta_n(\lambda_{2n-1})} \\
&= \left(-2v_n + \mathrm{i}\gamma_n \sqrt[+]{1 - \rho_n^2}\right) \cdot \left(-\rho_n + \mathrm{i}\sqrt[+]{1 - \rho_n^2}\right)^{\hat{\zeta}_n}.
\end{aligned}$$

The gradients of both factors have a limit as $q \to p$ and the product rule can be applied. For $q \in B_n$, we have $\mu_n = \tau_n$ and hence $v_n = 0$ as well as $\rho_n = 0$. Thus, in the limit the first factor vanishes, while the second converges to 1 by Lemma 9.2. As a consequence the identity

$$\begin{aligned}\partial z_n^+ &= \lim_{B_n \ni q \to p} \partial\left(\gamma_n \mathrm{e}^{\mathrm{i}\eta_n'}\right) \\ &= \lim_{B_n \ni q \to p} \partial\left(-2v_n + \mathrm{i}\gamma_n \sqrt[+]{1-\rho_n^2}\right) \\ &= 2(\partial\tau_n - \partial\mu_n) + \mathrm{i} \lim_{B_n \ni q \to p} \partial\gamma_n \\ &= 2(\partial\tau_n - \partial\mu_n) + \mathrm{i} \lim_{B_n \ni q \to p} f_{2n}^2 - f_{2n-1}^2\end{aligned}$$

holds.

By what we have just proven, $\lim_{B_n \ni q \to p} f_{2n}^2 - f_{2n-1}^2$ exists. By the next lemma below, f_{2n} and f_{2n-1} converge individually to solutions of the corresponding differential equation and thus to periodic eigenfunctions of p, again denoted f_{2n} and f_{2n-1}. They are orthonormal and satisfy $\langle f_{2n}, g_n \rangle > 0 > \langle f_{2n-1}, g_n \rangle$.

For the limiting eigenfunctions we therefore must have

$$\begin{aligned}f_{2n} &= \alpha h_n + \beta g_n, \\ f_{2n-1} &= \beta h_n - \alpha g_n,\end{aligned} \tag{9.5}$$

with $\alpha, \beta > 0$ and $\alpha^2 + \beta^2 = 1$, and g_n and h_n as stated in the theorem. Moreover, inserting this representation into equation (9.6), we obtain $\beta/\alpha = \alpha/\beta$, or $\alpha = \beta$. Thus we have

$$\begin{aligned}f_{2n}^2 - f_{2n-1}^2 &= 2g_n h_n, \\ 2\partial\tau_n = f_{2n}^2 + f_{2n-1}^2 &= g_n^2 + h_n^2,\end{aligned}$$

while $2\partial\mu_n = 2g_n^2$ as usual. Altogether,

$$\partial z_n^+ = h_n^2 - g_n^2 + 2\mathrm{i} g_n h_n.$$

Clearly, this gradient has mean value zero and is in $\mathcal{H}^2$.

With the asymptotic formulas $g_n = \sqrt{2}\sin \pi n x + O(1/n)$ and, by orthogonality and chosen normalization, $h_n = \sqrt{2}\cos \pi n x + O(1/n)$, the stated asymptotic formula for ∂z_n^+ then follows with standard trigonometric identities. The formula for ∂z_n^- is obtained analogously. Moreover, at $q = 0$ these formulas hold without the error terms. □

Lemma 9.6. *As $q \in B_n$ tends to $p \in L_0^2 \cap D_n$, the periodic eigenfunctions f_{2n} and f_{2n-1} of q do converge to normalized eigenfunctions of p, denoted by the same symbols, such that in the limit $\langle f_{2n}, g_n \rangle > 0 > \langle f_{2n-1}, g_n \rangle$ and*

$$\frac{\langle f_{2n}, g_n \rangle}{\langle f_{2n-1}, g_n \rangle} = -\frac{f_{2n}(0)}{f_{2n-1}(0)}. \tag{9.6}$$

Proof. As q tends to p, the initial values of the normalized eigenfunctions f_{2n} and f_{2n-1} remain bounded in $\mathbb{R}^2$, since each eigenfunction is represented in terms of its initial data and the fundamental solution y_1, y_2 of $-y'' + qy = \lambda y$, and the angle between y_1 and y_2 as well as their L^2-norms are locally uniformly bounded away from zero. Hence we can always choose a convergent subsequence of these initial values. As the fundamental solution depends analytically on λ and q, the corresponding eigenfunctions then converge uniformly to some eigenfunctions of p, which we temporarily denote by $\bar{f}_{2n}$ and $\bar{f}_{2n-1}$. We show that these limiting eigenfunctions are indeed uniquely determined. This will establish the convergence of the *entire* sequence.

We observed that $\lim_{B_n \ni q \to p} f_{2n}^2 - f_{2n-1}^2$ exists. By the analyticity of τ_n, also $\lim_{B_n \ni q \to p} f_{2n}^2 + f_{2n-1}^2$ exists. Hence, also the individual limits of f_{2n}^2 and f_{2n-1}^2 exist, which fixes $\bar{f}_{2n}$ and $\bar{f}_{2n-1}$ uniquely up to sign. This sign is also fixed, once we show that

$$\bar{f}_{2n}(0) \neq 0,$$

since $\bar{f}_{2n}(0) \geq 0$ by normalization. The same argument applies to $\bar{f}_{2n-1}(0)$.

To establish that $\bar{f}_{2n}(0)$ does not vanish we observe that

$$\begin{aligned}(\lambda_{2n} - \mu_n) \langle f_{2n}, g_n \rangle &= \langle \lambda_{2n} f_{2n}, g_n \rangle - \langle f_{2n}, \mu_n g_n \rangle \\ &= f_{2n} g_n' \big|_0^1 \\ &= f_{2n}(0) \left((-1)^n g_n'(1) - g_n'(0) \right)\end{aligned}$$

by the differential equation and partial integration. As we will show in a moment,

$$\lim_{B_n \ni q \to p} \frac{(-1)^n g_n'(1) - g_n'(0)}{\lambda_{2n} - \mu_n} = \kappa_n > 0, \tag{9.7}$$

that is, the limit exists and is positive. Hence, in the limit we have

$$\langle \bar{f}_{2n}, g_n \rangle = \kappa_n \bar{f}_{2n}(0).$$

But then $\bar{f}_{2n}(0)$ can not be zero, since otherwise, $\bar{f}_{2n}$ would be a nontrivial multiple of g_n, and the left hand side could not be zero.

By an analogous calculation,

$$(\lambda_{2n-1} - \mu_n) \langle f_{2n-1}, g_n \rangle = f_{2n-1}(0) \left((-1)^n g_n'(1) - g_n'(0) \right).$$

Taking the quotient with the similar equation for f_{2n} and observing that by construction $\lambda_{2n} - \mu_n = \mu_n - \lambda_{2n-1}$ for $q \in B_n$, we also obtain (9.6).

It remains to prove (9.7), or equivalently,

$$\lim_{B_n \ni q \to p} \frac{(-1)^n y_2'(1, \mu_n) - 1}{\lambda_{2n} - \mu_n} > 0.$$

Use $\sqrt[*]{\Delta^2(\mu_n) - 4} = y_1(1, \mu_n) - y_2'(1, \mu_n)$ to get for $q \in B_n$

$$\begin{aligned} 2y_2'(1, \mu_n) &= \Delta(\mu_n) - \sqrt[*]{\Delta^2(\mu_n) - 4} \\ &= \Delta(\mu_n) + (-1)^n \sqrt[+]{\Delta^2(\mu_n) - 4}. \end{aligned}$$

As $2 = (-1)^n \Delta(\lambda_{2n})$, it thus remains to prove that

$$\lim_{B_n \ni q \to p} \frac{1}{\lambda_{2n} - \mu_n} \left((-1)^n (\Delta(\mu_n) - \Delta(\lambda_{2n})) + \sqrt[+]{\Delta^2(\mu_n) - \Delta^2(\lambda_{2n})} \right) > 0.$$

Concerning the first term in the sum,

$$\frac{\Delta(\mu_n) - \Delta(\lambda_{2n})}{\mu_n - \lambda_{2n}} \to \dot{\Delta}(\lambda_{2n}) = 0,$$

as we are at a double eigenvalue. Concerning the second term in the sum, write

$$\begin{aligned} \Delta^2(\mu_n) - \Delta^2(\lambda_{2n}) &= -2 \int_{\mu_n}^{\lambda_{2n}} \Delta(\lambda) \dot{\Delta}(\lambda) \, \mathrm{d}\lambda \\ &= -2 \int_{\mu_n}^{\lambda_{2n}} \Delta(\lambda) \int_{\dot{\lambda}_n}^{\lambda} \ddot{\Delta}(\mu) \, \mathrm{d}\mu \, \mathrm{d}\lambda, \end{aligned}$$

where $\dot{\lambda}_n$ is the unique root of $\dot{\Delta}$ in the n-th gap. With $\dot{\lambda}_n = \tau_n + o(\gamma_n^2)$ it follows that

$$\frac{\Delta^2(\mu_n) - \Delta^2(\lambda_{2n})}{(\mu_n - \lambda_{2n})^2} \to -\Delta(\lambda_{2n}) \ddot{\Delta}(\lambda_{2n}) > 0,$$

which proves our claim. □

Let us now define the cartesian coordinates

$$\begin{aligned} x_n &= \sqrt{2I_n} \cos \theta_n = \xi_n \gamma_n \frac{\mathrm{e}^{\mathrm{i}\theta_n} + \mathrm{e}^{-\mathrm{i}\theta_n}}{4}, \\ y_n &= \sqrt{2I_n} \sin \theta_n = \xi_n \gamma_n \frac{\mathrm{e}^{\mathrm{i}\theta_n} - \mathrm{e}^{-\mathrm{i}\theta_n}}{4\mathrm{i}}. \end{aligned}$$

Expressed in terms of $z_n^{\pm}$ this is equivalent with

$$\begin{aligned} x_n &= \frac{\xi_n}{4} \left(z_n^+ \mathrm{e}^{\mathrm{i}\beta_n} + z_n^- \mathrm{e}^{-\mathrm{i}\beta_n} \right), \\ y_n &= \frac{\xi_n}{4\mathrm{i}} \left(z_n^+ \mathrm{e}^{\mathrm{i}\beta_n} - z_n^- \mathrm{e}^{-\mathrm{i}\beta_n} \right). \end{aligned}$$

The latter expressions apply to all complex potentials in W, while the former expressions apply to potentials in $W \setminus D_n$ only.

Theorem 9.7. *The functions x_n and y_n are real analytic on W with gradients having mean value zero. At a point in $L_0^2 \cap D_n$ their gradients are*

$$\begin{pmatrix} \partial x_n \\ \partial y_n \end{pmatrix} = \frac{\xi_n}{2} \begin{pmatrix} \cos\beta_n & -\sin\beta_n \\ \sin\beta_n & \cos\beta_n \end{pmatrix} \begin{pmatrix} h_n^2 - g_n^2 \\ 2h_n g_n \end{pmatrix}.$$

They are in $\mathcal{H}^2$ and satisfy the asymptotic estimates

$$\begin{pmatrix} \partial x_n \\ \partial y_n \end{pmatrix} = \frac{1}{\sqrt{\pi n}} \begin{pmatrix} \cos 2\pi n x \\ \sin 2\pi n x \end{pmatrix} + O\left(\frac{\log n}{n^{3/2}}\right)$$

locally uniformly on L_0^2 in the sense that at each point q only those n with $\gamma_n(q) = 0$ are taken into account. At $q = 0$ the above formulas hold without the error terms for all $n \geq 1$.

Proof. Analyticity of x_n and y_n follows from Theorems 7.3 and 8.5 and what we have shown above. On $L_0^2 \smallsetminus D_n$ their gradients are given in terms of ∂I_n and $\partial\theta_n$, hence have mean value zero by Theorems 7.1 and again 8.5.

As to their gradients on $L_0^2 \cap D_n$, we have

$$x_n + \mathrm{i} y_n = \frac{\xi_n}{2} z_n^+ \mathrm{e}^{\mathrm{i}\beta_n}.$$

Since z_n^+ vanishes on $L_0^2 \cap D_n$,

$$\partial x_n + \mathrm{i}\partial y_n = \frac{\xi_n}{2} \mathrm{e}^{\mathrm{i}\beta_n} \partial z_n^+.$$

Inserting $\partial z_n^+ = h_n^2 - g_n^2 + 2\mathrm{i} g_n h_n$ and decomposing into the real and imaginary part we obtain the representations for ∂x_n and ∂y_n above. The asymptotics follow with Theorem 9.5 and

$$\xi_n = \frac{1}{\sqrt{\pi n}} + O\left(\frac{\log n}{n^{3/2}}\right), \qquad \beta_n = O\left(\frac{1}{n}\right) \tag{9.8}$$

from Theorems 7.3 and 8.5. At $q = 0$ we have

$$\partial x_n + \mathrm{i}\partial y_n = \frac{1}{\sqrt{4\pi n}} \partial z_n^+ = \frac{1}{\sqrt{\pi n}} (\cos 2\pi n x + \mathrm{i} \sin 2\pi n x)$$

for all $n \geq 1$ by Theorems 7.3 and 9.5 and Lemma 8.2. □

Summarizing our construction so far we may associate with every potential q in L_0^2 infinitely many coordinates

$$\Omega(q) = (\boldsymbol{x}(q), \boldsymbol{y}(q)),$$

where

$$\boldsymbol{x}(q) = (x_1(q), x_2(q), \dots),$$
$$\boldsymbol{y}(q) = (y_1(q), y_2(q), \dots).$$

In view of the asymptotic estimates of Theorem 9.4 and equation (9.8),

$$|x_n| + |y_n| = O\left(\frac{|\gamma_n| + |\mu_n - \tau_n|}{\sqrt{n}}\right)$$

locally uniformly on W. Thus, Ω maps L_0^2 into the space $\mathscr{h}_{1/2} = \ell_{1/2}^2 \times \ell_{1/2}^2$.

Theorem 9.8. *The map*

$$\begin{aligned} \Omega\colon L_0^2 &\to \mathscr{h}_{1/2} \\ q &\mapsto \Omega(q) = (\boldsymbol{x}(q), \boldsymbol{y}(q)) \end{aligned}$$

is real analytic and extends analytically to all of W. Its Jacobian at $q = 0$ is boundedly invertible, its inverse is the weighted Fourier transform

$$\begin{aligned} d_0\Omega^{-1}\colon\ \mathscr{h}_{1/2} &\to L_0^2 \\ (\boldsymbol{x}, \boldsymbol{y}) &\mapsto \sum_{n\geq 1} \sqrt{2\pi n}\,\big(x_n e_n + y_n e_{-n}\big), \end{aligned}$$

where $e_n = \sqrt{2}\cos 2\pi n x$ and $e_{-n} = \sqrt{2}\sin 2\pi n x$ for $n \geq 1$.

Proof. The map Ω is analytic as a map from L_0^2 into $\mathscr{h}_{1/2}$ by Theorem A.5 in view of the analyticity of its components and its local boundedness. Its Jacobian at 0 is given by

$$d_0\Omega(h) = \big((d_0 x_n(h))_{n\geq 1}, (d_0 y_n(h))_{n\geq 1}\big),$$

with $d_0 x_n(h) = \langle \partial x_n, h\rangle = \langle e_n, h\rangle/\sqrt{2\pi n}$ by Theorem 9.7, and similarly for $d_0 y_n$. This proves the theorem. □

Thus, Ω is a local diffeomorphism at $q = 0$. To show that Ω is a global diffeomorphism, we need to establish some orthogonality relations.

10 Orthogonality Relations

In this section we establish orthogonality relations among the gradients of I_n, θ_n, x_n and y_n. They will enable us to show that Ω is a global diffeomorphism and indeed a symplectomorphism. These orthogonality relations are nothing but the Poisson brackets between I_n, θ_n, x_n and y_n. To avoid issues of regularity and anti-symmetry, however, we introduce the notation

$$[\Phi, \Psi] := \langle \partial\Phi, \partial_x \partial\Psi\rangle,$$

where we assume Φ and Ψ to be differentiable functions on L_0^2 such that

$$\partial\Phi \in L^2, \qquad \partial\Psi \in \mathcal{H}^1.$$

We refer to $[\Phi, \Psi]$ as the *bracket of Φ and Ψ*.

To start, we consider products of solutions of $-y'' + qy = \lambda y$. In this case, regularity is not an issue.

Lemma 10.1. *Let F and G be linear combinations of products of solutions of $-y'' + qy = \lambda y$ for $\lambda = \alpha$ and $\lambda = \beta$, respectively, with coefficients depending only on q. If $q \in \mathcal{H}^1$ and $FG|_0^1 = 0$, then*

$$4(\alpha - \beta)\,\langle F, G' \rangle = \left(F''G - F'G' + FG''\right)\Big|_0^1.$$

Remark. We assume q to be in $\mathcal{H}^1$ in order that the boundary values of F'' and G'' make sense.

Proof. Consider the operator $L = -\frac{1}{2}D^3 + qD + Dq$, where $D = \partial_x$. One verifies that for $q \in \mathcal{H}^1$, the product $Y = y_i y_j$ of any two functions of the fundamental solution satisfies the differential equation

$$LY = 2\lambda DY.$$

It follows by partial integration that

$$\begin{aligned} &2\alpha\,\langle DF, G\rangle + 2\beta\,\langle F, DG\rangle \\ &\quad = \langle LF, G\rangle + \langle F, LG\rangle \\ &\quad = -\tfrac{1}{2}\left(F''G - F'G' + FG'' - 4qFG\right)\Big|_0^1. \end{aligned}$$

By another partial integration,

$$\begin{aligned} &2\alpha\,\langle DF, G\rangle + 2\beta\,\langle F, DG\rangle \\ &\quad = 2(\alpha - \beta)\,\langle DF, G\rangle + 2\beta FG\Big|_0^1 \\ &\quad = (\alpha - \beta)\,(\langle DF, G\rangle - \langle F, DG\rangle) + (\alpha + \beta)FG\Big|_0^1 \end{aligned}$$

by symmetrization. Combining the two identities,

$$\begin{aligned} &2(\alpha - \beta)\left(\langle F, G'\rangle - \langle F', G\rangle\right) = \\ &\qquad\qquad \left(F''G - F'G' + FG'' + 2(\alpha + \beta - 2q)FG\right)\Big|_0^1, \end{aligned}$$

from which the result follows. □

We first apply this lemma to the discriminant Δ. For each λ consider

$$\Delta_\lambda \overset{\text{def}}{=} \Delta(\lambda, \cdot)$$

as a function on L_0^2. By Proposition B.3, its gradient is

$$\partial\Delta_\lambda = m_2 y_1^2 + (m_2' - m_1) y_1 y_2 - m_1' y_2^2, \tag{10.1}$$

where all functions are evaluated at λ and q, and where the convenient notation

$$\begin{pmatrix} m_1 & m_2 \\ m_1' & m_2' \end{pmatrix}(\lambda, q) = \begin{pmatrix} y_1 & y_2 \\ y_1' & y_2' \end{pmatrix}(1, \lambda, q)$$

for the entries of the Floquet matrix for $-y'' + qy = \lambda y$ at $x = 1$ is used. Hence, $\partial\Delta_\lambda$ is a linear combination of y_1^2, $y_1 y_2$ and y_2^2 and belongs to $\mathcal{H}^2$. Moreover, by Proposition B.3 one also has the representation

$$\partial\Delta_\lambda = y_2(1, \lambda, T_t q), \qquad T_t q = q(\cdot + t). \tag{10.2}$$

This shows that $\partial\Delta_\lambda$ is 1-periodic, and that $y_2(1, \lambda, T_t q)$ is represented by products of the fundamental solution.

Lemma 10.2. (i) *For all complex α and β,*

$$[\Delta_\alpha, \Delta_\beta] = 0.$$

(ii) *More generally, if Φ is a differentiable function on L_0^2, which at every point q depends only on the periodic spectrum of q, then*

$$[\Phi, \Delta_\beta] = 0$$

for all complex β.

Proof. (i) In view of the representation of $\partial\Delta_\lambda$ in equation (10.1) the bracket $[\Delta_\alpha, \Delta_\beta]$ is continuous on L_0^2. It therefore suffices to prove the claim for smooth q. But then $\partial\Delta_\lambda$ is smoothly 1-periodic in t by (10.2), and so all boundary terms in Lemma 10.1 cancel each other. Consequently, $[\Delta_\alpha, \Delta_\beta] = 0$ for $\alpha \neq \beta$. The case $\alpha = \beta$ is trivial.

(ii) Fix β, and consider $X = \partial_x \partial\Delta_\beta$ as a vector field on L_0^2. As

$$L_X \Delta_\alpha \overset{\text{def}}{=} \langle \partial\Delta_\alpha, X \rangle = [\Delta_\alpha, \Delta_\beta] = 0$$

for all α, we conclude that the function Δ and hence the periodic spectrum is constant along the flow lines of X. But then also Φ is constant along these flow lines, as Φ is assumed to depend only on the periodic spectrum. Therefore, also the Lie-derivative of Φ with respect to X vanishes, so

$$L_X \Phi = [\Phi, \Delta_\beta] = 0$$

for all β. □

Lemma 10.3. *On L_0^2,*

$$[\mu_n, \Delta_\lambda] = \frac{1}{2} \frac{m_2(\lambda)}{\dot{m}_2(\mu_n)} \frac{\sqrt[*]{\Delta^2(\mu_n) - 4}}{\lambda - \mu_n}$$

for all $n \geq 1$ and all $\lambda \neq \mu_n$.

Proof. By continuity, it suffices to prove the lemma for q in $\mathcal{H}^1$. By Proposition B.7, we have

$$\partial\mu_n = \frac{m_1(\mu_n)}{\dot{m}_2(\mu_n)} y_2^2(\cdot, \mu_n).$$

Thus, with equation (10.2),

$$[\mu_n, \Delta_\lambda] = \frac{m_1(\mu_n)}{\dot m_2(\mu_n)} \big\langle y_2^2(\cdot\,,\mu_n), \partial_x y_2(1,\lambda, T_x q) \big\rangle .$$

Applying Lemma 10.1 with $F = y_2^2(\cdot\,,\mu_n)$ and observing that all boundary terms of F, F' and F'' vanish except the contribution from $(y_2')^2$ we obtain

$$\begin{aligned} 2(\mu_n - \lambda)\,[\mu_n, \Delta_\lambda] &= \frac{m_1(\mu_n)}{\dot m_2(\mu_n)} \big(y_2'(x,\mu_n)^2 y_2(1,\lambda,T_x q) \big)\Big|_0^1 \\ &= \frac{m_1(\mu_n)}{\dot m_2(\mu_n)} \big(m_2'(\mu_n)^2 m_2(\lambda) - m_2(\lambda) \big) \\ &= \frac{m_2(\lambda)}{\dot m_2(\mu_n)} \big(m_2'(\mu_n) - m_1(\mu_n) \big), \end{aligned}$$

since $m_1(\mu_n) m_2'(\mu_n) = 1$ by the Wronskian identity. The claim now follows with $m_1(\mu_n) - m_2'(\mu_n) = \sqrt[*]{\Delta^2(\mu_n) - 4}$ by (6.2). □

Lemma 10.4. $2\,[\theta_n, \Delta_\lambda] = \psi_n(\lambda)$ *for all* $n \ge 1$ *and all* λ *on* $L_0^2 \smallsetminus D_n$.

Proof. We first consider the bracket $\big[\beta_k^n, \Delta_\lambda\big]$ for $k \ne n$ under the assumption that $\lambda_{2k-1} < \mu_k < \lambda_{2k}$. Recall that

$$\beta_k^n = \int_{\lambda_{2k-1}}^{\mu_k^*} \frac{\psi_n(\lambda)}{\sqrt{\Delta^2(\lambda) - 4}}\, d\lambda .$$

The lower bound of the integral and the integrand are spectral invariants, hence invariant under the flow of the vector field $X = \partial_x \partial \Delta_\lambda$. Thus we have

$$\begin{aligned} \big[\beta_k^n, \Delta_\lambda\big] = L_X \beta_k^n &= \frac{\psi_n(\mu_k)}{\sqrt[*]{\Delta^2(\mu_k) - 4}} L_X \mu_k \\ &= \frac{\psi_n(\mu_k)}{\sqrt[*]{\Delta^2(\mu_k) - 4}} [\mu_k, \Delta_\lambda] \\ &= \frac{1}{2} \frac{m_2(\lambda)}{\dot m_2(\mu_k)} \frac{\psi_n(\mu_k)}{\lambda - \mu_k} \end{aligned}$$

by the previous lemma. This holds indeed on all of L_0^2, since the first and the last expression are continuous on L_0^2 as long as $\lambda \ne \mu_k$, and since $\lambda_{2k-1} < \mu_k < \lambda_{2k}$ holds on an open and dense subset of L_0^2.

The same expression holds for $[\eta_n, \Delta_\lambda]$ on $L_0^2 \smallsetminus D_n$, since the formula for η_n is analogous to the formula for β_k^n modulo 2π, and additive constants vanish upon differentiation.

Now consider $\theta_n = \eta_n + \sum_{k\neq n} \beta_k^n$. By Lemma 8.4, this series converges locally uniformly. By Cauchy's theorem, the same holds true for its gradient. Thus we obtain

$$\begin{aligned} 2\,[\theta_n, \Delta_\lambda] &= 2\,[\eta_n, \Delta_\lambda] + 2\sum_{k\neq n} [\beta_k^n, \Delta_\lambda] \\ &= \sum_{k\geq 1} \frac{m_2(\lambda)}{\dot m_2(\mu_k)} \frac{\psi_n(\mu_k)}{\lambda - \mu_k} = \psi_n(\lambda) \end{aligned}$$

by Proposition D.11. □

Proposition 10.5.

$$[I_n, I_m] = 0, \qquad [\theta_n, I_m] = -\delta_{nm}$$

on L_0^2 and $L_0^2 \smallsetminus D_n$, respectively. Moreover,

$$[x_n, I_m] = \delta_{nm} y_n, \qquad [y_n, I_m] = -\delta_{nm} x_n,$$

on L_0^2. Finally, $[\Phi, I_m] = 0$ for every differentiable function Φ on L_0^2, which at every point q depends only on the periodic spectrum of q.

Proof. By Theorem 7.1 and the representation (10.1) of $\partial\Delta_\lambda$,

$$\partial I_m = -\frac{1}{\pi}\int_{\Gamma_m} \frac{\partial\Delta_\lambda}{\sqrt[c]{\Delta^2(\lambda) - 4}}\,\mathrm{d}\lambda$$

is in $\mathcal{H}^2$. The first identity as well as $[\Phi, I_m] = 0$ then follow with Lemma 10.2 by integration. The second identity follows from

$$\begin{aligned} [\theta_n, I_m] &= -\frac{1}{\pi}\int_{\Gamma_m} \frac{[\theta_n, \Delta_\lambda]}{\sqrt[c]{\Delta^2(\lambda) - 4}}\,\mathrm{d}\lambda \\ &= -\frac{1}{2\pi}\int_{\Gamma_m} \frac{\psi_n(\lambda)}{\sqrt[c]{\Delta^2(\lambda) - 4}}\,\mathrm{d}\lambda = -\delta_{nm}, \end{aligned}$$

using Lemma 10.4 for the second identity and (8.1) for the last identity. Finally, on $L_0^2 \smallsetminus D_n$ the definitions $x_n = \sqrt{2I_n}\cos\theta_n$ and $y_n = \sqrt{2I_n}\sin\theta_n$ give

$$[x_n, I_m] = \frac{1}{\sqrt{2I_n}}\cos\theta_n\,[I_n, I_m] - \sqrt{2I_n}\sin\theta_n\,[\theta_n, I_m] = y_n\delta_{nm},$$

and similarly $[y_n, I_m] = -x_n\delta_{nm}$. Since both sides are continuous on L_0^2, this holds on all of L_0^2. □

To determine the brackets among x_n, x_m, y_n, y_m at points in $D_n \cap D_m$, we first consider the functions $z_n^\pm$. Recall that $\partial z_n^\pm = h_n^2 - g_n^2 \pm 2\mathrm{i} g_n h_n$ on $L_0^2 \cap D_n$ by Theorem 9.5, where g_n denotes the normalized eigenfunction for the Dirichlet eigenvalue μ_n, and h_n that periodic eigenfunction orthonormal to g_n which is positive at 0. In particular, the gradients $\partial z_n^\pm$ are in $\mathcal{H}^2$.

Lemma 10.6. *On $L_0^2 \cap D_n$,*

$$\left[z_n^+, z_n^-\right] = -4\mathrm{i}\, g_n' h_n\big|_0 \neq 0,$$

while on $L_0^2 \cap D_n \cap D_m$ with $n \neq m$, all brackets among $z_n^\pm$ and $z_m^\pm$ vanish.

Proof. By Theorem 9.5,

$$\left[z_n^\pm, z_m^\pm\right] = \left\langle h_n^2 - g_n^2 \pm 2\mathrm{i} g_n h_n, (h_m^2 - g_m^2 \pm 2\mathrm{i} g_m h_m)'\right\rangle.$$

As both sides are continuous in q, it suffices to consider q in $\mathcal{H}^1$. For $n \neq m$, the right hand side vanishes by Lemma 10.1, since each product is periodic. For $n = m$, we obtain

$$\begin{aligned}
\frac{1}{2\mathrm{i}}\left[z_n^+, z_n^-\right] &= \left\langle g_n h_n, (h_n^2 - g_n^2)'\right\rangle - \left\langle h_n^2 - g_n^2, (g_n h_n)'\right\rangle \\
&= \int_0^1 \left(g_n h_n^2 h_n' - g_n^2 g_n' h_n - g_n' h_n^3 + g_n^3 h_n'\right) \mathrm{d}x \\
&= \int_0^1 W(g_n, h_n)(g_n^2 + h_n^2)\, \mathrm{d}x \\
&= 2W(g_n, h_n) = -2g_n' h_n\Big|_0,
\end{aligned}$$

where $W(g_n, h_n) = g_n h_n' - g_n' h_n$ denotes the constant Wronskian of g_n and h_n. The function h_n does not vanish at 0, since it is linearly independent of g_n. □

This lemma immediately carries over to a corresponding result for the functions x_n and y_n.

Proposition 10.7. *On $L_0^2 \cap D_n$,*

$$[x_n, y_n] = \frac{\xi_n^2}{2}\, g_n' h_n\big|_0 \neq 0,$$

while on $L_0^2 \cap D_n \cap D_m$ with $n \neq m$, all brackets among x_n, x_m, y_n, y_m vanish.

Proof. Recall from the preceding section that

$$\begin{aligned}
x_n &= \frac{\xi_n}{4}\left(z_n^+ \mathrm{e}^{\mathrm{i}\beta_n} + z_n^- \mathrm{e}^{-\mathrm{i}\beta_n}\right), \\
y_n &= \frac{\xi_n}{4\mathrm{i}}\left(z_n^+ \mathrm{e}^{\mathrm{i}\beta_n} - z_n^- \mathrm{e}^{-\mathrm{i}\beta_n}\right).
\end{aligned}$$

Hence, ∂x_n and ∂y_n are in $\mathcal{H}^2$ on D_n. As $z_n^\pm$ vanishes on D_n, the preceding lemma gives

$$[x_n, y_n] = \frac{\xi_n^2}{16\mathrm{i}}\left([z_n^-, z_n^+] - [z_n^+, z_n^-]\right) = \frac{\xi_n^2}{2} g_n' h_n\Big|_0.$$

The other identities follow analogously. □

Eventually we will see that $[x_n, y_n] = 1$. But the far simpler result stated in Proposition 10.7 is sufficient to show in the next section that Ω is a global diffeomorphism.

11 The Diffeomorphism Property

We are going to show that

$$\Omega\colon\ L_0^2 \to \mathscr{h}_{1/2}$$

is a global diffeomorphism onto $\mathscr{h}_{1/2}$. This is done in two steps. First we show that Ω is a local diffeomorphism not only at 0 by Theorem 9.8, but at every point in L_0^2. Then we show that Ω is globally one-to-one and onto.

To establish Ω as a local diffeomorphism it is more convenient to consider the map

$$\begin{aligned}\Phi = d_0\Omega^{-1} \circ \Omega\colon\ L_0^2 &\to L_0^2 \\ q &\mapsto \sum_{n\geq 1} \sqrt{2\pi n}\,\bigl(x_n(q)e_n + y_n(q)e_{-n}\bigr)\end{aligned}$$

with

$$e_n = \sqrt{2}\cos 2\pi n x, \qquad e_{-n} = \sqrt{2}\sin 2\pi n x \tag{11.1}$$

for $n \geq 1$. Clearly, Ω is a local diffeomorphism if and only if Φ is a local diffeomorphism.

Consider the Jacobian of Φ at q,

$$\begin{aligned}d_q\Phi\colon\ L_0^2 &\to L_0^2 \\ h &\mapsto d_q\Phi(h) = \sum_{n\neq 0} \langle h, d_n(q)\rangle e_n,\end{aligned}$$

where for $n \geq 1$,

$$d_n(q) = \sqrt{2\pi n}\,\partial x_n, \qquad d_{-n}(q) = \sqrt{2\pi n}\,\partial y_n. \tag{11.2}$$

To verify that $d_q\Phi$ is a linear isomorphism we make use of the following general result.

First two definitions. A system of vectors (v_n) in a Hilbert space is said to be *linearly independent*, if none of the vectors v_n is contained in the closed linear span of all the other vectors v_m, $m \neq n$. A system of vectors (v_n) is a *basis* for a Hilbert space H, if there exists a Hilbert space $\mathscr{h}$ of sequences $\boldsymbol{\alpha} = (\alpha_n)$ such that the correspondence

$$\boldsymbol{\alpha} \mapsto \sum_{n\geq 1} \alpha_n v_n$$

is a linear isomorphism between $\mathscr{h}$ and H.

Proposition 11.1. *Let $(e_n)_{n\in N}$, $N \subset \mathbb{Z}$, be an orthonormal basis of a Hilbert space H. Suppose $(d_n)_{n\in N}$ is another sequence of vectors in H that* (i) *either spans or is linearly independent and* (ii) *satisfies*

$$\sum_{n\in N} \|d_n - e_n\|^2 < \infty.$$

Then $(d_n)_{n\in N}$ is also a basis of H, and the map

$$h \mapsto \sum_{n\in N} \langle h, d_n\rangle e_n$$

is a linear isomorphism of H.

Moreover, if the vectors d_n, $n \in N$, depend continuously on a parameter such that (i) *holds for every parameter,* (ii) *holds on a dense subset of parameters, and the above map depends continuously in the operator norm on the parameter, then the conclusion holds for every parameter.*

Proof. Define an operator A on H by

$$Ah = \sum_{n\in N} \langle h, e_n\rangle d_n.$$

By item (ii) A is a well defined, bounded operator, which maps e_n into d_n. Since

$$\sum_{n\in N} \|(A-I)e_n\|^2 = \sum_{n\in N} \|d_n - e_n\|^2 < \infty,$$

$A - I$ is Hilbert-Schmidt, hence A is a compact perturbation of the identity.

If the system $(d_n)_{n\in N}$ is linearly independent, then A is one-to-one, for if $Ah = 0$, then $\langle h, e_n\rangle = 0$ for all $n \in N$ and hence $h = 0$. If, on the other hand, $(d_n)_{n\in N}$ spans, then the range of A is dense in H. But the range of a compact perturbation of the identity is closed, so A is onto. So in either case, A is boundedly invertible by the Fredholm alternative. Hence $(d_n)_{n\in N}$ is also a basis.

Next, we note that

$$\langle h, d_n\rangle = \langle h, Ae_n\rangle = \langle A^*h, e_n\rangle.$$

Since also A^* is boundedly invertible, the map $h \mapsto (\langle A^*h, e_n\rangle)_{n\in N} = (\langle h, d_n\rangle)_{n\in N}$ is a linear isomorphism between H and ℓ^2. Consequently, the map

$$h \mapsto \sum_{n\in N} \langle h, d_n\rangle e_n$$

is a linear isomorphism of H.

Finally, suppose the system $(d_n)_{n\in N}$ depends on a parameter as stipulated. Then $A - I$ is compact for a dense set of parameters. By assumption, A^* is continuous in the parameter, hence $A - I$ is compact for all parameters, and the argument applies as before. □

We apply the preceding proposition to the system of vectors $(d_n)_{n\neq 0}$ defined by (11.2) in the Hilbert space L_0^2 and compare them to the orthonormal basis $(e_n)_{n\neq 0}$ defined by (11.1). The point q is considered as a parameter on which the d_n depend continuously.

Lemma 11.2. *For q in the dense subset of finite gap potentials in L_0^2,*

$$\sum_{n\neq 0} \|d_n - e_n\|^2 < \infty.$$

Proof. This is an immediate consequence of Theorem 9.7. For the density of finite gap potentials in L_0^2, see Theorem B.16. □

We go on to show that the system of vectors $(d_n)_{n\neq 0}$ is linearly independent at every point q in L_0^2. We fix q, and let

$$A(q) = \{n \in \mathbb{N} : \gamma_n(q) > 0\}.$$

The argument is given in two steps.

Lemma 11.3. *At every point q in L_0^2, the vectors $(d_n)_{n\neq 0}$ are linearly independent in L_0^2, if the vectors $(\partial I_n)_{n\in A(q)}$ are linearly independent in L_0^2.*

Proof. Fix q, and let $A = A(q)$. Suppose $(\alpha_n)_{n\neq 0}$ are real numbers such that

$$f = \sum_{n\neq 0} \alpha_n d_n = 0.$$

For $m \in A$, Proposition 10.5 gives

$$\begin{aligned} 0 = \langle f, \partial_x \partial I_m \rangle &= \sqrt{2\pi m}\, (\alpha_m\, [x_m, I_m] + \alpha_{-m}\, [y_m, I_m]) \\ &= \sqrt{2\pi m}\, (\alpha_m y_m - \alpha_{-m} x_m). \end{aligned}$$

It follows that the 2-vector (α_m, α_{-m}) is parallel to the 2-vector (x_m, y_m). Hence $(\alpha_m, \alpha_{-m}) = a_m(\cos\theta_m, \sin\theta_m)$ with $a_m^2 = \alpha_m^2 + \alpha_{-m}^2$, and

$$\begin{aligned} \frac{1}{\sqrt{2\pi m}} (\alpha_m d_m + \alpha_{-m} d_{-m}) &= \alpha_m \partial x_m + \alpha_{-m} \partial y_m \\ &= a_m (\cos\theta_m\, \partial x_m + \sin\theta_m\, \partial y_m) \\ &= \frac{a_m}{\sqrt{2 I_m}} \partial I_m \end{aligned}$$

by inserting the definitions of x_m and y_m.

Consider now $m \notin A$, so that $\gamma_m = 0$. In view of the above representation and Propositions 10.7 and 10.5 we find that

$$\begin{aligned} 0 = \langle f, \partial_x \partial y_m \rangle &= \sqrt{2\pi m}\, \alpha_m\, [x_m, y_m], \\ 0 = \langle f, \partial_x \partial x_m \rangle &= \sqrt{2\pi m}\, \alpha_{-m}\, [y_m, x_m]. \end{aligned}$$

Hence $\alpha_{\pm m} = 0$ for $m \notin A$. We thus conclude that

$$0 = \sum_{n\neq 0} \alpha_n d_n = \sum_{n\in A} \frac{a_n}{\sqrt{2 I_n}} \partial I_n.$$

If the vectors $(\partial I_n)_{n\in A}$ are linearly independent, then the $a_n = \alpha_n^2 + \alpha_{-n}^2$ and hence also the $\alpha_{\pm n}$ must vanish, since both are real. This proves the claim. □

Lemma 11.4. *At every point q in L_0^2 the vectors $(\partial I_n)_{n\in A(q)}$ are linearly independent in L_0^2.*

Proof. Fix q, and let $A = A(q)$. Suppose

$$\sum_{n\in A} \alpha_n \partial I_n = 0$$

with some real coefficients α_n. We have to show that they are all zero. This would be straightforward if we were allowed to take inner products with $\partial_x \partial\theta_m$ and make use of the orthogonality relations of Proposition 10.5. However, we do not know yet that the gradient of θ_m is in $\mathcal{H}^1$, nor can we assume that $\sum_{n\in A} \alpha_n \partial I_n$ converges in $\mathcal{H}^1$. Therefore, we can not apply Proposition 10.5 right away.

Instead we redo part of the argument that led to these orthogonality relations. Define the functions

$$h_k^m = \begin{cases} -\dfrac{\psi_m(\mu_k)}{\sqrt[*]{\Delta^2(\mu_k)-4}}\,\partial\mu_k & \text{for} \quad \lambda_{2k-1} < \mu_k < \lambda_{2k}, \\[2ex] \dfrac{\psi_m(\mu_k)}{\dot{\Delta}(\mu_k)}\, y_1 y_2\big|_{\mu_k} & \text{for} \quad \mu_k \in \{\lambda_{2k-1}, \lambda_{2k}\}. \end{cases}$$

In the case $\lambda_{2k-1} < \mu_k < \lambda_{2k}$, one obtains with Lemma 10.3

$$\begin{aligned} \big\langle \partial\Delta_\lambda, \partial_x h_k^m \big\rangle &= \frac{\psi_m(\mu_k)}{\sqrt[*]{\Delta^2(\mu_k)-4}}\,[\mu_k, \Delta_\lambda] \\ &= \frac{1}{2}\frac{m_2(\lambda)}{\dot{m}_2(\mu_k)}\frac{\psi_m(\mu_k)}{\lambda-\mu_k}. \end{aligned}$$

In the case $\mu_k \in \{\lambda_{2k-1}, \lambda_{2k}\}$, we have with equation (10.2)

$$\big\langle \partial\Delta_\lambda, \partial_x h_k^m \big\rangle = -\frac{\psi_m(\mu_k)}{\dot{\Delta}(\mu_k)} \big\langle\, y_1 y_2\big|_{\mu_k}, \partial_x y_2(1,\lambda,T_x q)\big\rangle.$$

To evaluate the right hand side, we first assume $q \in \mathcal{H}^1$ and apply Lemma 10.1 with $F = y_1 y_2|_{\mu_k}$ and $G = y_2(1,\lambda,T_x q)$. Since $y_2 = 0$ and, by the Wronskian identity, $y_1 y_2' = 1$ at the boundary of $[0,1]$, and since G is periodic, we have

$$F'G'\big|_0^1 = 0, \qquad FG''\big|_0^1 = 0.$$

Moreover, since $y_2'' = 0$ at the boundary by the differential equation,

$$\begin{aligned} F''G\big|_0^1 &= 2y_1' y_2' G\big|_0^1 \\ &= 2y_1'(1,\mu_k)y_2'(1,\mu_k)y_2(1,\lambda) \\ &= 2m_1'(\mu_k)m_2'(\mu_k)m_2(\lambda), \end{aligned}$$

where for the second equality we used that $y_1'(0) = 0$. Hence we obtain

$$\langle \partial \Delta_\lambda, \partial_x h_k^m \rangle = \frac{\psi_m(\mu_k)}{\dot{\Delta}(\mu_k)} \frac{1}{4(\lambda - \mu_k)} F''G\Big|_0^1$$
$$= \frac{1}{2} \frac{\psi_m(\mu_k)}{\lambda - \mu_k} \frac{m_1'(\mu_k) m_2'(\mu_k)}{\dot{\Delta}(\mu_k)} m_2(\lambda)$$

for $q \in \mathcal{H}^1$ by Lemma 10.1, and then for general q in L_0^2 by continuity.

Differentiating the Wronskian identity $m_1 m_2' - m_1' m_2 = 1$ with respect to λ and using that $m_2 = 0$ and $m_1 = m_2' = (-1)^k$ when μ_k is a periodic eigenvalue, one finds that

$$m_1'(\mu_k)\dot{m}_2(\mu_k) = m_2'(\mu_k)\dot{\Delta}(\mu_k) = \dot{\Delta}(\mu_k)/m_2'(\mu_k),$$

or

$$\frac{m_1'(\mu_k) m_2'(\mu_k)}{\dot{\Delta}(\mu_k)} = \frac{1}{\dot{m}_2(\mu_k)}.$$

Together with the preceding identity we then arrive at the same result as in the case $\lambda_{2k-1} < \mu_k < \lambda_{2k}$.

By Theorem 7.1 we then get

$$\langle \partial I_n, \partial_x h_k^m \rangle = \frac{1}{2\pi} \int_{\Gamma_n} \frac{m_2(\lambda)}{\dot{m}_2(\mu_k)} \frac{\psi_m(\mu_k)}{\lambda - \mu_k} \frac{\mathrm{d}\lambda}{\sqrt[c]{\Delta^2(\lambda) - 4}}.$$

By standard asymptotic estimates,

$$\frac{\psi_m(\mu_k)}{\dot{m}_2(\mu_k)} = O\left(\frac{m}{|m^2 - k^2|} \right), \qquad k \neq m.$$

We may therefore take the sum over all k to obtain

$$\sum_{k \geq 1} \langle \partial I_n, \partial_x h_k^m \rangle = \frac{1}{2\pi} \int_{\Gamma_n} \left(\sum_{k \geq 1} \frac{m_2(\lambda)}{\dot{m}_2(\mu_k)} \frac{\psi_m(\mu_k)}{\lambda - \mu_k} \right) \frac{\mathrm{d}\lambda}{\sqrt[c]{\Delta^2(\lambda) - 4}}$$
$$= \frac{1}{2\pi} \int_{\Gamma_n} \frac{\psi_m(\lambda)}{\sqrt[c]{\Delta^2(\lambda) - 4}} \mathrm{d}\lambda$$
$$= \delta_{nm},$$

using Proposition D.11 in the second line and Theorem 8.1 in the third line.

We can now prove the lemma. Suppose $\sum_{n \in A} \alpha_n \partial I_n = 0$. Taking the inner product with $\partial_x h_k^m$ and summing over all k we obtain

$$0 = \sum_{n \in A} \alpha_n \sum_{k \geq 1} \langle \partial I_n, \partial_x h_k^m \rangle$$
$$= \sum_{n \in A} \alpha_n \delta_{nm} = \alpha_m.$$

Thus, all coefficients α_n vanish. □

As the last three lemmas hold and the Jacobian $d_q\Phi$ depends continuously on q, the vectors $(d_n)_{n\neq 0}$ satisfy the assumptions of Proposition 11.1. Thus, $d_q\Phi$ is a linear isomorphism at every point q in L_0^2, and we conclude the following intermediate result.

Proposition 11.5. *The map $\Omega\colon L_0^2 \to \hbar_{1/2}$ is a local diffeomorphism everywhere.*

We now show that this map is one-to-one and onto. This is a topological argument and boils down to verifying that the map Ω is *proper*. That is, the preimage of compact sets is compact.

Lemma 11.6. *The map $\Omega\colon L_0^2 \to \hbar_{1/2}$ is proper.*

Proof. It suffices to show that for any sequence (q_m) in L_0^2 with the property that $\Omega(q_m)$ converges strongly in $\hbar_{1/2}$ there exists a subsequence that converges strongly in L_0^2.

So let (q_m) be such a sequence. If $\Omega(q_m)$ converges in $\hbar_{1/2}$, then also the norms $\|\Omega(q_m)\|_{\hbar_{1/2}}$ converge. By Proposition E.1 we know that

$$\begin{aligned}\|\Omega(q_m)\|_{\hbar_{1/2}}^2 &= \sum_{k\geq 1} k\left(x_k^2(q_m)+y_k^2(q_m)\right)\\ &= \sum_{k\geq 1} 2k\, I_k(q_m) = \frac{1}{2\pi}\,\|q_m\|_{L^2}^2. \qquad (11.3)\end{aligned}$$

Hence, (q_m) is bounded in L_0^2 and thus admits a weakly convergent subsequence again denoted by (q_m).

Let q be the weak limit of (q_m). By Lemma 7.2, each I_k is a compact function on L_0^2, hence

$$I_k(q_m) \to I_k(q)$$

for each $k \geq 1$. With (11.3) we then conclude that also

$$\|q_m\|_{L^2} \to \|q\|_{L^2}.$$

This together with the weak convergence of q_m to q implies that (q_m) converges to q strongly in L_0^2. □

Proposition 11.7. *The map $\Omega\colon L_0^2 \to \hbar_{1/2}$ is globally one-to-one and onto.*

Proof. Consider the set

$$M = \left\{ z \in \hbar_{1/2} \colon \#\Omega^{-1}(z) = 1 \right\}.$$

M is open and closed, since Ω is a local diffeomorphism everywhere and proper. M is not empty, since

$$\Omega^{-1}(\mathbf{0}) = \left\{ q \in L_0^2 \colon \gamma_n(q) = 0 \text{ for all } n \right\} = \left\{ 0 \in L_0^2 \right\},$$

$q \equiv 0$ being the only real valued potential with $\Omega(q) = \mathbf{0}$ by Proposition E.1. It follows that $M = \hbar_{1/2}$, so Ω is globally one-to-one and onto. □

Thus we have established the following result.

Theorem 11.8. *The map* $\Omega\colon L_0^2 \to \hbar_{1/2}$ *is a global real analytic diffeomorphism.*

The map Ω also respects certain subsets of L_0^2 in a natural way. Recall that

$$\mathcal{G}_A = \left\{ q \in L_0^2 \colon \gamma_k(q) > 0 \text{ iff } k \in A \right\}$$

denotes the set of A-gap potentials in L_0^2, and that

$$\hbar_A = \left\{ (\boldsymbol{x}, \boldsymbol{y}) \in \hbar_0 \colon x_k^2 + y_k^2 > 0 \text{ iff } k \in A \right\}.$$

Theorem 11.9. *For each* $A \subset \mathbb{N}$,

$$\mathcal{G}_A = \Omega^{-1}(\hbar_A).$$

In particular, $\mathcal{G}_A$ *is a real analytic submanifold of* L_0^2.

Proof. This is an immediate consequence of Theorem 7.3, which implies that for any $n \geq 1$, we have $\gamma_n = 0$ iff $I_n = 0$ iff $x_n^2 + y_n^2 = 0$. □

Remark. It follows that also the union of all A-gap potentials with $A = \{1, \dots, M\}$, the so called *M-gap potentials*, is dense in L_0^2.

In addition, Ω maps each isospectral set $\mathrm{Iso}(q) \subset L_0^2$ onto a corresponding torus

$$\mathrm{Tor}(\boldsymbol{I}) = \left\{ (\boldsymbol{x}, \boldsymbol{y}) \colon x_n^2 + y_n^2 = 2I_n \text{ for } n \geq 1 \right\} \subset \hbar_{1/2},$$

where $\boldsymbol{I} = \boldsymbol{I}(q) = (I_n(q))_{n\geq 1}$.

Theorem 11.10. *The diffeomorphism* Ω *maps each isospectral set* $\mathrm{Iso}(q)$ *in* L_0^2 *one-to-one onto the torus* $\mathrm{Tor}(\boldsymbol{I}(q))$ *in* $\hbar_{1/2}$.

Proof. Fix q and $\boldsymbol{I} = \boldsymbol{I}(q)$. As any $p \in \mathrm{Iso}(q)$ has the same periodic spectrum as q, it also has the same Δ-function as q in view of its product representation. The actions I_n are defined entirely in terms of this Δ-function. Consequently, $\boldsymbol{I}(p) = \boldsymbol{I}$ and $\Omega(p) \in \mathrm{Tor}(\boldsymbol{I})$. Thus,

$$\Omega(\mathrm{Iso}(q)) \subset \mathrm{Tor}(\boldsymbol{I}).$$

Conversely, take any point $\boldsymbol{w} = (\boldsymbol{x}, \boldsymbol{y})$ in $\mathrm{Tor}(\boldsymbol{I})$. Set $A = \{n \colon I_n \neq 0\}$, and for any $n \in A$ consider the curve $\boldsymbol{w}^n$ in $\mathrm{Tor}(\boldsymbol{I})$, which moves around the n-th circle $\{x_n^2 + y_n^2 = 2I_n\}$ with unit speed, while all other coordinates stay put. Its preimage under Ω is a smooth curve q^n in L_0^2 with nonvanishing velocity vector $\dot{q}^n$ everywhere.

This velocity vector is perpendicular to ∂I_k for $k \geq 1$, to $\partial\theta_k$ for $n \neq k \in A$, and to ∂x_k, ∂y_k for $k \notin A$, since all these coordinates are constant along this curve. By comparison with Proposition 10.5 it follows that in every point,

$$\dot{q}^n = \alpha \partial_x \partial I_n$$

with some real α. Again with Proposition 10.5 this implies that

$$\frac{\mathrm{d}}{\mathrm{d}t}\Delta \circ q^n = L_{\dot{q}^n}\Delta = \alpha\,[\Delta, I_n] = 0.$$

That is, the Δ-function is invariant along the curve q^n. Consequently, all potentials along this curve have the same periodic spectrum, and so

$$\operatorname{image} q^n \subset \operatorname{Iso}(q).$$

Since this holds for each $n \in A$, we conclude by induction that

$$\Omega^{-1}(\operatorname{Tor}(\boldsymbol{I})) \subset \operatorname{Iso}(q).$$

This completes the proof. □

Finally, Theorem 11.8 is sharpened to the extent that Ω respects higher regularity. This was first proven in [6] with a quite different approach. Denote by Ω_N the restriction of Ω to $\mathcal{H}_0^N$. In particular, $\Omega_0 = \Omega$.

Theorem 11.11. *For each $N \geq 0$, the map Ω_N is a global, real analytic diffeomorphism between $\mathcal{H}_0^N$ and $h_{N+1/2}$.*

Remark. Thus, the union of all M-gap potentials is also dense in $\mathcal{H}_0^N$ for any $N \geq 0$.

Proof. For $N = 0$, this is the content of the preceding theorem. So fix $N \geq 1$. We already noticed in section 9 that

$$|x_n| + |y_n| = O\left(\frac{|\gamma_n| + |\mu_n - \tau_n|}{\sqrt{n}}\right)$$

locally uniformly on W. Moreover, there exists a complex neighbourhood W_N of $\mathcal{H}_0^N$ in $\mathcal{H}_{0,\mathbb{C}}^N$ such that

$$\sum_{n\geq 1} n^{2N}\left(|\gamma_n|^2 + |\mu_n - \tau_n|^2\right) = O(1)$$

locally uniformly by Proposition B.9. Hence, Ω_N maps $W_N \cap W$ into $h_{N+1/2,\mathbb{C}}$ and is locally bounded. Since each coordinate function is real analytic, the entire map

$$\Omega_N\colon\ \mathcal{H}_0^N \to h_{N+1/2}$$

is real analytic by Theorem A.5.

This map is clearly one-to-one, since Ω is one-to-one. To show that Ω_N is onto, let $(\boldsymbol{x}, \boldsymbol{y}) \in h_{N+1/2}$. Then $q = \Omega^{-1}(\boldsymbol{x}, \boldsymbol{y})$ is well defined in L_0^2. As q is real valued, we may apply Theorem 7.3 to obtain for all large n that

$$|x_n|^2 + |y_n|^2 = 2I_n \geq \frac{1}{8\pi}\frac{\gamma_n^2}{n}.$$

Hence,

$$\sum_{n\geq 1} n^{2N}\gamma_n^2 < \infty,$$

and consequently $q \in \mathcal{H}_0^N$, as q is real [84]. This shows that Ω_N is onto.

It remains to show that Ω_N is a local diffeomorphism everywhere. As before, this is done by considering the map

$$\Phi_N = d_0\Omega_N^{-1} \circ \Omega_N\colon\ \mathcal{H}_0^N \to \mathcal{H}_0^N,$$

where $d_0\Omega_N^{-1} = (d_0\Omega^{-1})\big|\hbar_{N+1/2}$ is the restriction of the inverse discrete Fourier transform $d_0\Omega^{-1}$ to $\hbar_{N+1/2}$. Clearly, $\Phi_N = \Phi|\mathcal{H}_0^N$, and

$$d_q\Phi_N\colon\ \mathcal{H}_0^N \to \mathcal{H}_0^N, \quad h \mapsto \sum_{n\neq 0} \langle h, d_n(q)\rangle e_n$$

is the restriction of $d_q\Phi$ to $\mathcal{H}_0^N$, where the d_n are given by (11.2). Thus, $d_q\Phi_N$ is one-to-one. Its deviation from the identity is the map

$$A_q\colon\ h \mapsto A_q h = \sum_{n\neq 0} \langle h, d_n(q) - e_n\rangle e_n,$$

and it suffices to show that A_q is compact at any finite gap potential q, which are dense in $\mathcal{H}_0^N$ – see appendix B. By continuity, A_q is then compact everywhere, and $d_q\Phi_N = I + A_q$ is a linear isomorphism by the Fredholm alternative.

To show that A_q is a compact operator on $\mathcal{H}_0^N$, we consider the N-th derivative of $A_q h$,

$$\partial_x^N A_q h = \sum_{n\neq 0} (2\pi|n|)^N \langle h, d_n(q) - e_n\rangle \tilde{e}_n,$$

where $\tilde{e}_n$ is either $\pm e_n$ or $\pm e_{-n}$ with the choice of sign depending on N. In view of Lemma 11.12 below,

$$\sum_{n\neq 0} |n|^{2N} \sup_{h\in\mathcal{H}_0^N,\, \|h\|_{\mathcal{H}^N}\leq 1} |\langle h, d_n(q) - e_n\rangle|^2 < \infty,$$

when q is a finite gap potential. Hence, A_q is a bounded operator on $\mathcal{H}_0^N$, and for each $\varepsilon > 0$ there exists an integer M such that

$$\Bigg\| \sum_{|n|\geq M} \langle h, d_n(q) - e_n\rangle e_n \Bigg\|_{\mathcal{H}^N} \leq \varepsilon$$

uniformly for $\|h\|_{\mathcal{H}^N} \leq 1$. From this, it follows that A_q maps the unit ball in $\mathcal{H}^N$ into a relatively compact subset of $\mathcal{H}^N$. Hence, A_q is compact. □

Lemma 11.12. *At a finite gap potential, the estimate*

$$|\langle h, d_n - e_n \rangle| = O\left(\frac{\log n}{n^{N+3/2}} \right) \|h\|_{\mathcal{H}^N}$$

holds uniformly for $h \in \mathcal{H}_0^N$ *for each* $N \geq 0$.

Proof. The proof consists in verifying the statement for $N = 0$ and $N = 1$ and in proving an induction step. We start with the latter and assume that $h \in \mathcal{H}_0^N$ with $N \geq 2$.

Given q, consider n sufficiently large so that $\gamma_n = 0$. It follows from Theorem 9.5 that ∂x_n and ∂y_n are linear combinations of squares of eigenfunctions of q for the eigenvalue $\lambda_{2n-1} = \lambda_{2n}$. Hence they are smooth functions of x, and as in the proof of Lemma 10.1 we have

$$Ld_n = 2\lambda_{2n} Dd_n,$$

where $L = -\frac{1}{2}D^3 + qD + Dq$. As the mean values of ∂x_n and ∂y_n vanish on D_n by Theorem 9.7, we thus have

$$d_n = \frac{1}{2\lambda_{2n}} D_0^{-1} L d_n,$$

where D_0^{-1} denotes the inverse of the restriction of D to $\mathcal{H}_0^1$.

It follows that

$$\begin{aligned} \langle d_n, h \rangle &= \frac{1}{2\lambda_{2n}} \left\langle D_0^{-1} L d_n, h \right\rangle \\ &= \frac{1}{2\lambda_{2n}} \left\langle d_n, L D_0^{-1} h \right\rangle \\ &= -\frac{1}{4\lambda_{2n}} \left\langle d_n, h^* \right\rangle \end{aligned}$$

with

$$h^* = h'' - 4qh - 2q' D_0^{-1} h \in \mathcal{H}^{N-2}.$$

On the other hand, by partial integration,

$$\langle e_n, h \rangle = -\frac{1}{4\pi^2 n^2} \left\langle e_n, h'' \right\rangle.$$

Hence we obtain

$$\begin{aligned} \langle d_n - e_n, h \rangle = &-\frac{1}{4\lambda_{2n}} \left\langle d_n - e_n, h^* \right\rangle \\ &+ \left(\frac{1}{4\pi^2 n^2} - \frac{1}{4\lambda_{2n}} \right) \left\langle e_n, h^* \right\rangle \\ &+ \frac{1}{4\pi^2 n^2} \left\langle e_n, h'' - h^* \right\rangle. \end{aligned}$$

By the induction hypotheses,

$$|\langle d_n - e_n, h^* \rangle| = O\left(\frac{\log n}{n^{N-1/2}}\right) \|h^*\|_{\mathcal{H}^{N-2}},$$

and furthermore

$$\left|\frac{1}{\lambda_{2n}} - \frac{1}{\pi^2 n^2}\right| = O\left(\frac{1}{n^4}\right),$$

$$|\langle e_n, h^* \rangle| = O\left(\frac{1}{n^{N-2}}\right) \|h^*\|_{\mathcal{H}^{N-2}},$$

$$|\langle e_n, h'' - h^* \rangle| = O\left(\frac{1}{n^N}\right) \|h\|_{\mathcal{H}^N},$$

where the latter estimate holds as $h'' - h^*$ is in $\mathcal{H}^N$, and q is smooth as a finite gap potential. As $\|h^*\|_{\mathcal{H}^{N-2}} = O(\|h\|_{\mathcal{H}^N})$, the claimed result follows.

It remains to verify the statement for $N = 0$ and $N = 1$. The case $N = 0$ is an immediate consequence of Theorem 9.7. The case $N = 1$ is proved similarly to the induction step. Performing another integration by parts, one gets

$$\langle d_n, h \rangle = \frac{1}{4\lambda_{2n}} \langle d_n', D_0^{-1} h^* \rangle, \qquad \langle e_n, h \rangle = \frac{1}{4\pi^2 n^2} \langle e_n', h' \rangle.$$

From here on, one proceeds as before to estimate $\langle d_n - e_n, h \rangle$. Using asymptotic formulas for the derivatives of ∂x_n and ∂y_n, which are obtained from their representations in Theorem 9.7 and standard asymptotic estimates of the derivatives of the fundamental solution one sees that $d_n' - e_n' = O(\log n / \sqrt{n})$. This proves the claim for $N = 1$ as well. □

The last two theorems allow us to conclude that *each* Hamiltonian in the KdV hierarchy, and more generally in the KdV algebra, is a function of the actions I_n alone, when expressed in terms of the coordinates $(\boldsymbol{x}, \boldsymbol{y})$.

Theorem 11.13. *Each Hamiltonian in the KdV hierarchy, expressed in terms of the coordinates* $(\boldsymbol{x}, \boldsymbol{y})$ *in a sufficiently regular subspace of* $h_{1/2}$, *is a function of the actions alone.*

Proof. Each KdV Hamiltonian H is defined on some subspace $\mathcal{H}_0^N \subset L_0^2$ with N sufficiently large. Then $H \circ \Omega^{-1}$ is defined on $\hbar_{N+1/2}$ by Theorem 11.11.

Moreover, each KdV Hamiltonian H is a spectral invariant due to its representation in terms of the asymptotic expansion of the Δ-function for $\lambda \to -\infty$. Hence, $\hat{H} = H \circ \Omega^{-1}$ is constant on each torus

$$\operatorname{Tor}(\boldsymbol{I}) = \left\{ (\boldsymbol{x}, \boldsymbol{y}) : x_n^2 + y_n^2 = 2I_n \text{ for } n \geq 1 \right\} \subset \hbar_{N+1/2}$$

by Theorem 11.10. But this means that $\hat{H}$ is a function of $I_1, I_2, \ldots$ alone. □

It remains to show that Ω preserves the Poisson bracket in order to establish $(\boldsymbol{x}, \boldsymbol{y})$ as Birkhoff coordinates. This is done in the following section.

12 The Symplectomorphism Property

Finally we show that $\Omega\colon L_0^2 \to \hbar_{1/2}$ is not only a diffeomorphism between these two spaces, but that it also preserves the associated Poisson brackets. It turns out, however, that it is more convenient to work with symplectic structures rather than Poisson structures, since this way we can avoid the issue of establishing the regularity of the gradient $\partial\theta_n$ *everywhere* on $L_0^2 \setminus D_n$. We therefore show equivalently that Ω is a symplectomorphism.

The symplectic form on $\hbar_{1/2}$ induced by the standard Poisson bracket is

$$\omega_0 = \sum_{n\geq 1} \mathrm{d}x_n \wedge \mathrm{d}y_n .$$

The symplectic form on L_0^2 induced by the Gardner bracket is $\omega = \langle \partial_x^{-1} \cdot\, , \cdot \rangle$, as the associated Poisson structure ∂_x is nondegenerate on $\mathcal{H}_0^1$ – see section 2 on page 22.

Theorem 12.1. *The diffeomorphism $\Omega\colon L_0^2 \to \hbar_{1/2}$ is symplectic with respect to the symplectic forms ω on L_0^2 and ω_0 on $\hbar_{1/2}$.*

The proof of this theorem requires that we establish the regularity and behavior of $\partial\theta_n$ at *certain* finite gap potentials. This is done in the following three statements.

Lemma 12.2. *If $\lambda_{2k-1} = \mu_k < \lambda_{2k}$, then for any n,*

$$\partial\beta_k^n = \frac{\psi_n(\mu_k)}{\dot{\Delta}(\mu_k)} g_k h_k ,$$

with $\beta_k^k = \eta_k$, where g_k denotes the k-th normalized Dirichlet eigenfunction and h_k the unique solution of $-y'' + qy = \mu_k y$ orthogonal to g_k with Wronskian $W(g_k, h_k) = 1$. In particular, $\partial\beta_k^n$ is in $\mathcal{H}^2$.

Proof. As to the gradient, there is no difference between the β_k^n for $k \neq n$ and η_k. So it suffices to consider the former.

The gradient of β_k^n exists by analyticity. To compute it at a potential p with $\lambda_{2k-1} = \mu_k < \lambda_{2k}$, we use a trick of McKean & Vaninsky [91] and approach p by isospectral potentials q with $\sqrt[*]{\Delta^2(\mu_k) - 4} > 0$ and $\lambda_{2k-1} < \mu_k < \tau_k$. As on page 71, we can write

$$\beta_k^n = \int_{\lambda_{2k-1}}^{\mu_k^*} \frac{\psi_n(\lambda)}{\sqrt{\Delta^2(\lambda) - 4}} \,\mathrm{d}\lambda = \int_0^{\mu_k - \lambda_{2k-1}} \frac{\psi_n(\lambda_{2k-1} + z)}{\sqrt{z}\sqrt{D(z)}} \,\mathrm{d}z,$$

where $D(z) = (\Delta^2(\lambda_{2k-1} + z) - 4)/z$ is analytic and bounded away from zero near $z = 0$ for q near p. Taking the gradient with respect to q and undoing the substitution,

$$\begin{aligned}\partial\beta_k^n &= \frac{\psi_n(\mu_k)}{\sqrt[*]{\Delta^2(\mu_k) - 4}} (\partial\mu_k - \partial\lambda_{2k-1}) \\ &\quad + \int_0^{\mu_k - \lambda_{2k-1}} \frac{\partial}{\partial q} \left(\frac{\psi_n(\lambda_{2k-1} + z)}{\sqrt{D(z)}} \right) \frac{\mathrm{d}z}{\sqrt{z}} .\end{aligned}$$

The gradient under the integral is bounded, so the integral vanishes for $\mu_k \to \lambda_{2k-1}$. Hence, at p,

$$\partial \beta_k^n = \lim_{q \to p} \frac{\psi_n(\mu_k)}{\sqrt[*]{\Delta^2(\mu_k) - 4}} (\partial \mu_k - \partial \lambda_{2k-1}).$$

In particular, the latter limit exists.

To compute this limit we follow [91]. Recall that

$$\partial \lambda_{2k-1} = -\frac{\partial \Delta}{\dot{\Delta}}\bigg|_{\lambda_{2k-1}}, \qquad \partial \mu_k = -\frac{\partial m_2}{\dot{m}_2}\bigg|_{\mu_k}.$$

To write the gradient of μ_k in a useful way, differentiate the Wronskian identity $m_1' m_2 = m_1 m_2' - 1$ with respect to λ and q and use $m_2 = 0$ at μ_k to obtain

$$\begin{aligned} m_1' \dot{m}_2 &= \dot{m}_1 m_2' + m_1 \dot{m}_2' = m_2' \dot{\Delta} + (m_1 - m_2') \dot{m}_2', \\ m_1' \partial m_2 &= \partial m_1 m_2' + m_1 \partial m_2' = m_2' \partial \Delta + (m_1 - m_2') \partial m_2'. \end{aligned} \tag{12.1}$$

Substituting these expressions into $\partial \mu_k = -\partial m_2 / \dot{m}_2$ we then have

$$\partial \mu_k - \partial \lambda_{2k-1} = \frac{\partial \Delta}{\dot{\Delta}}\bigg|_{\lambda_{2k-1}} - \frac{m_2' \partial \Delta + (m_1 - m_2') \partial m_2'}{m_2' \dot{\Delta} + (m_1 - m_2') \dot{m}_2'}\bigg|_{\mu_k}.$$

Cross multiplying and dividing by $\sqrt[*]{\Delta^2(\mu_k) - 4} = (m_1 - m_2')\big|_{\mu_k}$, which is of the order of $\sqrt{\mu_k - \lambda_{2k-1}}$, we arrive in the limit at

$$\begin{aligned} \lim_{q \to p} \frac{\partial \mu_k - \partial \lambda_{2k-1}}{\sqrt[*]{\Delta^2(\mu_k) - 4}} &= \frac{\dot{m}_2' \partial m_1 - \dot{m}_1 \partial m_2'}{\dot{\Delta}^2 m_2'} \\ &= \frac{1}{\dot{\Delta}} \frac{\dot{m}_2' \partial m_1 - \dot{m}_1 \partial m_2'}{m_1' \dot{m}_2}, \end{aligned}$$

where we used that $m_2' \dot{\Delta} = m_1' \dot{m}_2$ at $\mu_k = \lambda_{2k-1}$ by (12.1).

By standard identities, $\partial m_1 = m_2 y_1^2 - m_1 y_1 y_2$ and $\partial m_2' = m_2' y_1 y_2 - m_1' y_2^2$, see appendix B. Thus,

$$\dot{m}_2' \partial m_1 - \dot{m}_1 \partial m_2' = m_1' \left(-\dot{m}_2 y_1 y_2 + \dot{m}_1 y_2^2\right),$$

again using (12.1). Hence altogether we obtain

$$\partial \beta_k^n = \frac{\psi_n(\mu_k)}{\dot{\Delta}(\mu_k)} \left(-y_1 y_2 + \frac{\dot{m}_1}{\dot{m}_2} y_2^2\right) = \frac{\psi_n(\mu_k)}{\dot{\Delta}(\mu_k)} y_0 y_2$$

with $y_0 = -y_1 + (\dot{m}_1 / \dot{m}_2) y_2$. Since $\partial \beta_k^n$ has mean value zero, y_0 and y_2 are orthogonal to each other, and the result follows by setting $h_k = \|y_2\|\, y_0$. □

Recall from Theorem 11.9 that

$$\mathcal{G}_A = \{u \in L_0^2 \colon \gamma_k(u) > 0 \text{ iff } k \in A\}$$

denotes the submanifold of A-gap potentials in L_0^2, for any $A \subset \mathbb{N}$. Within $\mathcal{G}_A$ we single out a submanifold of "normalized" potentials, namely

$$\mathcal{G}_A^{\mathrm{o}} = \{u \in \mathcal{G}_A \colon \theta_k(u) = 0 \text{ for } k \in A\}.$$

Equivalently, these A-gap potentials are characterized by $\mu_k = \lambda_{2k-1}$ for $k \in A$.

Lemma 12.3. *At a finite gap potential in $\mathcal{G}_A^{o}$ the series*

$$\partial\theta_n = \partial\eta_n + \sum_{k \neq n} \partial\beta_k^n, \qquad n \in A,$$

converges in $\mathcal{H}^1$ to the L^2-gradient of the function θ_n.

Proof. Fix $n \in A$. For $k \in A$, $\partial\beta_k^n$ is in $\mathcal{H}^2$ by the preceding lemma. For $k \notin A$ approximate a given $q \in \mathcal{G}_A^{\mathrm{o}}$ by finite gap potentials in $\mathcal{G}_{A\cup\{k\}}^{\mathrm{o}}$ with $\lambda_{2k-1} = \mu_k$. By the same lemma and l'Hospital's rule,

$$\partial\beta_k^n = \frac{\dot{\psi}_n(\mu_k)}{\ddot{\Delta}(\mu_k)} g_k h_k.$$

The product expansions for ψ_n and $\dot{\Delta}$ – see Proposition B.13 – give

$$\dot{\psi}_n(\mu_k) = -\frac{2}{\pi n}\frac{1}{k^2\pi^2}\prod_{m\neq k,n}\frac{\sigma_m^n - \mu_k}{m^2\pi^2}$$

$$\ddot{\Delta}(\mu_k) = \frac{1}{k^2\pi^2}\prod_{m\neq k}\frac{\dot{\lambda}_m - \mu_k}{m^2\pi^2}.$$

As $\sigma_m^n = \dot{\lambda}_m$ for $m \notin A$, we thus have

$$\frac{\dot{\psi}_n(\mu_k)}{\ddot{\Delta}(\mu_k)} = -\frac{2\pi n}{\dot{\lambda}_n - \mu_k}\prod_{\substack{m\in A\\ m\neq n}}\frac{\sigma_m^n - \mu_k}{\dot{\lambda}_m - \mu_k} = O\left(\frac{1}{k^2}\right)$$

for n fixed. Finally, $g_k h_k$ is in $\mathcal{H}^2$ for each k, and by the asymptotics for g_k from appendix B, the orthogonality of h_k to g_k and their Wronskian being one, we have

$$g_k = \sqrt{2}\sin\pi kx + O\left(\frac{1}{k}\right),$$

$$h_k = -\frac{1}{\sqrt{2}\pi k}\cos\pi kx + O\left(\frac{1}{k^2}\right).$$

Hence

$$g_k h_k = -\frac{1}{2\pi k}\sin 2\pi kx + O\left(\frac{1}{k^2}\right).$$

A corresponding estimate holds for the first derivative. From this the final result follows. □

Proposition 12.4. *At a finite gap potential in $\mathcal{G}^o_A$,*

$$[\theta_n, \theta_m] = 0, \qquad [I_n, \theta_m] = \delta_{nm},$$

for $n, m \in A$. Moreover,

$$[x_k, \theta_m] = 0, \qquad [y_k, \theta_m] = 0,$$

for $k \notin A$ and $m \in A$.

Proof. By the preceding lemma the series $\partial\theta_n = \partial\eta_n + \sum_{k\neq n}\partial\beta^n_k$ converges in $\mathcal{H}^1$. Thus we have $[\theta_n, \theta_m] = 0$ if we show that

$$[\beta^n_k, \beta^m_l] = 0$$

for all $n, m \in A$ and all $k, l \geq 1$, with $\beta^n_n = \eta_n$. In view of the representation of the gradients of β^n_k and β^m_l in Lemma 12.2 this reduces to showing that

$$\langle g_k h_k, (g_l h_l)'\rangle = 0$$

for all k and l, since only the factors involving the ψ-function depend on n and m.

But this follows by a straightforward computation. There is nothing to do for $k = l$, so we may assume that $k \neq l$. Denoting by $W(g, h) = gh' - g'h$ the Wronskian between g and h, we have

$$\begin{aligned} 2\langle g_k h_k, (g_l h_l)'\rangle &= \langle g_k h_k, (g_l h_l)'\rangle - \langle (g_k h_k)', g_l h_l\rangle \\ &= \int_0^1 \left(g_k h_k g_l h_l' + g_k h_k g_l' h_l - g_k h_k' g_l h_l - g_k' h_k g_l h_l\right) \mathrm{d}x \\ &= \int_0^1 \left(g_k g_l W(h_k, h_l) + h_k h_l W(g_k, g_l)\right) \mathrm{d}x. \end{aligned}$$

Since $W(h_k, h_l)' = (\mu_k - \mu_l) h_k h_l$, and similarly for $W(g_k, g_l)'$, this equals

$$\begin{aligned} &\frac{1}{\mu_k - \mu_l}\int_0^1 \left(W(g_k, g_l)' W(h_k, h_l) + W(g_k, g_l) W(h_k, h_l)'\right) \mathrm{d}x \\ &\qquad = \frac{1}{\mu_k - \mu_l} W(g_k, g_l) W(h_k, h_l)\Big|_0^1 = 0. \end{aligned}$$

This proves that $[\theta_n, \theta_m] = 0$.

The second identity $[I_n, \theta_m] = \delta_{nm}$ follows from Proposition 10.5 by partial integration, using that $\partial\theta_m$ is in $\mathcal{H}^1$.

Finally, given $k \notin A$ consider first a potential in $\mathcal{G}^o_{A\cup\{k\}}$. The standard definitions $x_k = \sqrt{2I_k}\cos\theta_k$ and $y_k = \sqrt{2I_k}\sin\theta_k$ give

$$[x_k, \theta_m] = \frac{1}{\sqrt{2I_k}}\cos\theta_m\,[I_k, \theta_m] - \sqrt{2I_k}\sin\theta_m\,[\theta_k, \theta_m] = 0,$$

and similarly $[y_k, \theta_m] = 0$. By continuity, the same holds for potentials in $\mathcal{G}^o_A$. □

We are now in a position to show that $\Omega\colon L^2_0 \to \ell_{1/2}$ is symplectic with respect to ω on L^2_0 and ω_0 on $\ell_{1/2}$. More precisely, we show that

$$\tilde{\omega} \overset{\text{def}}{=} \Omega_*\omega = \omega_0$$

on $\ell_{1/2}$. By continuity, it suffices to verify this for the restrictions of $\tilde{\omega}$ and ω_0 to a family of submanifolds filling $\ell_{1/2}$ densely. We consider the family of spaces

$$g_N = \left\{ (x, y) \in \ell_0 : x_n^2 + y_n^2 > 0 \text{ iff } 1 \le n \le N \right\}.$$

By Theorem 11.9, their preimages under Ω are the manifolds $\mathcal{G}_N$ of N-gap potentials in L^2_0.

On g_N we may use standard angle-action coordinates $(I, \theta) = (I_1, \ldots, I_N, \theta_1, \ldots, \theta_N)$, in which the restriction of ω_0 is again a symplectic form, namely

$$\omega_0|g_N = \sum_{1\le n\le N} \mathrm{d}I_n \wedge \mathrm{d}\theta_n.$$

So we have to show that $\tilde{\omega}|g_N = \omega_0|g_N$ for each large N. Henceforth, we drop the '$|g_N$' from the notation.

Consider the standard tangent vectors

$$e_n = \frac{\partial}{\partial I_n}, \qquad f_n = \frac{\partial}{\partial \theta_n}, \qquad 1 \le n \le N,$$

which form a basis of the tangent space to g_N at any point of g_N and are supplemented to a basis of $\ell_{1/2}$ by the vectors $\partial/\partial x_k$ and $\partial/\partial y_k$ for $k \ge N+1$. Let

$$E_n = \Omega^* e_n, \qquad F_n = \Omega^* f_n, \qquad 1 \le n \le N,$$

be their preimages under Ω.

Lemma 12.5. *For* $1 \le n \le N$,

$$E_n = \partial_x \partial\theta_n, \qquad F_n = -\partial_x \partial I_n$$

everywhere on $\mathcal{G}^o_N$ *and on* $\mathcal{G}_N$*, respectively.*

Proof. Fix $1 \le n \le N$. Expressing $\Omega_* F_n = f_n$ in the coordinates I_n, θ_n and x_k, y_k one obtains

$$\begin{aligned} \langle \partial I_m, F_n \rangle &= 0, \\ \langle \partial \theta_m, F_n \rangle &= \delta_{mn}, \\ \langle \partial x_k, F_n \rangle &= 0, \\ \langle \partial y_k, F_n \rangle &= 0, \end{aligned}$$

for $1 \le m \le N$ and $k \ge N + 1$. In view of Lemma 10.5, exactly the same is true, when F_n is replaced by $-\partial_x \partial I_n$, which is also in L_0^2. Since Ω is a diffeomorphism, we thus must have $F_n = -\partial_x \partial I_n$.

The same reasoning applies to E_n, using Lemma 12.4 and the fact that on $\mathcal{G}_N^{\circ}$, $\partial\theta_n$ is in $\mathcal{H}^1$ for $1 \le n \le N$. □

Proof of Theorem 12.1. We can now prove that Ω is symplectic. Considering

$$\tilde{\omega} = \Omega_* \omega,$$

where $\omega = \langle \partial_x^{-1} \cdot\, , \cdot \rangle$ is the symplectic form on L_0^2, we have

$$\tilde{\omega}(\xi, \eta) = \omega(\Omega^* \xi, \Omega^* \eta),$$

where ξ, η are tangent vectors to $\mathcal{g}_N$ at $\Omega(q)$. We then find that for $1 \le m, n \le N$,

$$\begin{aligned} \tilde{\omega}(f_m, f_n) &= \omega(F_m, F_n) \\ &= \langle \partial_x^{-1} F_m, F_n \rangle \\ &= -\langle \partial I_m, F_n \rangle = 0 \end{aligned}$$

by Lemma 12.5, and similarly

$$\begin{aligned} \tilde{\omega}(e_n, f_m) &= \omega(E_n, F_m) \\ &= \langle \partial_x^{-1} E_n, F_m \rangle \\ &= \langle \partial \theta_n, F_m \rangle = \delta_{nm}. \end{aligned}$$

Hence for the symplectic form restricted to $\mathcal{g}_N$ we must have

$$\tilde{\omega} = \sum_{1 \le n \le N} \mathrm{d} I_n \wedge \mathrm{d}\theta_n + \sum_{1 \le m, n \le N} a_{mn}\, \mathrm{d} I_m \wedge \mathrm{d} I_n,$$

with coefficients a_{mn} depending smoothly on I and θ in view of Theorem 11.8.

However, since $\tilde{\omega}$ is closed,

$$L_{f_n} \tilde{\omega} = \mathrm{d}(\tilde{\omega}(f_n, \cdot)) = \mathrm{d}(\mathrm{d} I_n) = 0,$$

so the a_{mn} are independent of θ. It therefore suffices to consider points in $\mathcal{g}_N$ with

$\theta = 0$, which correspond to N-gap potentials in $\mathcal{G}_N^{\mathrm{o}}$. But then, by Proposition 12.4,

$$\begin{aligned}\tilde{\omega}(e_m, e_n) &= \omega(E_m, E_n)\\ &= \langle \partial_x^{-1} E_m, E_n \rangle\\ &= [\theta_m, \theta_n]\\ &= 0.\end{aligned}$$

That is, $a_{mn} = 0$ for $1 \le m, n \le N$. This proves the theorem. □

Thus, we finally completed the proof of Theorem 6.1. Having established Ω as a canonical diffeomorphism we know that $(\boldsymbol{x}, \boldsymbol{y})$ are Birkhoff coordinates for each Hamiltonian H in the KdV hierarchy, or more generally in the Poisson algebra of KdV, when restricted to the appropriate subspace of $\hbar_{1/2}$. That is, $H' = H \circ \Omega^{-1}$ is a function of the actions alone, and in fact a real analytic one – see the Addendum to Theorem 15.1. Its equations of motions are given by

$$\begin{aligned}\dot{x}_n &= \omega_n(\boldsymbol{I}) y_n,\\ \dot{y}_n &= -\omega_n(\boldsymbol{I}) x_n,\end{aligned}$$

where

$$\omega_n(\boldsymbol{I}) = \frac{\partial H'}{\partial I_n}(\boldsymbol{I})$$

with $\boldsymbol{I} = (I_n)_{n \ge 1} = \frac{1}{2}(x_n^2 + y_n^2)_{n \ge 1}$ are the associated frequencies. Thus, each KdV equation becomes an infinite-dimensional system of ordinary differential equations, describing the motion of infinitely many oscillators whose frequencies depend on their amplitudes in a nonlinear fashion.

For the convenience of reference we collect the results stated in Theorems 6.1, 9.8, 11.9 and 11.10 again in one single statement, using the notations established there.

Theorem 12.6. *There exists a diffeomorphism*

$$\Omega\colon\ L_0^2 \to \hbar_{1/2}$$

with the following properties.

(i) *Ω is one-to-one, onto, bi-analytic, and preserves the Poisson bracket.*

(ii) *For each $N \ge 0$, the restriction of Ω to $\mathcal{H}_0^N$, denoted by the same symbol, is a map*

$$\Omega\colon\ \mathcal{H}_0^N \to \hbar_{N+1/2},$$

which is one-to-one, onto, and bi-analytic as well.

(iii) *The coordinates $(\boldsymbol{x}, \boldsymbol{y})$ in $\hbar_{3/2}$ are global Birkhoff coordinates for the KdV equation. That is, the transformed KdV Hamiltonian $H \circ \Omega^{-1}$ depends only on $x_n^2 + y_n^2$, $n \ge 1$, with $(\boldsymbol{x}, \boldsymbol{y})$ being canonical coordinates in $\hbar_{3/2}$.*

(iv) *The same holds for any other Hamiltonian in the KdV hierarchy, considered on a subspace* $\mathcal{H}_0^N$ *with appropriate* N.

(v) *The Jacobian of* Ω *at* $q = 0$ *is boundedly invertible, its inverse is the weighted Fourier transform*

$$\begin{aligned} d_0\Omega^{-1}\colon\ \mathcal{h}_{1/2} &\to L_0^2 \\ (\boldsymbol{x}, \boldsymbol{y}) &\mapsto \sum_{n\geq 1} \sqrt{2\pi n}\,\big(x_n e_n + y_n e_{-n}\big). \end{aligned}$$

(vi) *For each finite index set* $A \subset \mathbb{N}$,

$$\mathcal{G}_A = \Omega^{-1}(\mathcal{h}_A).$$

(vii) Ω *maps each isospectral set* $\operatorname{Iso}(q)$ *in* L_0^2 *one-to-one onto the torus* $\operatorname{Tor}(\boldsymbol{I}(q))$ *in* $\mathcal{h}_{1/2}$.

(viii) $\bigcup_{M\geq 1} \mathcal{G}_{\{1,\dots,M\}}$ *is dense in* $\mathcal{H}_0^N$.

IV

Perturbed KdV Equations

13 The Main Theorems

In this chapter we study small perturbations of the KdV equation

$$u_t = -u_{xxx} + 6uu_x$$

on the real line with periodic boundary conditions. We consider this equation as an infinite dimensional, *integrable Hamiltonian* system and subject it to sufficiently small *Hamiltonian* perturbations. The aim is to show that large families of time-quasi-periodic solutions persist under such perturbations. This is true not only for this KdV equation, but in principle for all higher order KdV equations as well. As an example, the second equation in the KdV hierarchy will be considered in detail.

Background

To set the stage we introduce for any integer $N \geq 0$ the phase space

$$\mathcal{H}^N = \left\{ u \in L^2(S^1; \mathbb{R}) \colon \ \|u\|_N < \infty \right\}$$

of real valued functions on $S^1 = \mathbb{R}/\mathbb{Z}$, where

$$\|u\|_N^2 = \left|\hat{u}(0)\right|^2 + \sum_{k \in \mathbb{Z}} |k|^{2N} \left|\hat{u}(k)\right|^2$$

is defined in terms of the Fourier transform $\hat{u}$ of u, $u(x) = \sum_{k\in\mathbb{Z}} \hat{u}(k) e^{2\pi i k x}$. In particular, we have $\mathcal{H}^0 = L^2(S^1)$ with norm $\|\cdot\| = \|\cdot\|_0$. We endow $\mathcal{H}^N$ with the Poisson structure proposed by Gardner,

$$\{F, G\} = \int_{S^1} \frac{\partial F}{\partial u(x)} \frac{\mathrm{d}}{\mathrm{d}x} \frac{\partial G}{\partial u(x)} \, \mathrm{d}x,$$

where F, G are differentiable functions on $\mathcal{H}^N$ with L^2-gradients in $\mathcal{H}^1$. The Hamiltonian corresponding to KdV is then given by

$$H(u) = \int_{S^1} \left(\tfrac{1}{2}u_x^2 + u^3\right) \mathrm{d}x,$$

and

$$\frac{\partial u}{\partial t} = \frac{\mathrm{d}}{\mathrm{d}x} \frac{\partial H}{\partial u}$$

is the KdV equation written in Hamiltonian form.

The Poisson bracket $\{\,\cdot\,,\,\cdot\,\}$ is degenerate and admits the average

$$[u] = \int_{S^1} u(x)\, \mathrm{d}x$$

as a Casimir function. Moreover, the Poisson structure induces a trivial foliation with leaves

$$\mathcal{H}_c^N = \left\{u \in \mathcal{H}^N : [u] = c\right\}, \qquad c \in \mathbb{R}.$$

Instead of considering the restriction of the Hamiltonian H to each leaf $\mathcal{H}_c^N$, it is more convenient to choose a fixed phase space, $\mathcal{H}_0^N$, which is symplectomorphic to every other leaf $\mathcal{H}_c^N$ by translation. Writing $u = v + c$ with $[v] = 0$ and $c = [u]$ the Hamiltonian then takes the form

$$H(u) = H_c(v) + c^3$$

with

$$H_c(v) = \int_{S^1} \left(\tfrac{1}{2}v_x^2 + v^3\right) \mathrm{d}x + 6c \int_{S^1} \tfrac{1}{2}v^2\, \mathrm{d}x. \tag{13.1}$$

Here, $\int_{S^1} \frac{1}{2}v^2\, \mathrm{d}x$ is the zero-th Hamiltonian of the KdV hierarchy and corresponds to translation.

To describe the structure of the phase space we recall some facts from the spectral theory of Hill's equation, with more details given in appendix B. For u in $L_0^2 = \mathcal{H}_0^0$ consider the differential operator

$$L = -\frac{\mathrm{d}^2}{\mathrm{d}x^2} + u$$

on the interval $[0, 2]$ with periodic boundary conditions. Its *spectrum*, denoted by $\operatorname{spec}(u)$, is pure point and consists of an unbounded sequence of *periodic eigenvalues*

$$\lambda_0(u) < \lambda_1(u) \le \lambda_2(u) < \lambda_3(u) \le \lambda_4(u) < \dots\ .$$

Equality or inequality may occur in every place with a '$\le$'-sign, and one speaks of the *gaps* $(\lambda_{2n-1}(u), \lambda_{2n}(u))$ of the *potential* u and its *gap lengths*

$$\gamma_n(u) = \lambda_{2n}(u) - \lambda_{2n-1}(u), \qquad n \ge 1.$$

In the case of a double periodic eigenvalue, the gap is empty, and one speaks of a *collapsed gap*. Otherwise, the gap is said to be *open*.

The space L_0^2 naturally decomposes into the isospectral sets

$$\operatorname{Iso}(u) = \left\{ v \in L_0^2 \colon \operatorname{spec}(v) = \operatorname{spec}(u) \right\}.$$

Our particular interest is in the isospectral sets of so called *finite gap potentials* u, which are characterized by the fact that only a finite number of gaps are open. Such potentials are known to be real analytic and dense in any space $\mathcal{H}_0^N$, see [49, 84]. In particular, for every *finite* index set $A \subset \mathbb{N}$ we may consider the set of *A-gap potentials*

$$\mathcal{G}_A = \left\{ u \in L_0^2 \colon \gamma_n(u) > 0 \Leftrightarrow n \in A \right\}.$$

Clearly,

$$u \in \mathcal{G}_A \quad \Leftrightarrow \quad \operatorname{Iso}(u) \subset \mathcal{G}_A,$$

and $\mathcal{G}_A \subset \bigcap_{N \geq 1} \mathcal{H}_0^N$, as all A-gap potentials are real analytic.

It turns out that the isospectral sets of A-gap potentials are actually $|A|$-dimensional tori, which are uniquely parameterized by their positive gap lengths. As functions on L_0^2, however, the periodic eigenvalues and their associated gap lengths are not differentiable at points where they are double. To avoid this difficulty, we may instead consider nonnegative *actions*

$$I_n(u) \geq 0, \qquad n \geq 1,$$

which are also defined entirely in terms of the periodic spectrum of u and closely related to the gap lengths $\gamma_n(u)$. For instance, I_n vanishes precisely when γ_n vanishes. But unlike the latter, the former are real analytic on all of L_0^2. See section 7 for the details.

We then have the following picture. The set $\mathcal{G}_A$ of A-gap potentials is a real analytic submanifold of L_0^2, and there exists a bi-analytic diffeomorphism

$$\Omega_A \colon \ \mathcal{G}_A \to \mathbb{T}^A \times \mathbb{R}_+^A,$$

which maps each isospectral set $\mathcal{T} = \operatorname{Iso}(u)$ in $\mathcal{G}_A$ onto some torus $T_I = \mathbb{T}^A \times \{I\}$, where $I = (I_n)_{n \in A}$ are the nontrivial actions of u. See Theorems 6.1, 11.9 and 11.10 for the details.

Incidentally, if u is not a finite gap potential, then $\operatorname{Iso}(u)$ is homeomorphic to a product of infinitely many circles endowed with the product topology, one circle for each open gap. It is not a manifold in the usual sense, since $\operatorname{Iso}(u)$ is compact in L_0^2, whereas an infinite dimensional Hilbert manifold is never compact.

Returning to the KdV equation, the relevance of these results stems from the fact that, considered as functions on L_0^2, the periodic eigenvalues represent *integrals* for the evolution of the KdV equation, as follows from the Lax pair formalism for KdV. The same is then true for the gap lengths γ_n and the actions I_n, as they are defined in terms of the periodic spectrum, too. The latter are moreover real analytic, functionally independent and in involution on L_0^2.

Thus, if u^t is a solution curve of the KdV equation in some space $\mathcal{H}_0^N$ with initial value u^{o}, then

$$\operatorname{spec}(u^t) = \operatorname{spec}(u^{\text{o}})$$

for all t. So the entire solution is confined to $\operatorname{Iso}(u^{\text{o}})$. Consequently, the whole phase space decomposes into a collection of tori of varying dimension which are *invariant* under the KdV flow. In particular, each manifold $\mathcal{G}_A$ of A-gap potentials is completely foliated into invariant tori of the same dimension $|A|$, and the actions I_n, $n \in A$, form a complete set of independent integrals in involution on $\mathcal{G}_A$. Hence, by the Liouville-Arnold-Jost-Theorem, the flow on each torus is linear in suitable coordinates and completely characterized by $|A|$ fixed frequencies: it consists of quasi-periodic motions winding around the torus in phase space.

Analytically speaking, one can find a bi-analytic diffeomorphism

$$\Phi_A\colon\ \mathbb{T}^A \times \mathbb{R}_+^A \to \mathcal{G}_A$$

and a real analytic frequency map

$$\varphi_A\colon\ \mathbb{R}_+^A \to \mathbb{R}^A,$$

such that Φ_A maps each fiber $T_I = \mathbb{T}^A \times \{I\}$ onto an invariant torus $\mathcal{T}_I = \Phi_A(T_I)$, and such that for each $\theta \in \mathbb{T}^A$,

$$\Phi_A(\theta + t\varphi_A(I), I), \qquad t \in \mathbb{R},$$

is a smooth, even real analytic solution of the KdV equation.

On $\mathcal{G}_A$ the KdV equation is thus *completely integrable* in the classical sense as a system of $|A|$ degrees of freedom. In addition, all its invariant tori are *linearly stable* in the infinite dimensional ambient phase space, since all their Lyapunov exponents vanish. This follows from the fact that all nearby solutions are also confined to invariant tori.

Results

In the following theorem we are going to describe a perturbation theory for these A-gap solutions. Let $\Gamma \subset \mathbb{R}_+^A$ be a compact set of A-actions of positive Lebesgue measure, and let

$$\mathcal{T}_\Gamma = \bigcup_{I \in \Gamma} \mathcal{T}_I \subset \mathcal{G}_A, \qquad \mathcal{T}_I = \Omega_A^{-1}(T_I),$$

be the union of the corresponding tori in $\mathcal{G}_A$. Recall that $\mathcal{G}_A \subset \bigcap_{N \geq 1} \mathcal{H}_0^N$. With $\mathcal{H}_{0,\mathbb{C}}^N$ we denote the complexification of $\mathcal{H}_0^N$ with norm $\|\cdot\|_N$, and with

$$\|F\|_{N;U}^{\sup} = \sup_{u \in U} \|F(u)\|_N$$

the usual sup-norm of a function F on a domain $U \subset \mathcal{H}_{0,\mathbb{C}}^N$.

The following result was first proven by Kuksin for the case $c = 0$ [74].

Theorem 13.1. *Let $A \subset \mathbb{N}$ be a finite index set of cardinality m, $\Gamma \subset \mathbb{R}_+^A$ a compact subset of positive Lebesgue measure, and $N \geq 1$. Assume that the Hamiltonian K is real analytic in a complex neighbourhood V of $\mathcal{T}_\Gamma$ in $\mathcal{H}_{0,\mathbb{C}}^N$ and satisfies the regularity condition*

$$\frac{\partial K}{\partial u}\colon \ V \to \mathcal{H}_{0,\mathbb{C}}^N, \qquad \left\| \frac{\partial K}{\partial u} \right\|_{N;V}^{\sup} \leq 1.$$

Then, for any real c, there exists an $\varepsilon_0 > 0$ depending only on A, N, c and the size of V such that for $|\varepsilon| < \varepsilon_0$ the following holds. There exist

(i) *a nonempty Cantor set $\Gamma_\varepsilon \subset \Gamma$ with* $\operatorname{meas}(\Gamma - \Gamma_\varepsilon) \to 0$ *as $\varepsilon \to 0$,*

(ii) *a Lipschitz family of real analytic torus embeddings*

$$\Xi\colon \ \mathbb{T}^m \times \Gamma_\varepsilon \to V \cap \mathcal{H}_0^N,$$

(iii) *a Lipschitz map $\chi : \Gamma_\varepsilon \to \mathbb{R}^m$,*

such that for each $(\theta, I) \in \mathbb{T}^m \times \Gamma_\varepsilon$, the curve $u(t) = \Xi(\theta + \chi(I)t, I)$ is a quasi-periodic solution of

$$\frac{\partial u}{\partial t} = \frac{\mathrm{d}}{\mathrm{d}x}\left(\frac{\partial H_c}{\partial u} + \varepsilon \frac{\partial K}{\partial u} \right)$$

winding around the invariant torus $\Xi(\mathbb{T}^m \times \{I\})$. Moreover, each such torus is linearly stable.

Remark 1. The set Γ_ε depends on the perturbation εK. Its measure estimate, however, depends only on its size ε and is independent of other properties of K.

Remark 2. The "size of V" refers to the diameter of a complex neighbourhood around a real domain. See section 17 for details.

For further remarks see section 1 on page 9.

Besides the KdV equation there is a whole hierarchy of so called higher order KdV equations, which are all in involution with each other – see appendix C. They also admit the same set of integrals as the KdV equation. Consequently, they all possess the same families of invariant tori. The difference is only in the frequencies of the quasi-periodic motions on each of these tori. Therefore, similar results should also hold for the higher order KdV equations.

As an example we consider the second KdV equation, which reads

$$\partial_t u = \partial_x^5 u - 10u\partial_x^3 u - 20\partial_x u \partial_x^2 u + 30u^2 \partial_x u.$$

Its Hamiltonian is

$$H^2(u) = \int_{S^1} \left(\tfrac{1}{2}u_{xx}^2 + 5uu_x^2 + \tfrac{5}{2}u^4\right) \mathrm{d}x,$$

which is defined on $\mathcal{H}^2 = H^2(S^1; \mathbb{R})$. Again, with $u = v + c$ and $[v] = 0$,

$$H^2(u) = H_c^2(v) + \frac{5}{2}c^4.$$

Here

$$H_c^2(v) = H^2(v) + 10cH^1(v) + 30c^2H^0(v),$$

with $H^1 = \int_{S^1} \left(\frac{1}{2}v_x^2 + v^3\right) \mathrm{d}x$ the KdV Hamiltonian, and $H^0 = \int_{S^1} \frac{1}{2}v^2 \,\mathrm{d}x$ the Hamiltonian of translation. We study this Hamiltonian on $\mathcal{H}_0^N$ with $N \geq 2$, considering c as a real parameter.

A version of the following result was first stated by Kuksin [74].

Theorem 13.2. *Let $A \subset \mathbb{N}$ be a finite index set, $\Gamma \subset \mathbb{R}_+^A$ a compact subset of positive Lebesgue measure, and $N \geq 2$. Assume that the Hamiltonian K is real analytic in a complex neighbourhood V of $\mathcal{T}_\Gamma$ in $\mathcal{H}_{0,\mathbb{C}}^N$ and satisfies the regularity condition*

$$\frac{\partial K}{\partial u} \colon \; V \to \mathcal{H}_{0,\mathbb{C}}^{N-2}, \qquad \left\| \frac{\partial K}{\partial u} \right\|_{N-2;V}^{\sup} \leq 1.$$

If $c \notin \mathcal{E}_A^2$, where the exceptional set $\mathcal{E}_A^2$ is an at most countable subset of the real line not containing 0 and with at most $|A|$ accumulation points, then the same conclusions as in Theorem 13.1 hold for the system with Hamiltonian $H_c^2 + \varepsilon K$.

Remark. A more detailed description of the set $\mathcal{E}_A^2$ is given in appendix J. It is likely that the theorem is true for *all* $c \in \mathbb{R}$.

Outline of Proof

The proof of Theorems 13.1 and 13.2 is based on an infinite dimensional version of the KAM theory that is concerned with the persistence of finite dimensional invariant tori and is applicable to small perturbations of the KdV equation.

A prerequisite for developing such a perturbation theory is the existence of coordinates with respect to which the linearized equations along the unperturbed motions on the invariant tori reduce to constant coefficient form. Often, such coordinates are difficult to construct even locally. In the case of the KdV equation, however, such coordinates exist *globally*.

For any $r \geq 0$ we introduce the model space of real sequences

$$h_r = \ell_r^2 \times \ell_r^2$$

with elements $(\boldsymbol{x}, \boldsymbol{y})$, where

$$\ell_r^2 = \left\{ \boldsymbol{x} \in \ell^2(\mathbb{N}, \mathbb{R}) \colon \; \|\boldsymbol{x}\|_r^2 = \sum_{n \geq 1} n^{2r} \, |x_n|^2 < \infty \right\}.$$

We endow this space with the standard symplectic structure $\sum_{n \geq 1} \mathrm{d}x_n \wedge \mathrm{d}y_n$. In chapter III we construct a bi-analytic symplectomorphism

$$\Psi \colon \; h_{1/2} \to L_0^2,$$

such that the KdV Hamiltonian on the model space $\hbar_{3/2}$, again denoted H_c, is of the form

$$H_c = H_c(I_1, I_2, \ldots), \qquad I_n = \frac{1}{2}(x_n^2 + y_n^2).$$

The equations of motion are thus

$$\dot{x}_n = \omega_n(\boldsymbol{I})y_n, \qquad \dot{y}_n = -\omega_n(\boldsymbol{I})x_n,$$

with frequencies

$$\omega_n = \frac{\partial H_c}{\partial I_n}(\boldsymbol{I}), \qquad \boldsymbol{I} = (I_n)_{n\geq 1},$$

that are constant along each orbit. So each orbit is winding around some invariant torus

$$T_{\boldsymbol{I}} = \{(\boldsymbol{x}, \boldsymbol{y}) : x_n^2 + y_n^2 = 2I_n,\ n \geq 1\},$$

where the parameters $\boldsymbol{I} = (I_n)_{n\geq 1}$ are the actions of its initial data.

We are interested in a perturbation theory for families of *finite*-dimensional tori $T_{\boldsymbol{I}}$. So we fix an index set $A \subset \mathbb{N}$ of finite cardinality $m = |A|$, and consider tori with

$$I_n > 0 \quad \Leftrightarrow \quad n \in A.$$

The linearized equations of motion along any such torus have now constant coefficients and are determined by m *internal frequencies* $\omega = (\omega_n)_{n\in A}$ and infinitely many *external frequencies* $\Omega = (\omega_n)_{n\notin A}$. Both depend on the m-dimensional parameter

$$\xi = (I_n)_{n\in A},$$

since all other components of $\boldsymbol{I}$ vanish.

The KAM theorem for such families of finite dimensional tori requires a number of assumptions, among which the most notorious and unpleasant ones are the so called *nondegeneracy* and *nonresonance conditions*. In this case, they essentially amount to the following. First, the map

$$\xi \mapsto \omega(\xi)$$

from the parameters to the internal frequencies has to be a local homeomorphism, which is Lipschitz in both directions. This is known as *Kolmogorov's condition* in the classical theory. Second, the zero set of any of the frequency combinations

$$\langle k, \omega(\xi)\rangle + \langle l, \Omega(\xi)\rangle$$

has to be a set of measure zero in Π, for each $k \in \mathbb{Z}^m$ and $l \in \mathbb{Z}^\infty$ with $1 \leq |l| \leq 2$. This is sometimes called *Melnikov's condition.*

To verify these conditions for the KdV Hamiltonian we need some knowledge of its frequencies. To this end we compute the first coefficients of the Birkhoff normal form of the KdV Hamiltonian. Writing

$$u = \sum_{n \neq 0} \gamma_n q_n \mathrm{e}^{2\pi \mathrm{i} n x}$$

with weights $\gamma_n = \sqrt{2\pi\,|n|}$ and complex coefficients $q_{\pm n} = (x_n \mp \mathrm{i} y_n)/\sqrt{2}$, the KdV Hamiltonian becomes

$$H_c = \sum_{n \geq 1} \lambda_n \left|q_n\right|^2 + \sum_{k+l+m=0} \gamma_k \gamma_l \gamma_m q_k q_l q_m$$

on $\hbar_{3/2}$ with

$$\lambda_n = (2\pi n)^3 + 6c \cdot 2\pi n.$$

Thus, at the origin we have an elliptic equilibrium with characteristic frequencies $\lambda_1, \lambda_2, \ldots$.

We then construct a real analytic, symplectic coordinate transformation Φ in a neighbourhood of the origin, which transforms H_c into

$$H_c \circ \Phi = \frac{1}{2} \sum_{n \geq 1} \lambda_n (x_n^2 + y_n^2) - \frac{3}{4} \sum_{n \geq 1} (x_n^2 + y_n^2)^2 + \ldots\ .$$

The important fact about the non-resonant Birkhoff normal form is that its coefficient are uniquely determined, once the quadratic term is fixed, and do not depend on the normalizing transformation – see appendix G. Comparing the global and the local KdV Hamiltonians we may thus conclude that they agree up to terms of order four – that is, the local result provides us with the first terms of the Taylor series expansion of the globally integrable KdV Hamiltonian. We obtain

$$H_c = H_c(I_1, I_2, \ldots) = \sum_{n \geq 1} \lambda_n I_n - 3 \sum_{n \geq 1} I_n^2 + \ldots\ .$$

Consequently

$$\omega_n(\boldsymbol{I}) = \frac{\partial H_c}{\partial I_n}(\boldsymbol{I}) = \lambda_n - 6 I_n + \ldots,$$

where λ_n and ω_n also depend on c. By further computing some additional terms of order six in the expansion above, we gain sufficient control over the frequencies ω to verify all nondegeneracy and nonresonance conditions for any c. This allows us to apply KAM theory and eventually prove Theorems 13.1 and 13.2.

14 Birkhoff Normal Form

We begin by transforming the KdV Hamiltonians into their Birkhoff normal forms up to order four. Recall that after writing $u = v + c$ with $[v] = 0$, our phase space is

$\mathcal{H}_0^N$ endowed with the Poisson structure

$$\{F, G\} = \int_{S^1} \frac{\partial F}{\partial u(x)} \frac{\mathrm{d}}{\mathrm{d}x} \frac{\partial G}{\partial u(x)} \, \mathrm{d}x.$$

The KdV Hamiltonian is

$$H_c(v) = \int_{S^1} \left(\tfrac{1}{2} v_x^2 + v^3\right) \mathrm{d}x + 3c \int_{S^1} v^2 \, \mathrm{d}x, \tag{14.1}$$

where c is considered as a real parameter.

To write this Hamiltonian system more explicitly as an infinite dimensional system we introduce infinitely many coordinates $\boldsymbol{v} = (v_n)_{n \neq 0}$ by writing

$$v = \mathcal{F}(\boldsymbol{v}) \stackrel{\mathrm{def}}{=} \sum_{n \neq 0} \gamma_n v_n \mathrm{e}^{2\pi \mathrm{i} n x}, \tag{14.2}$$

where

$$\gamma_n = \sqrt{2\pi |n|}$$

are fixed positive weights. The sequence $\boldsymbol{v} = (v_n)_{n \neq 0}$ is an element of the Hilbert space $\mathcal{h}_r^\bullet$ of all *complex* valued sequences $\boldsymbol{w} = (w_n)_{n \neq 0}$ satisfying

$$\|\boldsymbol{w}\|_r^2 = \sum_{n \neq 0} |n|^{2r} |w_n|^2 < \infty, \qquad w_{-n} = \overline{w}_n.$$

Due to the choice of the weights, we have an *isomorphism* $\mathcal{F} : \mathcal{h}_{N+1/2}^\bullet \to \mathcal{H}_0^N$ for each $N \geq 1$.

The complex space $\mathcal{h}_r^\bullet$ is canonically identified with the real space $\mathcal{h}_r$ by setting

$$w_n = (x_n - \mathrm{i} y_n)/\sqrt{2}, \qquad w_{-n} = \overline{w}_n, \qquad n \geq 1.$$

The minus sign in the definition of w_n is chosen so that $\mathrm{d}w_n \wedge \mathrm{d}w_{-n} = \mathrm{i}\, \mathrm{d}x_n \wedge \mathrm{d}y_n$. A function on $\mathcal{h}_r^\bullet$ is said to be *real analytic*, if with this identification it is real analytic in x_n and y_n in the usual sense. The *complexification* of $\mathcal{h}_r^\bullet$ is the same space of sequences, but with the condition $w_{-n} = \overline{w}_n$ dropped.

The Hamiltonian expressed in the new coordinates $\boldsymbol{v}$ is determined by inserting the expansion (14.2) of v into the definition (14.1) of H_c. Using for simplicity the same symbol for the Hamiltonian as a function of $\boldsymbol{v}$ we obtain

$$H_c(\boldsymbol{v}) = \Lambda_c(\boldsymbol{v}) + G(\boldsymbol{v})$$

with

$$\Lambda_c(\boldsymbol{v}) = \sum_{n \geq 1} \left(\gamma_n^6 + 6c\gamma_n^2\right) |v_n|^2, \quad G(\boldsymbol{v}) = \sum_{\substack{k,l,m \neq 0 \\ k+l+m=0}} \gamma_k \gamma_l \gamma_m v_k v_l v_m.$$

Note that the first sum is taken over $n \geq 1$, not $n \neq 0$, which accounts for the "missing" factor $1/2$. The phase space $h_r^\bullet$ is endowed with the Poisson structure

$$\{F,G\} = \mathrm{i}\sum_{n\neq 0} \sigma_n \frac{\partial F}{\partial v_n}\frac{\partial G}{\partial v_{-n}},$$

where $\sigma_n = \operatorname{sgn}(n)$ is the sign of n, and the equations of motion in the new coordinates are given by

$$\dot{v}_n = \mathrm{i}\sigma_n \frac{\partial H_c}{\partial v_{-n}}, \qquad n \neq 0.$$

This is most easily seen by observing that

$$\frac{\partial F}{\partial v(x)} = \sum_{n\neq 0} \frac{\partial F}{\partial v_n}\frac{\partial v_n}{\partial v(x)} = \sum_{n\neq 0} \frac{\partial F}{\partial v_n}\cdot \gamma_n^{-1}\mathrm{e}^{-2\pi \mathrm{i} n x}$$

and calculating $\{F,G\}$ on $\mathcal{H}_0^N$.

Since the transformed Poisson structure is nondegenerate, it also defines a symplectic structure

$$\omega = \frac{1}{\mathrm{i}}\sum_{n\geq 1} \mathrm{d}v_n \wedge \mathrm{d}v_{-n}$$

on $h_r^\bullet$, according to which the above equations of motion are the usual Hamiltonian equations with Hamiltonian $H_c(\boldsymbol{v})$. The associated Hamiltonian vector field with Hamiltonian H is given by

$$X_H = \mathrm{i}\sum_{n\neq 0} \sigma_n \frac{\partial H}{\partial v_{-n}}\frac{\partial}{\partial v_n}.$$

The vector field of the quadratic Hamiltonian Λ_c takes values in $h_{r-3}^\bullet$ for $\boldsymbol{v}$ in $h_r^\bullet$, hence it is unbounded of order 3. Strictly speaking, it is not a genuine vector field. The vector field of the cubic Hamiltonian G is also unbounded, but only of order 1. More precisely, we have the following regularity property of X_G.

Lemma 14.1. *The Hamiltonian vector field X_G is real analytic as a map from $h_r^\bullet$ into $h_{r-1}^\bullet$ for each $r \geq \frac{3}{2}$. Moreover,*

$$\|X_G\|_{r-1} = O\big(\|\boldsymbol{v}\|_r^2\big).$$

We remark that here and in the following, $r - \frac{1}{2}$ is thought of being an integer. But everything holds for arbitrary $r \geq \frac{3}{2}$ as well.

Proof. We have $G(\boldsymbol{v}) = \sum_{k+l+m=0} \gamma_k\gamma_l\gamma_m v_k v_l v_m$, hence

$$\frac{\partial G}{\partial v_{-n}} = 3\gamma_n \sum_{k+l=n} \gamma_k\gamma_l v_k v_l = 3\gamma_n g_n,$$

where

$$g_n = \sum_{k+l=n} \gamma_k \gamma_l v_k v_l = \sum_k \gamma_k v_k \gamma_{n-k} v_{n-k},$$

and where all indices are nonzero integers. This restriction may be dropped by understanding that $v_0 = 0$ and $\gamma_0 = 0$. Defining $\boldsymbol{w} = (w_n)_n = (\gamma_n v_n)_n$ and $\boldsymbol{g} = (g_n)$ we see that $g_n = (\boldsymbol{w} * \boldsymbol{w})_n$, hence

$$\boldsymbol{g} = \boldsymbol{w} * \boldsymbol{w}.$$

For $\boldsymbol{v} \in \mathscr{h}_r^\bullet$ we have $\boldsymbol{w} \in \mathscr{h}_{r-\sigma}^\bullet$ with $\sigma = \frac{1}{2}$. Moreover, the space $\mathscr{h}_{r-\sigma}^\bullet$ is a Banach algebra for $r - \sigma > \frac{1}{2}$, see [76]. Hence we have

$$\|\boldsymbol{g}\|_{r-\sigma} = \|\boldsymbol{w} * \boldsymbol{w}\|_{r-\sigma} \le c\,\|\boldsymbol{w}\|_{r-\sigma}^2 \le c\,\|\boldsymbol{v}\|_r^2,$$

and consequently

$$\|X_G\|_{r-1} \le c\,\|\boldsymbol{g}\|_{r-\sigma} \le c\,\|\boldsymbol{v}\|_r^2.$$

The analyticity of X_G follows with Theorem A.5 from the analyticity of each of its components and its local boundedness as a map from $\mathscr{h}_r^\bullet$ into $\mathscr{h}_{r-1}^\bullet$. □

The next theorem is the main result of this section.

Theorem 14.2. *There exists a real analytic symplectic coordinate transformation* $\boldsymbol{v} = \Phi(\boldsymbol{w})$ *defined in a neighbourhood of the origin in* $\mathscr{h}_{3/2}^\bullet$ *which transforms each Hamiltonian* H_c, $c \in \mathbb{R}$, *into its Birkhoff normal form up to order four. More precisely,*

$$H_c \circ \Phi = \Lambda_c - B + K_c$$

with

$$B = 3\sum_{n\ge 1} |w_n|^4, \qquad \|X_{K_c}\|_{1/2} = O\big(\|\boldsymbol{w}\|_{3/2}^4\big).$$

Moreover, for each $r \ge 3/2$, *the restriction of* Φ *to some neighbourhood of the origin in* $\mathscr{h}_r^\bullet$ *defines a similar coordinate transformation in* $\mathscr{h}_r^\bullet$, *so that*

$$\|X_{K_c}\|_{r-1} = O\big(\|\boldsymbol{w}\|_r^4\big).$$

Note that B happens to be independent of c and is a sum of terms each of which depends only on $|w_n|^2$. Thus the modes are *uncoupled* up to order four.

Before giving the proof of the theorem we mention that the transformation Φ not only normalizes the KdV Hamiltonian H_c up to order four, but indeed puts *each* Hamiltonian in the KdV hierarchy into its Birkhoff normal form up to order four. This is explained in appendix G. As an example we consider the second KdV equation at the end of this section. Another consequence is that the term K_c is actually independent of c.

Moreover, the construction of the transformation Φ allows us to determine the Birkhoff coefficients *up to order 6*. This is also explained in appendix G, and the somewhat lengthy calculations can be found in appendix H.

We turn to the proof of Theorem 14.2. To simplify notation we drop the subscript c and thus consider the Hamiltonian $H = \Lambda + G$, with

$$\Lambda = \sum_{n \geq 1} \lambda_n |v_n|^2, \qquad \lambda_n = \tilde{n}^3 + 6c\tilde{n}.$$

Here and later, we use the convenient short hand $\tilde{n} = 2\pi n$ for integers n. The coordinate transformation Φ is constructed in two steps:

$$\Phi = \Psi \circ \Xi,$$

where Ψ eliminates the third order term G and replaces it by higher order terms, and Ξ normalizes the resulting fourth order term. Each of these transformations is obtained as the time-1-map of the flow of some real analytic Hamiltonian vector field on $h_r^\bullet$, whose Hamiltonian has to be chosen properly.

Consider the first step, and let

$$\Psi = X_F^1 = X_F^t\big|_{t=1}.$$

Assuming for the moment that X_F^t is defined for $0 \leq t \leq 1$ in some neighbourhood of the origin in $h_r^\bullet$ we can use Taylor's formula to expand around $t = 0$:

$$\begin{aligned} H \circ \Psi &= \Lambda \circ X_F^t\big|_{t=1} + G \circ X_F^t\big|_{t=1} \\ &= \Lambda + \{\Lambda, F\} + \int_0^1 (1-t)\,\{\{\Lambda, F\}, F\} \circ X_F^t \,\mathrm{d}t \\ &\quad + G + \int_0^1 \{G, F\} \circ X_F^t \,\mathrm{d}t. \end{aligned}$$

If we can solve the equation

$$\{\Lambda, F\} + G = 0$$

with a homogeneous Hamiltonian F of order three, then we obtain

$$\begin{aligned} H \circ \Psi &= \Lambda + \int_0^1 t\,\{G, F\} \circ X_F^t \,\mathrm{d}t \\ &= \Lambda + \frac{1}{2}\{G, F\} - \frac{1}{2}\int_0^1 (t^2 - 1)\,\{\{G, F\}, F\} \circ X_F^t \,\mathrm{d}t \end{aligned} \tag{14.3}$$

by partial integration. Here, $\{G, F\}$ is homogeneous of order four, and the integral only contains terms of order five or more.

To solve $\{\Lambda, F\} + G = 0$ we make the ansatz $F = \sum F_{klm} v_k v_l v_m$ and note that

$$\{\Lambda, F\} = -\mathrm{i} \sum_{k,l,m} (\lambda_k + \lambda_l + \lambda_m) F_{klm} v_k v_l v_m,$$

where we extend the definition of $\lambda_n = \tilde{n}^3 + 6c\tilde{n}$ to *all* n. Since G contains only monomials $v_k v_l v_m$ with $k + l + m = 0$ and $k, l, m \neq 0$, also F need only contain monomials of this kind. Under this condition,

$$\lambda_k + \lambda_l + \lambda_m = 8\pi^3(k^3 + l^3 + m^3)$$

is *independent* of c. Moreover, we make the following observation.

Lemma 14.3. *Suppose k, l, m are nonzero integers with $k + l + m = 0$. Then*

$$k^3 + l^3 + m^3 = 3klm \neq 0.$$

Indeed, if $k + l + m = 0$, then $m = -k - l$, and

$$k^3 + l^3 + m^3 = -3k^2 l - 3kl^2 = -3kl(k + l) = 3klm \neq 0.$$

Hence it is justified to define F by setting

$$\mathrm{i}F_{klm} = \begin{cases} \dfrac{G_{klm}}{\lambda_k + \lambda_l + \lambda_m} & \text{for } k + l + m = 0, \\ 0 & \text{otherwise.} \end{cases}$$

Then, at least formally, we have $\{\Lambda, F\} + G = 0$.

The nonzero coefficients of F are more explicitly

$$\mathrm{i}F_{klm} = \frac{1}{8\pi^3}\frac{\gamma_k\gamma_l\gamma_m}{k^3 + l^3 + m^3} = \frac{1}{3}\frac{\gamma_k\gamma_l\gamma_m}{2\pi k \cdot 2\pi l \cdot 2\pi m} = \frac{1}{3}\frac{1}{\tilde{\gamma}_k\tilde{\gamma}_l\tilde{\gamma}_m},$$

with $\tilde{\gamma}_n = \sigma_n\gamma_n = \operatorname{sgn}(n)\sqrt{2\pi\,|n|}$, not to be confused with $\tilde{n}$ for $2\pi n$. It follows, in exactly the same manner as in the proof of Lemma 14.1, that X_F defines a real analytic vector field on $h_r^\bullet$ of order -1 with

$$\|X_F\|_{r+1} = O\big(\|v\|_r^2\big)$$

for each $r \geq \frac{3}{2}$.

A fortiori, X_F is a real analytic vector field on all of $h_r^\bullet$ for every $r \geq \frac{3}{2}$. It follows that if X_G is of order 1, then also $X_{\{G,F\}} = [X_G, X_F]$ is of order 1. Moreover, in a small neighbourhood of the origin in $h_r^\bullet$, the flow X_F^t exists for $0 \leq t \leq 1$ and defines a local diffeomorphism $\Psi = X_F^1$ with fixed point 0 and regular Jacobian $D\Psi$, which is also an isomorphism of $h_{r-1}^\bullet$. It follows from (14.3) that the transformed vector field

$$\Psi^* X_H - \Lambda = D\Psi^{-1} X_H \circ \Psi - \Lambda$$

is again analytic and unbounded of order 1. Moreover, by construction, $H \circ \Psi$ is normalized up to terms of order three. This completes the first step of the normalization procedure.

In the second step we normalize the resulting fourth order term $\frac{1}{2}\{G, F\}$ in (14.3). This term is easily calculated. Recall that

$$G = \sum \gamma_k\gamma_l\gamma_m v_k v_l v_m, \qquad F = \frac{1}{3i}\sum \frac{v_k v_l v_m}{\tilde{\gamma}_k\tilde{\gamma}_l\tilde{\gamma}_m}. \tag{14.4}$$

So we have

$$
\begin{aligned}
\{G,F\} &= \mathrm{i}\sum_{j\neq 0}\sigma_j \frac{\partial G}{\partial v_j}\frac{\partial F}{\partial v_{-j}} \\
&= 3\sum_{j\neq 0}\sigma_j \sum_{k,l:\,k+l=-j} \gamma_j\gamma_k\gamma_l v_k v_l \cdot \sum_{m,n:\,m+n=j} \frac{v_m v_n}{\tilde{\gamma}_{-j}\tilde{\gamma}_m\tilde{\gamma}_n} \\
&= -3 \sum_{\substack{k+l+m+n=0\\ k+l\neq 0}} \frac{\gamma_k\gamma_l}{\tilde{\gamma}_m\tilde{\gamma}_n} v_k v_l v_m v_n .
\end{aligned}
$$

We decompose the last sum into its contribution to the Birkhoff normal form and the rest, to be transformed away in a moment. The former consists of all terms for which $k+m=0$ or $k+n=0$, whereas terms with $k+l=0$ do not occur. If for example we have $k+m=0$, then $k+l+m+n=0$ leads to $l+n=0$, and the corresponding term reduces to

$$
\frac{\gamma_k\gamma_l}{\tilde{\gamma}_m\tilde{\gamma}_n} v_k v_l v_m v_n = \sigma_{kl}\,|v_k|^2\,|v_l|^2,
$$

where σ_{kl} is the sign of kl. The same contribution is obtained for $k+n=0$ and $l+m=0$. Hence the normal form part of $\{G,F\}$ is

$$
-3\left(2\sum_{k,l:k\neq l}\sigma_{kl}\,|v_k|^2\,|v_l|^2 + \sum_{k\neq 0}|v_k|^4\right) = -3\sum_{k\neq 0}|v_k|^4,
$$

since the first sum on the left hand side is zero as its terms annihilate each other. Thus we obtain

$$
\begin{aligned}
\frac{1}{2}\{G,F\} &= -3\sum_{k>0}|v_k|^4 - \frac{3}{2}\sum_{\substack{k+l+m+n=0\\ k+l,k+m,k+n\neq 0}} \frac{\gamma_k\gamma_l}{\tilde{\gamma}_m\tilde{\gamma}_n} v_k v_l v_m v_n \\
&= -B - Q,
\end{aligned}
\tag{14.5}
$$

where B is the term stated in the theorem.

The complete Hamiltonian at this stage is

$$
H\circ\Psi = \Lambda - B - Q - \frac{1}{2}\int_0^1 (t^2-1)\,\{\{G,F\},F\}\circ X_F^t\,\mathrm{d}t.
$$

It remains to eliminate Q by another coordinate transformation Ξ. In complete analogy to the first step we let

$$
\Xi = X_F^t\big|_{t=1}, \qquad F = \sum_{k,l,m,n} F_{klmn} v_k v_l v_m v_n .
$$

We need to solve the equation

$$
\{\Lambda, F\} = -\mathrm{i}\sum_{k,l,m,n}(\lambda_k+\lambda_l+\lambda_m+\lambda_n)F_{klmn} v_k v_l v_m v_n = Q.
$$

To this end it suffices that F contains only those monomials which also appear in Q. But for $k+l+m+n=0$ we have

$$\lambda_k+\lambda_l+\lambda_m+\lambda_n=8\pi^3(k^3+l^3+m^3+n^3).$$

Moreover, the following holds.

Lemma 14.4. *Suppose k, l, m, n are nonzero integers with $k+l+m+n=0$, but $k+l\neq 0$, $k+m\neq 0$ and $k+n\neq 0$. Then*

$$k^3+l^3+m^3+n^3=3(k+l)(k+m)(k+n)\neq 0.$$

Indeed, with $n=-(k+l+m)$ we have

$$k^3+l^3+m^3+n^3=-3(k^2l+kl^2+k^2m+km^2+l^2m+lm^2+2klm),$$

and the expression in parentheses equals $(k+l)(k+m)(l+m)$ by a straightforward computation.

Thus we may define

$$F_{klmn}=\frac{3}{2}\frac{\mathrm{i}}{\lambda_k+\lambda_l+\lambda_m+\lambda_n}\frac{\gamma_k\gamma_l}{\tilde{\gamma}_m\tilde{\gamma}_n}$$

for the relevant indices k, l, m, n, and $F_{klmn}=0$ otherwise. At least formally, the transformation $\Xi=X_F^1$ then eliminates the term Q.

To establish the regularity of the vector field X_F we write

$$\begin{aligned}F_{klmn}&=\frac{3}{16\pi^3}\frac{\mathrm{i}}{k^3+l^3+m^3+n^3}\frac{\gamma_k\gamma_l}{\tilde{\gamma}_m\tilde{\gamma}_n}\\&=\frac{3}{64\pi^5}\frac{\mathrm{i}}{k^3+l^3+m^3+n^3}\frac{\gamma_k\gamma_l\gamma_m\gamma_n}{mn}\end{aligned}$$

and observe that

$$(k^3+l^3+m^3+n^3)mn\geq\max\left(k^2,l^2,m^2,n^2\right).$$

Indeed, if k is the biggest integer in absolute value, then we write the left hand side as $3(k+l)(k+m)(k+n)mn$ and use

$$|(k+m)m|,\ |(k+n)n|\geq|k-1|\geq\frac{2}{3}|k|.$$

If, however, m is the biggest integer in absolute value, then we write the left hand side as $3(m+k)(m+l)(m+n)mn$ and use

$$|(m+n)n|\geq|m-1|\geq\frac{2}{3}|m|.$$

By symmetry in k, l and m, n, this covers all cases. It follows that for

$$\partial F/\partial v_j = \sum_{k+l+m=-j} (F_{jklm} + \cdots + F_{klmj}) v_k v_l v_m$$

we have the estimate

$$\left| \frac{\partial F}{\partial v_j} \right| \le \frac{3}{4\pi^3} \frac{\gamma_j}{j^2} \sum_{k+l+m=-j} \gamma_k \gamma_l \gamma_m \, |v_k| \, |v_l| \, |v_m| \le \frac{g_j}{\gamma_j^3},$$

where g_j stands for the entire sum. Thus,

$$\|X_F\|_{r+1} \le \|\boldsymbol{g}\|_{r-1/2},$$

and $\boldsymbol{g} = (g_n)_{n \ne 0}$ is the three-fold convolution of $\boldsymbol{w} = (\gamma_n \, |v_n|)_{n \ne 0}$. As in the proof of Lemma 14.1, we have $\|\boldsymbol{g}\|_{r-1/2} \le c \, \|\boldsymbol{w}\|_{r-1/2}^3 \le c \, \|\boldsymbol{v}\|_r^3$ for $r \ge \frac{3}{2}$. So finally,

$$\|X_F\|_{r+1} = O\big(\|\boldsymbol{v}\|_r^3\big)$$

for $\boldsymbol{v} \in \mathscr{h}_r^\bullet$ and $r \ge \frac{3}{2}$. This establishes the regularity of the vector field X_F, and finishes the proof of Theorem 14.2.

As we explain in appendix G we obtain the same result by first calculating the normal form for H_c at $c = 0$ and then adding the quadratic term $6c \sum_n \gamma_n^2 \, |v_n|^2$ to the result, as the latter is invariant under Φ. Hence, it is no accident that the divisors in Lemmas 14.3 and 14.4 are independent of c.

We already mentioned that the transformation Φ of Theorem 14.2 not only normalizes H_c up to order four, but puts *every* Hamiltonian in the KdV hierarchy into its own Birkhoff normal form up to order four – see appendix G. This is due to the fact that all these Hamiltonians are in involution. As an example we consider the second KdV equation

$$\partial_t u = \partial_x^5 u - 10 u \partial_x^3 u - 20 \partial_x u \partial_x^2 u + 30 u^2 \partial_x u,$$

whose Hamiltonian

$$H^2(u) = \int_{S^1} \left(\tfrac{1}{2} u_{xx}^2 + 5 u u_x^2 + \tfrac{5}{2} u^4\right) \mathrm{d}x$$

is defined on $\mathscr{H}^2$. With $u = v + c$, $[v] = 0$, we can write

$$H^2(u) = H_c^2(v) + \frac{5}{2} c^4$$

with

$$H_c^2(v) = H^2(v) + 10 c H^1(v) + 30 c^2 H^0(v),$$

where $H^1 = \int_{S^1} \left(\frac{1}{2} v_x^2 + v^3\right) \mathrm{d}x$ is the KdV Hamiltonian, and $H^0 = \int_{S^1} \frac{1}{2} v^2 \, \mathrm{d}x$ is the Hamiltonian of translation.

Theorem 14.5. *The transformation* Φ *of Theorem 14.2 also takes each Hamiltonian* H_c^2, $c \in \mathbb{R}$, *into its Birkhoff normal form of order four:*

$$H_c^2 \circ \Phi = \Lambda_c^2 - B_c^2 + K_c^2,$$

where, with $\tilde{n} = 2\pi n$,

$$\Lambda_c^2 = \sum_{n \geq 1} \left(\tilde{n}^5 + 10c\tilde{n}^3 + 30c^2\tilde{n}\right) |w_n|^2,$$

$$B_c^2 = \sum_{n \geq 1} \left(15\tilde{n}^2 + 30c\right) |w_n|^4 - 10 \sum_{k,l \geq 1} (3 - 2\delta_{kl}) \, |\tilde{k}\tilde{l}| \, |w_k|^2 \, |w_l|^2,$$

and K_c *defines a real analytic vector field of order 3 satisfying*

$$\left\| X_{K_c} \right\|_{r-3} = O\left(\|\boldsymbol{w}\|_r^4\right)$$

for $\boldsymbol{w} \in h_r^\bullet$ *for each* $r \geq \frac{5}{2}$.

For the second KdV Hamiltonian, the fourth order term B_c^2 does depend on c. Its first sum is the contribution of the term $\frac{1}{2}\{G, F\}$ for this case, and the second term is the normal form part arising from the quartic term $\frac{5}{2}\int_{S^1} v^4 \, \mathrm{d}x$. The calculations are given in appendix G.

15 Global Coordinates and Frequencies

The Birkhoff normal form theorem in section 14 provides symplectic coordinates in a neighbourhood of the origin in $h_r^\bullet$ in which the KdV Hamiltonian is a classically integrable system up to order four. These coordinates are completely sufficient, if we were only interested in *small* solutions u and their behavior under Hamiltonian perturbations of the equation. To study large solutions as well, however, we employ the global Birkhoff coordinates constructed in chapter III. The results from Theorem 12.6 relevant in this context may be formulated as follows.

Theorem 15.1. *There exists a canonical transformation*

$$\Psi\colon \; h_{1/2}^\bullet \to \mathcal{H}_0^0$$

with the following properties.

(i) Ψ *is one-to-one, onto, bi-analytic and symplectic with respect to the symplectic structures* $-\mathrm{i}\sum_{n\geq 1} \mathrm{d}w_n \wedge \mathrm{d}w_{-n}$ *and* $\langle \partial_x^{-1} \cdot \,, \cdot \rangle$, *respectively.*

(ii) *For each* $N \geq 0$, *the restriction of* Ψ *to* $h_{N+1/2}^\bullet$, *denoted by the same symbol, is a map*

$$\Psi\colon \; h_{N+1/2}^\bullet \to \mathcal{H}_0^N,$$

which is one-to-one, onto, and bi-analytic as well.

(iii) *The coordinates* $\boldsymbol{w}$ *on* $\hbar^{\bullet}_{3/2}$ *are global Birkhoff coordinates for KdV:*

$$H_c \circ \Psi = \hat{H}_c(|\boldsymbol{w}|^2), \qquad |\boldsymbol{w}|^2 = (w_n w_{-n})_{n\geq 1}.$$

The same holds for all higher KdV equations when considered on $\hbar^{\bullet}_{N+1/2}$ *with an appropriate* N.

(iv) *This transformation satisfies* $\Psi(0) = 0$ *and* $d_0\Psi = \mathcal{F}$, *where* $\mathcal{F}$ *is a weighted inverse Fourier transform as defined in* (14.2).

Let us introduce the actions

$$\boldsymbol{I} = (I_n)_{n\geq 1}, \qquad I_n = w_n w_{-n}.$$

If $\boldsymbol{w} \in \hbar^{\bullet}_{3/2}$, then $\boldsymbol{I}$ is an element of the Banach space ℓ^1_3 consisting of all real sequences $\boldsymbol{u} = (u_n)_{n\geq 1}$ with

$$|\boldsymbol{u}|_{\ell^1_3} = \sum_{n\geq 1} n^3\,|u_n| < \infty.$$

The map $A\colon \hbar^{\bullet}_{3/2} \to \ell^1_3$, $\boldsymbol{w} \mapsto \boldsymbol{I}$, is real analytic and onto the closed *positive cone*

$$\mathbb{P}\ell^1_3 = \left\{\boldsymbol{u} \in \ell^1_3 \colon u_n \geq 0 \text{ for all } n \geq 1\right\}.$$

Of course, this map is not one-to-one.

Addendum to Theorem 15.1. *There exists a real analytic Hamiltonian* $\check{H}_c$ *on the closed positive cone* $\mathbb{P}\ell^1_3$ *such that*

$$\hat{H}_c(|\boldsymbol{w}|^2) = \check{H}_c(\boldsymbol{I}).$$

Proof. The function $\check{H}_c$ is pointwise well defined on $\mathbb{P}\ell^1_3$, since $\hat{H}_c$ is a function of the amplitudes $|\boldsymbol{w}|^2$ and thus constant on the fibers of A over $\mathbb{P}\ell^1_3$.

To prove analyticity consider first the origin in ℓ^1_3. We may expand $\hat{H}_c$ into its Taylor series in $\boldsymbol{w}$, replace $w_n w_{-n}$ by I_n everywhere for all n, and re-index the coefficients. Thus, we obtain a Taylor series expansion of $\check{H}_c$ around the origin. At any other point $\boldsymbol{I} \in \mathbb{P}\ell^1_3$, we can do the same by expanding $\hat{H}_c$ around $\boldsymbol{w} = (w_n)_{n\neq 0}$ with $w_n = w_{-n} = \sqrt{I_n}$. Thus, there exists a Taylor series expansion of $\check{H}_c$ around every point in $\mathbb{P}\ell^1_3$. □

From now on we drop the accents and write again H_c for the Hamiltonian as a function of $\boldsymbol{I}$.

Theorem 15.1 states that Ψ puts H_c into a *global* Birkhoff normal form on all of $\hbar^{\bullet}_{3/2}$. So does the transformation Φ of Theorem 14.2 locally at the origin up to order four. By the last item of Theorem 15.1, the change between these two coordinate systems is of the form "identity + higher order terms", hence by its uniqueness the normal form up to order four is not affected – compare Theorem G.1. Theorem 14.2 therefore provides us with the Taylor series expansion of H_c up to order two.

Corollary 15.2. *The transformed KdV Hamiltonian H_c expanded at $\boldsymbol{I} = 0$ is of the form*

$$H_c(\boldsymbol{I}) = \sum_{n\geq 1} \lambda_n I_n - 3 \sum_{n\geq 1} I_n^2 + \dots,$$

where $\lambda_n = \tilde{n}^3 + 6c\tilde{n}$ with $\tilde{n} = 2\pi n$, and the dots stand for higher order terms in $\boldsymbol{I}$.

Before continuing our investigation of H_c, we consider the second KdV Hamiltonian. By Theorem 15.1, Ψ takes H_c^2 into a function of $|\boldsymbol{w}|^2$ on $\ell^2_{5/2}$ as well. As above, this gives rise to a Hamiltonian $\check{H}_c^2$ of the actions $\boldsymbol{I}$, which is real analytic on the closed positive cone $\mathbb{P}\ell^1_5$ defined analogously to $\mathbb{P}\ell^1_3$. Theorem 14.5 then provides the first two terms of its Taylor series expansion.

Theorem 15.3. *The transformation Ψ of Theorem 15.1 transforms the second KdV Hamiltonian H_c^2 into*

$$\hat{H}_c^2(|\boldsymbol{w}|^2) = \check{H}_c^2(\boldsymbol{I}) = \sum_{n\geq 1} \lambda_n^2 I_n - \frac{1}{2} \sum_{k,l\geq 1} C_{kl} I_k I_l + \dots,$$

where $\check{H}_c^2$ is real analytic on $\mathbb{P}\ell^1_5$ and

$$\lambda_n^2 = \tilde{n}^5 + 10c\tilde{n}^3 + 30c^2\tilde{n}, \qquad C_{kl} = \begin{cases} -60\tilde{k}\tilde{l}, & k \neq l \\ 10\tilde{k}^2 + 60c, & k = l \end{cases}.$$

The dots stand for higher order terms in $\boldsymbol{I}$.

Given the integrable Hamiltonian H_c as a function of $\boldsymbol{I}$ we may now define its *frequencies* $\boldsymbol{\omega}$ in the usual fashion:

$$\boldsymbol{\omega} = (\omega_n)_{n\geq 1}, \qquad \omega_n = \frac{\partial H_c}{\partial I_n}.$$

This definition does not require the use of angular variables. Since $\boldsymbol{\omega}$ is just the differential of H_c, we obtain a real analytic map from $\mathbb{P}\ell^1_3$ into the dual space of ℓ^1_3. That is,

$$\boldsymbol{\omega}\colon \mathbb{P}\ell^1_3 \to \ell^\infty_{-3}$$

is real analytic, where ℓ^∞_β is the Banach space of all real sequences $\boldsymbol{\lambda} = (\lambda_1, \lambda_2, \dots)$ with

$$|\boldsymbol{\lambda}|_\beta = \sup_{n\geq 1} n^\beta \, |\lambda_n| < \infty.$$

This follows from the general Cauchy estimate for analytic maps between Banach spaces, see Lemma A.2. Moreover, according to Corollary 15.2 we have the expansion $\omega_n = \lambda_n - 6I_n + O_2(\boldsymbol{I})$, or

$$\boldsymbol{\omega} = \boldsymbol{\lambda} - 6\boldsymbol{I} + \dots \quad . \tag{15.1}$$

Note that $\boldsymbol{\lambda} = (\lambda_n)_{n\geq 1}$ itself is an element of ℓ^∞_{-3}, so the above target space for $\boldsymbol{\omega}$ can not be chosen smaller.

To obtain a sharper statement, we consider the $\boldsymbol{I}$-dependent part

$$\tilde{\boldsymbol{\omega}} = \boldsymbol{\omega} - \boldsymbol{\lambda} = -6\boldsymbol{I} + \dots$$

of the frequencies $\boldsymbol{\omega}$. By Corollary F.5 they satisfy the asymptotic estimate

$$|\tilde{\omega}_n| = |\omega_n - \lambda_n| = O(n)$$

uniformly on bounded subsets of the phase space, hence uniformly on bounded subsets of $\mathbb{P}\ell^1_3$. This estimate also extends to small complex neighbourhoods around each point in $\mathbb{P}\ell^1_3$. So we have a map

$$\tilde{\boldsymbol{\omega}} \colon \mathbb{P}\ell^1_3 \to \ell^\infty_{-1},$$

which is componentwise real analytic and locally bounded on complex neighbourhoods. Since the norm of the target space is a weighted sup-norm, this implies that the whole map is real analytic – see Theorem A.3.

Theorem 15.4. *The frequency map*

$$\tilde{\boldsymbol{\omega}} \colon \quad \boldsymbol{I} \mapsto \tilde{\boldsymbol{\omega}}(\boldsymbol{I}) = \boldsymbol{\omega}(\boldsymbol{I}) - \boldsymbol{\lambda}$$

of the KdV Hamiltonian is real analytic as a map from $\mathbb{P}\ell^1_3$ *to* ℓ^∞_{-1}.

In the next section we are going to describe an abstract infinite dimensional KAM theorem. To be applicable to perturbations of the KdV equation, however, the frequencies $\boldsymbol{\omega}$ need to satisfy certain nondegeneracy conditions as functions of $\boldsymbol{I}$. In particular, since our focus is on *finite gap solutions*, their dependence on *finitely* many non-zero actions is important. These properties are now established.

Let $A \subset \mathbb{N}$ be an arbitrary *finite* index set, and let $Z = \mathbb{N} - A$ be its complement. According to this decomposition of $\mathbb{N}$, we write

$$\boldsymbol{I} = (\boldsymbol{I}_A, \boldsymbol{I}_Z), \qquad \boldsymbol{I}_A = (I_n)_{n\in A}, \quad \boldsymbol{I}_Z = (I_n)_{n\in Z},$$

and similarly in other cases. We are going to restrict ourselves to the subspace

$$\ell_A = \left\{ \boldsymbol{u} \in \ell^1_3 \colon \boldsymbol{u}_Z = 0 \right\} \simeq \mathbb{R}^A.$$

Of course, $\boldsymbol{\omega}$ and $\tilde{\boldsymbol{\omega}}$ are also real analytic maps from the positive cone $\mathbb{P}\ell_A \subset \ell_A$ into ℓ^∞_{-3} and ℓ^∞_{-1}, respectively.

For integer sequences $\boldsymbol{k} \in \mathbb{Z}^\infty$, let $\boldsymbol{k} \cdot \boldsymbol{\omega} = \sum_{n\geq 1} k_n \omega_n$ and $|\boldsymbol{k}| = \sum_{n\geq 1} |k_n|$. We need to know that for a certain set of $\boldsymbol{k}$, the frequency combinations $\boldsymbol{k} \cdot \boldsymbol{\omega}$ do not vanish identically when considered as functions of $\boldsymbol{I}_A$. Moreover, the finite dimensional frequency map $\boldsymbol{I}_A \mapsto \boldsymbol{\omega}_A$ has to have an almost everywhere regular Jacobian

$$Q_A = (Q_{kl})_{k,l\in A} = \left(\frac{\partial \omega_k}{\partial I_l} \right)_{k,l\in A}.$$

The following proposition and its addendum provide the essential facts. Recall that the frequencies also depend on the real parameter c, which we do not indicate explicitly.

Proposition 15.5. *For every c and every finite index set $A \subset \mathbb{N}$ the following holds on $\mathbb{P}\ell_A$.*

(i) *The map $\boldsymbol{I}_A \mapsto \boldsymbol{\omega}_A$ is nondegenerate in the sense that* $\det Q_A \not\equiv 0$.
(ii) $\boldsymbol{k} \cdot \boldsymbol{\omega} \not\equiv 0$ *for all* $\boldsymbol{k} \in \mathbb{Z}^\infty$ *with* $\boldsymbol{k}_A \neq 0$ *and* $|\boldsymbol{k}_Z| \leq 2$.
(iii) $\boldsymbol{k} \cdot \boldsymbol{\omega} \not\equiv 0$ *for all* $\boldsymbol{k} \in \mathbb{Z}^\infty$ *with* $\boldsymbol{k}_A = 0$ *and* $1 \leq |\boldsymbol{k}_Z| \leq 2$, *provided that*

$$c \notin \mathcal{E}_A = \left\{ -\tfrac{2}{3}\pi^2 \left(i^2 \pm ij + j^2 \right) : i, j \in Z \right\}.$$

Remark 1. In fact, Krichever proved that the map $\boldsymbol{I}_A \mapsto \boldsymbol{\omega}_A$ is a local diffeomorphism everywhere, by using the representation of the frequencies by periods of certain Abelian differentials – see [9]. In the case $c = 0$, the second statement is proven in [11] using Schottky uniformization.

Remark 2. $\mathcal{E}_A$ is a discrete subset of the negative real axis, and there are no exceptional values for $c > -\frac{2}{3}\pi^2$. Anyway, the Addendum below will show, that the condition $c \notin \mathcal{E}_A$ can be dropped.

Proof of Proposition 15.5. Recall that we have $\omega_n = \lambda_n - 6I_n + O_2(\boldsymbol{I})$. Thus,

$$Q_A = -6E_A + O(\boldsymbol{I}),$$

where E_A is the identity matrix of dimension $|A|$. This proves the first item. Also,

$$\boldsymbol{k} \cdot \boldsymbol{\omega} = \boldsymbol{k} \cdot \boldsymbol{\lambda} - 6\boldsymbol{k}_A \cdot \boldsymbol{I}_A + O_2(\boldsymbol{I}_A),$$

since $\boldsymbol{I}_Z = 0$ on ℓ_A. For $\boldsymbol{k}_A \neq 0$, the right hand side can not vanish identically in $\boldsymbol{I}_A$, so also the second item is proven.

For $\boldsymbol{k}_A = 0$, however, the right hand side is not explicitly under control as a function of $\boldsymbol{I}_A$. Therefore, we instead restrict the parameter c so that

$$\boldsymbol{k} \cdot \boldsymbol{\omega}|_{\boldsymbol{I}=0} = \boldsymbol{k} \cdot \boldsymbol{\lambda} = \boldsymbol{k}_Z \cdot \boldsymbol{\lambda}_Z \neq 0.$$

For instance, consider the case where $\boldsymbol{k}_Z \cdot \boldsymbol{\lambda}_Z = \lambda_i \pm \lambda_j$ for distinct $i, j \in Z$. With $\lambda_n = \tilde{n}^3 + 6c\tilde{n}$ we obtain the condition

$$0 \neq \tilde{\imath}^3 \pm \tilde{\jmath}^3 + 6c(\tilde{\imath} \pm \tilde{\jmath}) = (\tilde{\imath}^2 \mp \tilde{\imath}\tilde{\jmath} + \tilde{\jmath}^2)(\tilde{\imath} \pm \tilde{\jmath}) + 6c(\tilde{\imath} \pm \tilde{\jmath}),$$

or

$$-6c \neq \tilde{\imath}^2 \pm \tilde{\imath}\tilde{\jmath} + \tilde{\jmath}^2 = 4\pi^2(i^2 \pm ij + j^2).$$

The other cases lead to conditions which are contained in this one. This proves also the third item. □

The proof of the preceding proposition is fairly short and simple. With *more* effort we can show that the restriction $c \notin \mathcal{E}_A$ can be dropped from part (iii). The following is proven in appendix I, based on a result by Kramer.

Addendum to Proposition 15.5. *For any real parameter c and any finite index set $A \subset \mathbb{N}$ one has $\boldsymbol{k} \cdot \boldsymbol{\omega} \not\equiv 0$ for all $\boldsymbol{k} \in \mathbb{Z}^\infty$ with $\boldsymbol{k}_A = 0$ and $1 \le |\boldsymbol{k}_Z| \le 2$.*

In the remainder of this section we discuss the frequencies of the second KdV Hamiltonian,

$$\boldsymbol{\omega}^2 = (\omega_n^2)_{n\ge 1}, \qquad \omega_n^2 = \frac{\partial H_c^2}{\partial I_n}.$$

According to Theorem 15.3 this defines a real analytic map

$$\boldsymbol{\omega}^2 \colon\ \mathbb{P}\ell_5^1 \to \ell_{-5}^\infty$$

with expansion

$$\boldsymbol{\omega}^2 = \boldsymbol{\lambda}^2 - \boldsymbol{C}\boldsymbol{I} + \dots, \qquad \boldsymbol{C} = (C_{ij})_{i,j\ge 1}.$$

Note that $\boldsymbol{\lambda}^2 = (\lambda_n^2)_{n\ge 1}$ itself belongs to ℓ_{-5}^∞.

As before, we consider the $\boldsymbol{I}$-dependent term $\tilde{\boldsymbol{\omega}}^2 = \boldsymbol{\omega}^2 - \boldsymbol{\lambda}^2$. By Theorem F.5 we have the asymptotic estimate

$$\left|\omega_n^2 - \lambda_n^2\right| = O(n^3)$$

uniformly on bounded subsets of $\mathbb{P}\ell_5^1$, and also on small complex neighbourhoods around each point. Hence we have the following analogue to Theorem 15.4.

Theorem 15.6. *The frequency map*

$$\tilde{\boldsymbol{\omega}}^2 \colon\ \mathbb{P}\ell_5^1 \to \ell_{-3}^\infty, \qquad \boldsymbol{I} \mapsto \tilde{\boldsymbol{\omega}}^2(\boldsymbol{I}) = \boldsymbol{\omega}^2(\boldsymbol{I}) - \boldsymbol{\lambda}^2$$

of the second KdV Hamiltonian is real analytic.

In appendix J we prove the following nondegeneracy properties of the second KdV Hamiltonian. For $A \subset \mathbb{N}$, let Q_A^2 be the Jacobian of the map $\boldsymbol{I}_A \mapsto \boldsymbol{\omega}_A^2$. Again, recall that the frequencies and the Jacobian depend on the real parameter c.

Proposition 15.7. *For every finite index set $A \subset \mathbb{N}$ the following holds on $\mathbb{P}\ell_A$.*

(i) *There exists an $|A|$-point set $\mathfrak{C}_A^2 \subset \mathbb{R}$ not containing 0 such that for $c \notin \mathfrak{C}_A^2$,*

$$\det Q_A^2 \not\equiv 0.$$

(ii) *There exists an at most countable subset $\mathcal{E}_A^2 \subset \mathbb{R}$ not containing 0 and accumulating at most at the points of $\mathfrak{C}_A^2$ such that for $c \notin \mathcal{E}_A^2$,*

$$\boldsymbol{k} \cdot \boldsymbol{\omega}^2 \not\equiv 0$$

for all $0 \ne \boldsymbol{k} \in \mathbb{Z}^\infty$ with $1 \le |\boldsymbol{k}_Z| \le 2$.

These statements are proven in appendix J by looking at the first two terms of the expansion of $\boldsymbol{\omega}^2$ at $\boldsymbol{I}_A = 0$. It is reasonable to expect that the set of excluded values for c can be considerably reduced or even eliminated, by looking at more terms of this expansion, or by using more global arguments.

16 The KAM Theorem

In this section we formulate an infinite dimensional KAM theorem that is applicable to small perturbations of KdV equations.

We begin by explaining its set up in an informal way. In the previous section we saw that on some Hilbert space $\mathscr{h}_r^\bullet$ of complex sequences $\boldsymbol{w} = (w_n)_{n\neq 0}$ with $w_{-n} = \overline{w}_n$ the KdV Hamiltonians take the form

$$H = H(|\boldsymbol{w}|^2), \qquad |\boldsymbol{w}|^2 = (|w_n|^2)_{n\geq 1}.$$

The same form results by truncating the Birkhoff normal forms of Theorems 14.2 and 14.5 to fourth order. In any of these cases, the equations of motion are

$$\dot{w}_n = \mathrm{i}\frac{\partial H}{\partial \overline{w}_n}, \qquad n \neq 0.$$

It is immediate that we have infinitely many nonnegative integrals of motion

$$I_n = |w_n|^2 \geq 0, \qquad n \geq 1,$$

all of which are functionally independent and in involution. So the equations of motion reduce to

$$\dot{w}_n = \mathrm{i}\omega_n(\boldsymbol{I})w_n,$$

where

$$\omega_n(\boldsymbol{I}) = \frac{\partial H}{\partial I_n}(\boldsymbol{I}), \qquad \boldsymbol{I} = (I_n)_{n\geq 1}.$$

It follows that each motion takes place on the product of circles $\prod_{n\geq 1}\{|w_n|^2 = I_n\}$, and the n-th coordinate circles around with fixed frequency $\omega_n(\boldsymbol{I})$. Their combined motion is in general not quasi-periodic, but *almost-periodic*, that is, the uniform limit of trigonometric polynomials [63]. The dimension of the underlying invariant torus equals the number of nonvanishing actions I_n, and the whole phase space decomposes into such tori. So in this sense the Hamiltonian system is completely integrable.

As up to now there is no genuine KAM theorem for the infinite dimensional tori in this system – and there might never be – we need to restrict our attention to *quasi-periodic* motions on *finite* dimensional tori. In the context of the KdV equation this corresponds to the evolution of finite gap solutions. Thus, we fix a *finite* index set $A \subset \mathbb{N}$ of cardinality $|A| = m$, let $Z = \mathbb{N} - A$ be its complement, and restrict ourselves to the finite dimensional space

$$\ell_A = \{\boldsymbol{w} \in \mathscr{h}_r^\bullet : \boldsymbol{w}_Z = 0\}.$$

Here we use the notation introduced in the previous section on page 130. This space is invariant and foliated into the invariant tori

$$T_{I_A} = \left\{ \boldsymbol{w}\colon\ |\boldsymbol{w}_A|^2 = \boldsymbol{I}_A,\ \boldsymbol{w}_Z = 0 \right\} \subset \hbar_r^\bullet, \qquad \boldsymbol{I}_A \in \mathbb{PR}^A.$$

That is, on T_{I_A} we have $|w_n|^2 = I_n$ for $n \in A$, and $w_n = 0$ otherwise. Their dimension is maximal, if $I_n > 0$ for all $n \in A$, which we will assume henceforth. That is, we assume that $\boldsymbol{I}_A \in \mathbb{R}_+^A$.

To formulate a KAM theory for such invariant tori it is convenient to introduce angle-action coordinates (x_A, y_A) in $\mathbb{T}^A \times \mathbb{R}^A$ locally around each torus T_{I_A} by writing

$$w_n = \sqrt{I_n + y_n}\,\mathrm{e}^{-\mathrm{i}x_n}, \qquad n \in A,$$

while keeping the coordinates w_n for $n \in Z$. The integrable Hamiltonian becomes

$$\begin{aligned} H &= H(\boldsymbol{I}_A + y_A, |\boldsymbol{w}_Z|^2) \\ &= H(\boldsymbol{I}_A, \boldsymbol{0}) + \sum_{n \in A} \omega_n(\boldsymbol{I}_A) y_n + \sum_{n \in Z} \omega_n(\boldsymbol{I}_A)\, |w_n|^2 + \dots, \end{aligned}$$

where

$$\omega_n(\boldsymbol{I}_A) = \frac{\partial H}{\partial I_n}(\boldsymbol{I}_A, \boldsymbol{0}), \qquad n \geq 1,$$

and the dots stand for the integral remainder of Taylor's formula of second order in y_A and $|\boldsymbol{w}_Z|^2$. The phase space coordinates are now (x_A, y_A, w_Z), the $\boldsymbol{I}_A$ are parameters, and the linearized equations of motion are

$$\dot{x}_n = \omega_n(\boldsymbol{I}_A), \qquad \dot{y}_n = 0, \qquad \dot{w}_n = \mathrm{i}\omega_n(\boldsymbol{I}_A) w_n,$$

for $n \in A$ and $n \in Z$, respectively. Thus, they reduce to constant coefficient form.

Geometrically speaking, in the new coordinates (x_A, y_A, w_Z) we have an *invariant torus* $\mathbb{T}^m \times \{0\} \times \{0\}$, on which the flow is given by m *internal frequencies* ω_n, $n \in A$. They depend on $\boldsymbol{I}_A$ and thus on the position of the torus in the given family. In the normal space, described by the coordinates w_Z, we have an *elliptic fixed point* at the origin, which is characterized by the *external frequencies* ω_n, $n \in Z$. Thus, for each $\boldsymbol{I}_A$, we have what we call an *invariant, rotational, linearly stable m-torus.*

It turns out that in order to prove Theorems 13.1 and 13.2 it is sufficient to develop a KAM theory, which is concerned with the persistence of a single finite dimensional torus, such as T_0, in an infinite dimensional system, which depends on sufficiently many *parameters.* To formulate a general KAM theorem we therefore start now with the following set up.

We consider small perturbations of a family of infinite dimensional integrable Hamiltonians

$$\begin{aligned} N &= N(x, y, u, v; \xi) \\ &= \sum_{1 \leq n \leq m} \omega_n(\xi) y_n + \frac{1}{2} \sum_{n \geq 1} \Omega_n(\xi)(u_n^2 + v_n^2). \end{aligned} \tag{16.1}$$

The phase space is

$$\mathcal{S}_p^m = \mathbb{T}^m \times \mathbb{R}^m \times \ell_p^2 \times \ell_p^2$$

with coordinates (x, y, u, v), where $\mathbb{T}^m = \mathbb{R}^m / 2\pi\mathbb{Z}^m$ denotes the usual m-torus, and

$$\ell_p^2 = \left\{ \boldsymbol{x} \in \ell^2(\mathbb{N}, \mathbb{R}) \colon \ \|\boldsymbol{x}\|_p^2 = \sum_{n \geq 1} n^{2p} \, |x_n|^2 < \infty \right\}.$$

The symplectic structure is $\sum_{1 \leq n \leq m} \mathrm{d}x_n \wedge \mathrm{d}y_n + \sum_{n \geq 1} \mathrm{d}u_n \wedge \mathrm{d}v_n$. The Hamiltonian N depends on parameters

$$\xi \in \Pi \subset \mathbb{R}^m,$$

where Π is a compact subset of $\mathbb{R}^m$ of positive Lebesgue measure. For example, Π may be a compact Cantor set of positive measure.

Note that this Hamiltonian N equals the Hamiltonian H above, after dropping the higher order terms and an irrelevant constant term, setting $\xi = \boldsymbol{I}_A$,

$$(\omega_1, \ldots, \omega_m) = \omega_A, \qquad (\Omega_n)_{n \geq 1} = \omega_Z,$$

and re-indexing the coordinates.

Our aim is to prove the persistence of the invariant torus

$$T_0 = \mathbb{T}^m \times \{0\} \times \{0\} \times \{0\}$$

of N at ξ together with its elliptic fixed point at $(u, v) = (0, 0)$ under sufficiently small Hamiltonian perturbations $H = N + P$ of N, for parameter values ξ in a large Cantor subset of Π. To this end we make the following three assumptions.

Assumption A: Frequency Asymptotics. There exist two real numbers $d > 1$ and $\delta < d - 1$ such that the following holds. First, the frequencies Ω_n are real valued functions of ξ of the form

$$\Omega_n(\xi) = \bar{\Omega}_n + \tilde{\Omega}_n(\xi),$$

where $\bar{\Omega}_n$ is independent of ξ and of the form $\bar{\Omega}_n = cn^d + \ldots$, where the dots stand for an expansion in lower order terms in n. Second, the functions

$$\xi \mapsto \frac{\tilde{\Omega}_n(\xi)}{n^\delta}, \qquad n \geq 1,$$

are *uniformly* Lipschitz on Π, or equivalently, the map

$$\tilde{\Omega} \colon \ \Pi \to \ell_{-\delta}^\infty, \quad \xi \mapsto \tilde{\Omega}(\xi) = (\tilde{\Omega}_n(\xi))_{n \geq 1}$$

is Lipschitz on Π.

For the proof of the abstract KAM theorem the assumption on $\bar{\Omega}_n$ is slightly relaxed. The one given here, however, is more transparent and suffices when applied to the KdV Hamiltonians.

We assume here that $d > 1$. The case $d = 1$ may also be handled, as it is done in [72, 109], but is somewhat more involved and not relevant for the application to the KdV equations. We therefore omit it for the sake of clarity.

Assumption B: Nondegeneracy. The map

$$\xi \mapsto \omega(\xi)$$

between Π and its image is a homeomorphism which is Lipschitz continuous in both directions. Moreover, for every $k \in \mathbb{Z}^m$ and $l \in \mathbb{Z}^\infty$ with $1 \le |l| \le 2$ the *resonance set*

$$\mathcal{R}_{kl} = \{\xi \in \Pi: \langle k, \omega(\xi)\rangle + \langle l, \Omega(\xi)\rangle = 0\}$$

has Lebesgue measure zero.

For integer vectors such as l, we always understand that $|l| = \sum_n |l_n|$.

We note that the zero measure condition is satisfied, if each frequency is a *real analytic* function of ξ, and

$$\langle k, \omega(\xi)\rangle + \langle l, \Omega(\xi)\rangle \not\equiv 0$$

on Π for all relevant integer vectors k and l. Moreover, assumption A implies that the measure of $\mathcal{R}_{kl}$ is under control for *almost all* k and l, so that this assumption together with assumption A is relevant only for *finitely many* resonance sets. This will be made explicit in section 22.

The third assumption is concerned with the perturbing Hamiltonian P and its Hamiltonian vector field,

$$X_P = (P_y, -P_x, P_v, -P_u)^T.$$

We use the notation $i_\xi X_P$ for X_P evaluated at ξ, and likewise in analogous cases. With $\mathcal{S}_{p,\mathbb{C}}$ we denote the complexification of the phase space $\mathcal{S}_p = \mathcal{S}_p^m$.

Assumption C: Regularity. There is a neighbourhood U_p of T_0 in $\mathcal{S}_{p,\mathbb{C}}$ such that P is defined on $U_p \times \Pi$, and its Hamiltonian vector field defines a map

$$X_P\colon\ U_p \times \Pi \to \mathcal{S}_{q,\mathbb{C}},$$

where q satisfies

$$p - q < d - 1.$$

Moreover, $i_\xi X_P$ is real analytic on U_p for each $\xi \in \Pi$, and $i_w X_P$ is uniformly Lipschitz on Π for each $w \in U_p$.

The essential requirement is the following. For each ξ, the vector field $i_\xi X_P$, considered as a map from a subset of $\mathcal{S}_p$ to $\mathcal{S}_q$, is of the *order* $p - q$. By the preceding assumption, this order must be strictly smaller than $d - 1$, where by assumption A, d is a lower bound for the order of the unperturbed vector field X_N. Hence, using the language of partial differential equations, the perturbed system has to be *semi-linear*.

In previous versions of the KAM theorem [72, 109], an additional requirement was

$$p - q \le 0.$$

That is, the perturbation was also required to be *bounded* as an operator. This is suitable for nonlinear Schrödinger and wave equations on a bounded interval, but not for the KdV equations considered here. This assumption was later removed by Kuksin in [74] so that the theory applies also to perturbed KdV equations.

To state the KAM theorem we need to introduce some domains and norms. For $s > 0$ and $r > 0$ we introduce the complex T_0-neighbourhoods

$$\begin{aligned} D(s,r) &= \{ |\mathrm{Im}\, x| < s \} \times \{ |y| < r^2 \} \times \{ \|u\|_p + \|v\|_p < r \} \\ &\subset \mathbb{C}^m \times \mathbb{C}^m \times \ell^2_{p,\mathbb{C}} \times \ell^2_{p,\mathbb{C}} \\ &= \mathcal{S}_{p,\mathbb{C}}. \end{aligned}$$

Here, $|z| = \max_n |z_n|$ for vectors in $\mathbb{C}^m$, and $\ell^2_{p,\mathbb{C}}$ is the complexification of ℓ^2_p. On $\mathcal{S}_{q,\mathbb{C}}$ we introduce for $W = (W_x, W_y, W_u, W_v)$ the *weighted norm*

$$\|W\|_{r,q} = |W_x| + \frac{1}{r^2} \left|W_y\right| + \frac{1}{r} \|W_u\|_q + \frac{1}{r} \|W_v\|_q .$$

For a map $W\colon U \times \Pi \to \mathcal{S}_{q,\mathbb{C}}$, such as the Hamiltonian vector field X_P, we then define the norms

$$\begin{aligned} \|W\|^{\mathrm{sup}}_{r,q;U\times\Pi} &= \sup_{(w,\xi)\in U\times\Pi} \|W(w,\xi)\|_{r,q}, \\ \|W\|^{\mathrm{lip}}_{r,q;U\times\Pi} &= \sup_{\substack{\xi,\zeta\in\Pi \\ \xi\neq\zeta}} \frac{\left\|\Delta_{\xi\zeta} W\right\|^{\mathrm{sup}}_{r,q;U}}{|\xi-\zeta|}, \end{aligned}$$

where $\Delta_{\xi\zeta} W = i_\xi W - i_\zeta W$, and

$$\|i_\xi W\|^{\mathrm{sup}}_{r,q;U} = \sup_{w\in U} \|W(w,\xi)\|_{r,q}.$$

In a completely analogous manner, the Lipschitz semi-norm of the map $\tilde{\Omega}\colon \Pi \to \ell^\infty_{-\delta}$ is defined as

$$|\tilde{\Omega}|^{\mathrm{lip}}_{-\delta;\Pi} = \sup_{\substack{\xi,\zeta\in\Pi \\ \xi\neq\zeta}} \frac{\left|\Delta_{\xi\zeta}\tilde{\Omega}\right|_{-\delta}}{|\xi-\zeta|}.$$

Note that $|\tilde{\Omega}|^{\mathrm{lip}}_{-\delta;\Pi} = |\Omega|^{\mathrm{lip}}_{-\delta;\Pi}$, since $\bar{\Omega} = \Omega - \tilde{\Omega}$ is independent of ξ.

We introduce one more constant. By assumptions A and B,

$$|\omega|^{\mathrm{lip}}_{\Pi} + |\Omega|^{\mathrm{lip}}_{-\delta;\Pi} \leq M < \infty.$$

Finally observe that if X_P satisfies assumption C, then it does so with the T_0-neighbourhoods $D(s,r)$ for all $s > 0$, $r > 0$ sufficiently small.

A complete proof of the following theorem will be given in chapter V.

Theorem 16.1. *Suppose N is a family of Hamiltonians of the form* (16.1) *defined on a phase space $\mathcal{S}_p^m$ and depending on parameters in Π so that assumptions A and B are satisfied. Then there exists a positive constant γ depending only on m, d, δ, the frequencies ω and Ω and the real number $s > 0$ such that for every perturbed Hamiltonian $H = N + P$ that satisfies assumption C and the smallness condition*

$$\varepsilon = \|X_P\|^{\text{sup}}_{r,q;D(s,r)\times\Pi} + \frac{\alpha}{M}\|X_P\|^{\text{lip}}_{r,q;D(s,r)\times\Pi} \le \alpha\gamma$$

for some $r > 0$ and $0 < \alpha < 1$, the following holds. There exist

(i) *a Cantor set $\Pi_\alpha \subset \Pi$ with* $\operatorname{meas}(\Pi \setminus \Pi_\alpha) \to 0$ *as $\alpha \to 0$,*

(ii) *a Lipschitz family of real analytic torus embeddings $\Phi\colon \mathbb{T}^m \times \Pi_\alpha \to \mathcal{S}_p$,*

(iii) *a Lipschitz map $\varphi\colon \Pi_\alpha \to \mathbb{R}^m$,*

such that for each $\xi \in \Pi_\alpha$ the map Φ restricted to $\mathbb{T}^m \times \{\xi\}$ is a real analytic embedding of a rotational torus with frequencies $\varphi(\xi)$ for the system with Hamiltonian $H = N + P$ at ξ. In other words,

$$t \mapsto \Phi(\theta + t\varphi(\xi), \xi), \qquad t \in \mathbb{R},$$

is a real analytic, quasi-periodic solution for the Hamiltonian $i_\xi H$ for every $\theta \in \mathbb{T}^m$ and $\xi \in \Pi_\alpha$.

Moreover, each embedding is real analytic on $D(s/2) = \{|\operatorname{Im} x| < s/2\}$, and

$$\|\Phi - \Phi_0\|^{\text{sup}}_{r,p;D(s/2)\times\Pi_\alpha} + \frac{\alpha}{M}\|\Phi - \Phi_0\|^{\text{lip}}_{r,p;D(s/2)\times\Pi_\alpha} \le \frac{c\varepsilon}{\alpha},$$
$$|\varphi - \omega|^{\text{sup}}_{\Pi_\alpha} + \frac{\alpha}{M}|\varphi - \omega|^{\text{lip}}_{\Pi_\alpha} \le c\varepsilon,$$

where

$$\Phi_0\colon \ \mathbb{T}^m \times \Pi \to T_0, \quad (x,\xi) \mapsto (x,0,0,0)$$

is the trivial embedding for each ξ, and c is a positive constant which depends on the same parameters as γ.

Remark 1. The role of the parameter α is the following. In applications the size of the perturbation usually depends on a small parameter which we may also call ε. One then wants to choose α as a function of this parameter, for example $\alpha = \varepsilon/\gamma$. This way one can control the size of $\Pi \setminus \Pi_\alpha$ in terms of the size of the perturbation. See [76, 110] for examples.

Remark 2. The complement of the Cantor set Π_α can be written as

$$\Pi \setminus \Pi_\alpha = \Xi_\alpha = \Xi_\alpha^1 \cup \Xi_\alpha^2,$$

where Ξ_α^1 and Ξ_α^2 may be considered as the *coarse* and the *fine structure* of Ξ_α, respectively. The latter depends on the perturbation P and satisfies

$$\operatorname{meas}(\Xi_\alpha^2) = O(\alpha).$$

The former consists of finitely many resonance zones defined in terms of the unperturbed frequencies, and

$$\operatorname{meas}(\Xi_\alpha^1) \to 0 \quad \text{as} \quad \alpha \to 0$$

by assumption B. However, it requires further assumptions on the frequencies to make the rate of convergence explicit.

17 Proof of the Main Theorems

We now give the proofs of the two theorems in section 13 concerning perturbations of the first two KdV equations. Since they are treated in a completely analogous way, we focus on the proof of Theorem 13.1.

Recall the set up of Theorem 13.1. Let $A \subset \mathbb{N}$ be a finite index set of cardinality m, let $\Gamma \subset \mathbb{R}_+^m$ be a closed bounded set of positive Lebesgue measure, and

$$\mathcal{T}_\Gamma = \bigcup_{J \in \Gamma} \mathcal{T}_J$$

the union of all A-gap potentials with A-gap lengths in Γ. Consider a perturbed KdV Hamiltonian

$$H = H_c + \varepsilon K,$$

where H_c is the KdV Hamiltonian in (13.1), and K is real analytic on some complex neighbourhood V of $\mathcal{T}_\Gamma$ in $\mathcal{H}_{0,\mathbb{C}}^N$. Since all finite gap potentials are in $\mathcal{H}_{0,\mathbb{C}}^N$, the unperturbed Hamiltonian H_c is real analytic on V as well, while K satisfies

$$\left\| \frac{\partial K}{\partial u} \right\|_{N;V}^{\sup} \leq 1$$

by assumption.

As a first step we apply the global symplectic coordinate transformation of Theorem 15.1,

$$\Psi\colon \ h_{N+1/2}^\bullet \to \mathcal{H}_0^N,$$

which introduces global Birkhoff coordinates for the KdV equation. We know by Theorems 11.9 and 11.10 that Ψ maps the space h_A diffeomorphically onto the manifold $\mathcal{G}_A$ of A-gap potentials in such a way that every torus $T_I \subset h_A$ is mapped diffeomorphically onto an isospectral torus $\mathcal{T}_I \subset \mathcal{G}_A$. Hence there exists a closed bounded subset $\Pi \subset \mathbb{R}_+^A$ such that

$$\Psi(T_\Pi) = \mathcal{T}_\Gamma.$$

Since Ψ is real analytic, there is also a complex neighbourhood U of T_Π in the complexification of $h_{N+1/2}^\bullet$, which is mapped bi-analytically onto the neighbourhood V of $\mathcal{T}_\Gamma$. If necessary, we choose V smaller. Hence we have the following diagram, where each arrow represents a bi-analytic diffeomorphism given by the map Ψ:

$$\begin{array}{ccccccc} T_I & \subset & T_\Pi & \subset & U & \subset & h^\bullet_{N+1/2,\mathbb{C}} \\ \downarrow & & \downarrow & & \downarrow & & \\ \mathcal{T}_J & \subset & \mathcal{T}_\Gamma & \subset & V & \subset & \mathcal{H}^N_{0,\mathbb{C}} \end{array} .$$

Now we consider the transformed Hamiltonian

$$\begin{aligned} H \circ \Psi &= H_c \circ \Psi + \varepsilon K \circ \Psi \\ &= \hat{H}_c + \varepsilon R \\ &= \hat{H}_c(|\boldsymbol{w}|^2) + \varepsilon R(\boldsymbol{w}, \bar{\boldsymbol{w}}), \end{aligned}$$

which is real analytic on $U \supset T_\Pi$.

We first look at the integrable Hamiltonian, which can be written as

$$\hat{H}_c(|\boldsymbol{w}|^2) = \check{H}_c(\boldsymbol{I}), \qquad \boldsymbol{I} \in \ell^1_{2N+1},$$

according to Theorem 15.1, where $\ell^1_{2N+1} \subset \ell^1_3$, as $N \geq 1$. Using Taylor's formula and the frequencies introduced in section 15 we can write

$$\begin{aligned} \check{H}_c(\boldsymbol{I}_0 + \boldsymbol{I}) &= \check{H}_c(\boldsymbol{I}_0) + \sum_{i \geq 1} \frac{\partial \check{H}_c}{\partial I_i}(\boldsymbol{I}_0)\, I_i \\ &\quad + \int_0^1 (1-t) \sum_{i,j \geq 1} \frac{\partial^2 \check{H}_c}{\partial I_i \partial I_j} I_i I_j \, \mathrm{d}t \\ &= \text{const} + \sum_{i \geq 1} \omega_i(\boldsymbol{I}_0) I_i + \sum_{i,j \geq 1} Q_{ij}(\boldsymbol{I}_0, \boldsymbol{I}) I_i I_j \end{aligned} \tag{17.1}$$

with

$$Q_{ij}(\boldsymbol{I}_0, \boldsymbol{I}) = \int_0^1 (1-t) \frac{\partial \omega_i}{\partial I_j}(\boldsymbol{I}_0 + t\boldsymbol{I})\, \mathrm{d}t. \tag{17.2}$$

Note that Q_{ij} is symmetric in i and j. Using the asymptotics of Theorem 15.4 and Cauchy's estimate we obtain

$$\left| \sum_{j \geq 1} Q_{ij}(\boldsymbol{I}_0, \boldsymbol{I}) I_j \right| \leq ci \, \|\boldsymbol{I}\|_{\ell^1_{2N+1}} \tag{17.3}$$

uniformly in $\boldsymbol{I}_0$ on some complex neighbourhood of T_Π and $|\boldsymbol{I}|_{\ell^1_{2N+1}}$ sufficiently small.

The perturbing Hamiltonian vector field in the original phase space is

$$X_K = \frac{\mathrm{d}}{\mathrm{d}x} \frac{\partial K}{\partial u}.$$

It is defined on V and of order 1. Since Ψ is a symplectic diffeomorphism of the two Hilbert scales $(h^{\bullet}_{N+1/2})_{N\ge 1}$ and $(\mathcal{H}_0^N)_{N\ge 1}$, as explained in appendix K, the vector field of the transformed Hamiltonian $R = K \circ \Psi$ is

$$X_R = \Psi^* X_K = D\Psi^{-1} X_K \circ \Psi,$$

and it is also of order 1. Since we may choose the domain V so that the inverse of the Jacobian of Ψ is uniformly bounded, we obtain

$$\|X_R\|^{\sup}_{N-1/2;U} \le C\,\|X_K\|^{\sup}_{N-1;V} \le C \left\| \frac{\partial K}{\partial u} \right\|^{\sup}_{N;V} \le C. \tag{17.4}$$

As a second step we introduce symplectic polar coordinates around the tori in the family T_Π. To simplify notation we henceforth assume that $A = \{1, \dots, m\}$. For each $\xi = (\xi_1, \dots, \xi_m) \in \Pi$ we then introduce new coordinates by setting

$$w_n = \sqrt{\xi_n + y_n}\, \mathrm{e}^{-\mathrm{i}x_n}, \qquad w_{-n} = \sqrt{\xi_n + y_n}\, \mathrm{e}^{\mathrm{i}x_n}, \qquad 1 \le n \le m,$$

and

$$w_{m+n} = \frac{1}{\sqrt{2}}(u_n - \mathrm{i}v_n), \qquad w_{-m-n} = \frac{1}{\sqrt{2}}(u_n + \mathrm{i}v_n), \qquad n \ge 1.$$

This transformation is real analytic and symplectic on

$$\begin{aligned} D(s,r) &= \{\,|\mathrm{Im}\, x| < s\,\} \times \{\,|y| < r^2\,\} \times \{\,\|u\|_p + \|v\|_p < r\,\} \\ &\subset \mathbb{C}^m \times \mathbb{C}^m \times \ell^2_{p,\mathbb{C}} \times \ell^2_{p,\mathbb{C}} \\ &= \mathcal{S}_{p,\mathbb{C}} \end{aligned}$$

for all $s > 0$ and $r > 0$ sufficiently small, where $p = N + 1/2$. In the following we may fix such an s arbitrarily, while we keep the freedom to choose r smaller.

Using the expansion of $\check{H}_c$ in (17.1) and setting $\boldsymbol{I}_0 = (\xi, 0)$ the integrable Hamiltonian in the new coordinates is, up to a constant depending only on ξ and dropping the accent, given by

$$H_c = N + Q = N(y, u, v, \xi) + Q(y, u, v, \xi),$$

where

$$N = \sum_{1 \le n \le m} \omega_n(\xi) y_n + \tfrac{1}{2} \sum_{n \ge 1} \Omega_n(\xi)(u_n^2 + v_n^2),$$

$$\Omega_n(\xi) = \omega_{m+n}(\xi),$$

and, with $I_i = y_i$ for $1 \le i \le m$ and $I_i = \frac{1}{2}(u_{i-m}^2 + v_{i-m}^2)$ for $i \ge m+1$,

$$Q = \sum_{i,j \ge 1} Q_{ij}(\xi, \boldsymbol{I}) I_i I_j.$$

As the notation indicates, N will play the role of the integrable normal form depending on the parameters ξ, which is perturbed by Q and R.

We check assumptions A, B and C of the KAM Theorem 16.1 for this normal form. Its external frequencies Ω_n may be written as

$$\Omega_n(\xi) = \bar{\Omega}_n + \tilde{\Omega}_n(\xi)$$

with $\bar{\Omega}_n = \Omega_n(0)$ and $\tilde{\Omega}_n(\xi) = \Omega_n(\xi) - \Omega_n(0)$. We have

$$\bar{\Omega}_n = 8\pi^3(m+n)^3 + 12\pi c(m+n)$$

and a map

$$\tilde{\Omega} = (\tilde{\Omega}_n)_{n\geq 1}\colon \quad \Pi \to \ell^\infty_{-1},$$

which by Theorem 15.4 is real analytic on some complex neighbourhood of Π. Hence that map is also Lipschitz by Cauchy's estimate. So assumption A is satisfied with $d = 3$ and $\delta = 1$.

To verify assumption B we recall that by Proposition 15.5 we have

$$\det\left(\frac{\partial \omega_i}{\partial \xi_j}\right)_{1\leq i,j\leq m} \not\equiv 0$$

on Π. Since this determinant is a real analytic function, it is non-zero almost everywhere on Π. In particular, for any given $\eta > 0$ we may excise from Π a relatively open subset Π_η with $\operatorname{meas}(\Pi_\eta) < \eta$ such that on $\Pi \setminus \Pi_\eta$ the above determinant is uniformly bounded away from zero. Moreover, we may cover $\Pi \setminus \Pi_\eta$ by finitely many closed subsets Π_ι, so that on each such subset the map $\xi \mapsto \omega(\xi)$ is a bi-analytic homeomorphism onto its image in $\mathbb{R}^m$. Henceforth it suffices to consider each such parameter set Π_ι one at a time.

On such a set we have

$$\langle k,\omega(\xi)\rangle + \langle l,\Omega(\xi)\rangle \not\equiv 0$$

for every $k \in \mathbb{Z}^m$ and $l \in \mathbb{Z}^\infty$ with $1 \leq |l| \leq 2$ by Proposition 15.5 and its Addendum. Since each such expression is analytic in ξ, its zero set is a set of measure zero. Thus, assumption B is satisfied for each subset Π_ι.

As a consequence of assumption A and B we also have

$$|\omega|^{\text{lip}}_{\Pi} + |\Omega|^{\text{lip}}_{-1;\Pi} \leq M < \infty, \qquad \left|\omega^{-1}\right|^{\text{lip}}_{\omega(\Pi\setminus\Pi_\eta)} \leq L < \infty.$$

It remains to check assumption C. There are two contributions to the perturbation P:

$$P = Q + \varepsilon R.$$

From the form of Q and the estimate of Q_{ij} as given by equations (17.2) and (17.3) we obtain with Cauchy's estimate

$$\|X_Q\|^{\text{sup}}_{r,p-1;D(s,r)\times\Pi_\iota} \leq cr^2.$$

Just observe that $I_i = y_i$ for $1 \leq i \leq m$ and $I_i = \frac{1}{2}(u_{i-m}^2 + v_{i-m}^2)$ for $i \geq m+1$, and that $|\boldsymbol{I}|_{\ell^1_{2N+1}} \leq cr^2$ on the domain $D(s,r) \times \Pi_\iota$.

For the second term in P we have to take into account the weight factors in the definition of the norm $\|\cdot\|_{r,p-1}$ to obtain from (17.4) the estimate

$$\|X_R\|^{\sup}_{r,p-1;D(s,r)\times\Pi_\iota} \leq \frac{c}{r^2}.$$

The same bounds hold for the respective Lipschitz semi-norms, since both vector fields are real analytic in ξ, the above estimates extend to some complex neighbourhood of Π_ι, and Cauchy's estimate applies. So altogether we obtain

$$\epsilon \overset{\text{def}}{=} \|X_P\|^{\sup}_{r,p-1;D(s,r)\times\Pi_\iota} + \frac{\alpha}{M}\|X_P\|^{\text{lip}}_{r,p-1;D(s,r)\times\Pi_\iota} \leq C\left(r^2 + \frac{\varepsilon}{r^2}\right)$$

for $\alpha \leq M$ and all small $r > 0$. In particular, we have verified that

$$X_P\colon\ U_p \times \Pi_\iota \to \mathcal{S}_{q,\mathbb{C}},$$

with $U_p = D(s,r) \subset \mathcal{S}_{p,\mathbb{C}}$ and $q = p-1$. Since $1 = p - q < d - 1 = 2$, X_P has the required regularity properties.

To meet the smallness condition of the KAM theorem for $P = Q + \varepsilon R$ choose now

$$r^2 = \sqrt{\varepsilon}, \qquad \alpha = \frac{2C}{\gamma}\sqrt{\varepsilon},$$

with ε so small that $\alpha < 1$. Here, C is taken from the preceding estimate, and γ is taken from the KAM theorem. We then obtain

$$\epsilon \leq 2C\sqrt{\varepsilon} = \gamma\alpha$$

as required.

The conclusions of Theorem 13.1 now follow immediately from the conclusions of the KAM theorem. We only comment on the measure theoretic statement. For each Π_ι we have

$$\operatorname{meas}\left(\Pi_\iota \smallsetminus \Pi_{\iota,\alpha}\right) \to 0 \quad \text{as} \quad \varepsilon \to 0.$$

We have only finitely many such patches Π_ι, which cover the original parameter domain Π up to a set of measure η. By first choosing η and then ε small enough we can arrange that

$$\operatorname{meas}\left(\Pi \smallsetminus \bigcup_\iota \Pi_{\iota,\alpha}\right) \to 0 \quad \text{as} \quad \varepsilon \to 0.$$

The proof of Theorem 13.1 is now complete.

The proof of Theorem 13.2 is completely analogous. The only differences are that we have $d = 5$, $\delta = 3$, and $q = p - 3$. Moreover, the perturbation R may depend on u_x.

V

The KAM Proof

18 Set Up and Summary of Main Results

In the following we give a complete proof of the infinite dimensional KAM theorem used in chapter IV to study small Hamiltonian perturbations of KdV equations. To make this presentation independent of chapter IV we begin by recalling the set up.

We consider small perturbations of a family of infinite dimensional integrable Hamiltonians, where each Hamiltonian admits an invariant n-torus, on which the flow is linear with fixed frequencies, while in the remaining coordinates there is an infinite dimensional elliptic equilibrium. The aim is to prove the persistence of this torus under such perturbations.

Extending the set up of the previous discussion in view of a wider class of applications, the underlying phase spaces will also allow for exponentially decaying sequences. It turns out that this does not complicate the construction and the pertaining estimates at all, but is rather a matter of notation.

More specifically the unperturbed family is given by the *normal form Hamiltonians*

$$\begin{aligned} N &= N(x, y, u, v; \xi) \\ &= \sum_{1 \le j \le n} \omega_j(\xi) y_j + \tfrac{1}{2} \sum_{j \ge 1} \Omega_j(\xi)(u_j^2 + v_j^2) \end{aligned} \tag{18.1}$$

on a phase space

$$\mathcal{S}^n_{p,a} = \mathbb{T}^n \times \mathbb{R}^n \times \ell^2_{p,a} \times \ell^2_{p,a}$$

with coordinates (x, y, u, v), depending on parameters

$$\xi \in \Pi \subset \mathbb{R}^n.$$

$\mathbb{T}^n$ is the usual n-torus $\mathbb{R}^n / 2\pi\mathbb{Z}^n$ with $1 \le n < \infty$, and $\ell^2_{p,a}$ denotes the Hilbert space of all real sequences $w = (w_1, w_2, \dots)$ with

$$\|w\|^2_{p,a} = \sum_{j \ge 1} j^{2p} \mathrm{e}^{2aj} |w_j|^2 < \infty.$$

The parameter set Π may be any compact subset of $\mathbb{R}^n$ of positive Lebesgue measure. For example, Π may be a compact *Cantor set* of positive measure.

We will assume that $p \geq 0$ and $a \geq 0$. In the following the parameters n and a are fixed. Therefore we drop them from the notation and for simplicity write

$$\mathcal{S}_p = \mathcal{S}^n_{p,a}, \qquad \ell^2_p = \ell^2_{p,a}, \qquad \|\cdot\|_p = \|\cdot\|_{p,a}.$$

The symplectic structure on $\mathcal{S}_p$ is $\sum_{1 \leq j \leq n} \mathrm{d}x_j \wedge \mathrm{d}y_j + \sum_{j \geq 1} \mathrm{d}u_j \wedge \mathrm{d}v_j$. The equations of motion for N are therefore

$$\begin{aligned} \dot{x}_j &= \omega_j(\xi), & \dot{u}_j &= \Omega_j(\xi)v_j, \\ \dot{y}_j &= 0, & \dot{v}_j &= -\Omega_j(\xi)u_j. \end{aligned}$$

For each parameter $\xi \in \Pi$ there is an invariant torus

$$T_0 = T_0^n = \mathbb{T}^n \times \{0\} \times \{0\} \times \{0\},$$

on which the flow is rotational with *internal frequencies* $\omega(\xi) = (\omega_1(\xi), \ldots, \omega_n(\xi))$. In the normal space described by the u, v-coordinates we have an infinite dimensional elliptic equilibrium at the origin, whose characteristic *external frequencies* are $\Omega(\xi) = (\Omega_1(\xi), \Omega_2(\xi), \ldots)$. Hence, T_0 is an *invariant, rotational, linearly stable n-torus.*

Our aim is to prove the persistence of this torus under small perturbations $N + P$ of the integrable Hamiltonian N for a large Cantor set of parameter values ξ. To this end we make the following three assumptions. Of these, assumption A* differs slightly from the corresponding assumption A in chapter IV in order to simplify minor technical points later on. See the remark following Theorem 18.1 below.

Assumption A: Frequency Asymptotics.* There exist reals $d > 1$ and $\delta < d - 1$ such that the following holds. First,

$$\left|\Omega_i - \Omega_j\right| \geq m\,|i - j|(i^{d-1} + j^{d-1}) \tag{18.2}$$

for all $i \neq j \geq 0$ uniformly on Π with some constant $m > 0$. Here, $\Omega_0 = 0$. Second, the functions

$$\xi \mapsto \frac{\Omega_j(\xi)}{j^\delta}$$

are uniformly Lipschitz on Π for $j \geq 1$.

Setting one of the indices to zero in (18.2) one has $\left|\Omega_j\right| \geq mj^d$ for $j \geq 1$. On the other hand, the second assumption implies $\left|\Omega_j(\xi) - \Omega_j(\zeta)\right| \leq cj^\delta$ for $j \geq 1$ uniformly on Π. So the ξ-dependent part of the external frequencies is only of order δ, not d.

We assume here that $d > 1$. The case $d = 1$ may also be handled, as it is done in [72, 109], but is somewhat more involved and not relevant for the application to the KdV equations. We therefore omit it.

Assumption B: Nondegeneracy.* The map $\xi \mapsto \omega(\xi)$ between Π and its image $\omega(\Pi)$ is a homeomorphism which is Lipschitz continuous together with its inverse. Moreover, for every $k \in \mathbb{Z}^n$ and $l \in \mathbb{Z}^\infty$ with $1 \leq |l| \leq 2$ the *resonance set*

$$\mathcal{R}_{kl} = \left\{ \xi \in \Pi \colon \langle k, \omega(\xi) \rangle + \langle l, \Omega(\xi) \rangle = 0 \right\}$$

has Lebesgue measure zero. Here, $|l| = \sum_{j \geq 1} |l_j|$.

Note that by assumption A*, the sets $\mathcal{R}_{kl}$ are empty for $k = 0$ and $1 \leq |l| \leq 2$.

The third assumption is concerned with the perturbing Hamiltonian P and its vector field, $X_P = (P_y, -P_x, P_v, -P_u)^T$. We use the notation $i_\xi X_P$ for X_P evaluated at ξ. With $\mathcal{S}_{p,\mathbb{C}}$ we denote the complexification of the phase space $\mathcal{S}_p$.

Assumption C: Regularity of Perturbation.* There is a neighbourhood U_p of T_0 in $\mathcal{S}_{p,\mathbb{C}}$ such that P is defined on $U_p \times \Pi$, and its Hamiltonian vector field defines a map

$$X_P \colon \; U_p \times \Pi \to \mathcal{S}_{q,\mathbb{C}},$$

where $q \geq 0$ satisfies $p - q < d - 1$. Moreover, $i_\xi X_P$ is real analytic on U_p for each $\xi \in \Pi$, and $i_w X_P$ is uniformly Lipschitz on Π for each $w \in U_p$.

Without loss of generality we will also assume that $\delta \geq 0$ is chosen so that

$$p - q \leq \delta < d - 1.$$

To state the KAM theorem we need to introduce some domains and norms. For $s > 0$ and $r > 0$ we introduce the complex T_0-neighbourhoods

$$\begin{aligned} D(s,r) &= \{ |\operatorname{Im} x| < s \} \times \{ |y| < r^2 \} \times \{ \|u\|_p + \|v\|_p < r \} \\ &\subset \mathbb{C}^n \times \mathbb{C}^n \times \ell^2_{p,\mathbb{C}} \times \ell^2_{p,\mathbb{C}} \\ &= \mathcal{S}_{p,\mathbb{C}}. \end{aligned}$$

Here, $|z| = \max_j |z_j|$ for vectors in $\mathbb{C}^n$, and $\ell^2_{p,\mathbb{C}}$ is the complexification of ℓ^2_p. On $\mathcal{S}_{q,\mathbb{C}}$ we introduce for $W = (W_x, W_y, W_u, W_v)$ the *weighted norm*

$$\|W\|_{r,q} = |W_x| + \frac{1}{r^2} \left| W_y \right| + \frac{1}{r} \|W_u\|_q + \frac{1}{r} \|W_v\|_q .$$

For a map $W \colon U \times \Pi \to \mathcal{S}_{q,\mathbb{C}}$, such as the Hamiltonian vector field X_P, we then define the norms

$$\begin{aligned} \|W\|^{\sup}_{r,q;U\times\Pi} &= \sup_{(w,\xi)\in U\times\Pi} \|W(w,\xi)\|_{r,q}, \\ \|W\|^{\mathrm{lip}}_{r,q;U\times\Pi} &= \sup_{\substack{\xi,\zeta\in\Pi \\ \xi\neq\zeta}} \frac{\left\| \Delta_{\xi\zeta} W \right\|^{\sup}_{r,q;U}}{|\xi - \zeta|}, \end{aligned}$$

where $\Delta_{\xi\zeta} W = i_\xi W - i_\zeta W$ and $\|i_\xi W\|^{\sup}_{r,q;U} = \sup_{w\in U} \|W(w,\xi)\|_{r,q}$.

In a completely analogous manner, the Lipschitz semi-norm of the external frequencies Ω is defined as

$$|\Omega|^{\text{lip}}_{-\delta;\Pi} = \sup_{\substack{\xi,\zeta\in\Pi \\ \xi\neq\zeta}} \frac{\left|\Delta_{\xi\zeta}\Omega\right|_{-\delta}}{|\xi-\zeta|} = \sup_{\substack{\xi,\zeta\in\Pi \\ \xi\neq\zeta}} \sup_{j\geq 1} \frac{j^{-\delta}\left|\Delta_{\xi\zeta}\Omega_j\right|}{|\xi-\zeta|}.$$

Here and later, $|\lambda|_\beta = \sup_{j\geq 1} j^\beta |\lambda_j|$ for sequences $\lambda = (\lambda_1, \lambda_2, \dots)$. Finally, we introduce two more constants. Assuming that assumptions A* and B* hold, there exist constants M and L with

$$|\omega|^{\text{lip}}_{\Pi} + |\Omega|^{\text{lip}}_{-\delta;\Pi} \leq M < \infty, \qquad \left|\omega^{-1}\right|^{\text{lip}}_{\omega(\Pi)} \leq L < \infty.$$

The bound L will only be needed in section 22.

Theorem 18.1 (KAM Theorem). *Suppose N is a family of Hamiltonians of the form* (18.1) *defined on a phase space $\mathcal{S}_p$ and depending on parameters in Π so that assumptions A* and B* are satisfied. Then there exists a positive constant γ depending only on n, d, δ, m, the frequencies ω and Ω and $s > 0$ such that for every perturbation $H = N + P$ of N that satisfies assumption C* and the smallness condition*

$$\varepsilon = \|X_P\|^{\text{sup}}_{r,q;D(s,r)\times\Pi} + \frac{\alpha}{M}\|X_P\|^{\text{lip}}_{r,q;D(s,r)\times\Pi} \leq \alpha\gamma$$

for some $r > 0$ and $0 < \alpha < 1$, the following holds. There exist

(i) *a Cantor set $\Pi_\alpha \subset \Pi$ with* $\text{meas}(\Pi \setminus \Pi_\alpha) \to 0$ *as* $\alpha \to 0$,

(ii) *a Lipschitz family of real analytic torus embeddings $\Phi\colon \mathbb{T}^n \times \Pi_\alpha \to \mathcal{S}_p$,*

(iii) *a Lipschitz map $\varphi\colon \Pi_\alpha \to \mathbb{R}^n$,*

such that for each $\xi \in \Pi_\alpha$ the map Φ restricted to $\mathbb{T}^n \times \{\xi\}$ is a real analytic embedding of a rotational torus with frequencies $\varphi(\xi)$ for the perturbed Hamiltonian H at ξ. In other words,

$$t \mapsto \Phi(\theta + t\varphi(\xi), \xi), \qquad t \in \mathbb{R},$$

is a real analytic, quasi-periodic solution for the Hamiltonian $i_\xi H$ for every $\theta \in \mathbb{T}^n$ and $\xi \in \Pi_\alpha$.

Moreover, each embedding is real analytic on $D(s/2) = \{|\text{Im}\, x| < s/2\}$, and

$$\|\Phi - \Phi_0\|^{\text{sup}}_{r,p;D(s/2)\times\Pi_\alpha} + \frac{\alpha}{M}\|\Phi - \Phi_0\|^{\text{lip}}_{r,p;D(s/2)\times\Pi_\alpha} \leq \frac{c\varepsilon}{\alpha},$$
$$|\varphi - \omega|^{\text{sup}}_{\Pi_\alpha} + \frac{\alpha}{M}|\varphi - \omega|^{\text{lip}}_{\Pi_\alpha} \leq c\varepsilon,$$

where

$$\Phi_0\colon\ \mathbb{T}^n \times \Pi \to T_0, \qquad (x, \xi) \mapsto (x, 0, 0, 0)$$

is the trivial embedding for each ξ, and c is a positive constant which depends on the same parameters as γ.

Remark 1. The smallness condition on γ is made more explicit through equations (19.8), (21.6) and (22.2).

Remark 2. For a more explicit description of the Cantor set Π_α we refer to section 22.

Remark 3. The formulation of the KAM theorem here and in section 16 is identical, only the assumptions A and A* differ slightly. However, assuming assumptions A and B in section 16 one verifies that for any given $\eta > 0$ one can remove from Π a subset Π_η with $\mathrm{meas}(\Pi_\eta) < \eta$, such that on $\Pi \setminus \Pi_\eta$ assumption A* is satisfied with some constant $m > 0$. Therefore, the KAM theorem of section 16 follows from the theorem given here.

We now give an outline of the proof of the theorem. We focus on the case $q < p$, that is, the case of unbounded perturbations. The case $q \geq p$ is much simpler and has been dealt with in [72, 109].

The proof of Theorem 18.1 employs the rapidly converging iteration scheme of Newton type to handle small divisor problems introduced by Kolmogorov, and involves an infinite sequence of coordinate transformations. At the ν-th step of the scheme, a Hamiltonian

$$H_\nu = N_\nu + P_\nu$$

is considered, which is a small perturbation of some normal form N_ν. A transformation Φ_ν is set up so that

$$H_\nu \circ \Phi_\nu = N_{\nu+1} + P_{\nu+1}$$

with another normal form $N_{\nu+1}$ and a much smaller error term $P_{\nu+1}$. For instance,

$$\| P_{\nu+1} \| \leq C_\nu \, \| P_\nu \|^\kappa$$

for some $\kappa > 1$. This transformation is found by linearizing the above equation. Repetition of this process leads to a sequence of transformations $\Phi_0, \Phi_1, \ldots$, whose infinite composition transforms the initial Hamiltonian H_0 into a normal form N_∞ up to a certain order.

To describe the construction in more detail, let us drop the index ν. First we write

$$H = N + P = N + R + (P - R),$$

where R is obtained from P by truncating its Fourier and Taylor series expansion in a suitable way.

The coordinate transformation Φ is written as the time-1-map of the flow X_F^t of a Hamiltonian vectorfield X_F:

$$\Phi = X_F^t \big|_{t=1}.$$

Then Φ is symplectic. Moreover, we may expand $H \circ \Phi = H \circ X_F^t \big|_{t=1}$ with respect to t at 0 using Taylor's formula. Recall that

$$\frac{\mathrm{d}}{\mathrm{d}t} G \circ X_F^t = \{G, F\} \circ X_F^t.$$

Thus we may write

$$\begin{aligned}(N+R)\circ\Phi &= N\circ X_F^t\big|_{t=1} + R\circ X_F^t\big|_{t=1}\\ &= N + \{N,F\} + \int_0^1 (1-t)\,\{\{N,F\},F\}\circ X_F^t\,\mathrm{d}t\\ &\qquad + R + \int_0^1 \{R,F\}\circ X_F^t\,\mathrm{d}t\\ &= N + R + \{N,F\} + \int_0^1 \{R + (1-t)\,\{N,F\},F\}\circ X_F^t\,\mathrm{d}t.\end{aligned}$$

The latter integral is of quadratic order in R and F and will be part of the new error term.

The point is to find F such that $N + R + \{N,F\} = N_+$ is again a normal form, where '+' is short for '$\nu + 1$'. Equivalently, setting $N_+ = N + \hat{N}$, this amounts to solving the linear equation

$$\{F,N\} + \hat{N} = R$$

for F and $\hat{N}$, when R is given. Suppose such a solution exists. Then

$$(1-t)\,\{N,F\} + R = (1-t)\hat{N} + tR,$$

and hence

$$H\circ\Phi = N_+ + P_+ = N_+ + Q + (P-R)\circ\Phi$$

with

$$Q = \int_0^1 \{(1-t)\hat{N} + tR, F\}\circ X_F^t\,\mathrm{d}t.$$

The term Q is of quadratic order in R, F and $\hat{N}$.

What makes this scheme more complicated than previous ones is the fact that the vector field X_R is *unbounded*, whereas the vector field X_F has to be bounded to generate a *bona fide* coordinate transformation. For most terms in F this presents no problem, because they are obtained from the corresponding terms in R by division with a *large divisor*. There is no such smoothing effect, however, for terms in R of the form

$$\sum_{j\geq 1} R_j(x;\xi)(u_j^2 + v_j^2).$$

So we include them in $\hat{N}$ and hence in the new normal form N_+. Subsequently, however, we have to deal with generalized normal forms

$$N = \langle\omega(\xi), y\rangle + \tfrac{1}{2}\sum_{j\geq 1}\Omega_j(x;\xi)(u_j^2 + v_j^2),$$

where now the coefficients Ω_j in general depend on the angular variables x.

The linearized equation $\{F, N\} + \hat{N} = R$ then leads to the following type of first order partial differential equation for functions on the torus $\mathbb{T}^n$:

$$-\mathrm{i}\partial_\omega u + \lambda u + b(x)u = f, \qquad x \in \mathbb{T}^n, \tag{18.3}$$

where $\partial_\omega = \sum_{j=1}^n \omega_j \partial_{x_j}$. To obtain the required estimates for its solution we need to make the following three assumptions.

Assumption U. The frequencies ω are diophantine in the sense that there are constants $\alpha > 0$, $\tau > n$ and $l > 0$ such that $|\lambda| \geq \alpha l$ and

$$|\langle k, \omega \rangle + \lambda| \geq \frac{\alpha l}{|k|^\tau},$$
$$|\langle k, \omega \rangle| \geq \frac{\alpha}{|k|^\tau},$$

for all $0 \neq k \in \mathbb{Z}^n$, where $|k| = |k_1| + \cdots + |k_n|$.

Assumption V. The function b is analytic on some strip $D(s) = \{\, |\mathrm{Im}\, x| < s \,\}$ around the torus $\mathbb{T}^n$ with $[b] = 0$ and

$$\|b\|_{s,\tau} \overset{\text{def}}{=} \sum_k |\hat{b}_k|\, |k|^\tau \mathrm{e}^{|k|s} \leq \gamma\alpha$$

for $b = \sum_k \hat{b}_k \mathrm{e}^{\mathrm{i}\langle k, x \rangle}$ with some $\gamma > 0$ and the same τ as before.

Assumption W. The function f is analytic on the same strip $D(s)$ with

$$\|f\|_s \overset{\text{def}}{=} \|f\|_{s,0} < \infty.$$

Lemma 18.2 (Kuksin [74]). *Under assumptions U, V, W, equation* (18.3) *has a unique solution u that is analytic on $D(s)$ and satisfies*

$$\|u\|_{s-\sigma} \leq \frac{c\mathrm{e}^{2\gamma}}{\alpha l \sigma^\tau} \|f\|_s, \qquad 0 < \sigma \leq s.$$

Moreover, if $\dfrac{\gamma}{\lambda} \leq \dfrac{\sigma}{5\,|\omega|}$ *with* $|\omega| = \max |\omega_i|$, *then also*

$$\|u\|_{s-\sigma} \leq \frac{c}{\alpha l \sigma^{\tau+n+3}} \|f\|_s, \qquad 0 < \sigma \leq s.$$

The constant c can be chosen so that it only depends on τ in the first case, and only on τ and n in the second case.

The second estimate is the difficult one. It is needed to obtain a uniform bound for $\|u\|_{s-\sigma}$ for solutions of a *family* of equations of type (18.3) with *no* uniform bound on γ, but only a uniform bound on γ/λ.

19 The Linearized Equation

We consider the linearized equation mentioned above,

$$\{F, N\} + \hat{N} = R.$$

Using convenient complex notation $z = (u - \mathrm{i}v)/\sqrt{2}$ and $\bar{z} = (u + \mathrm{i}v)/\sqrt{2}$, where, however, u and v are *complex*, the generalized normal form reads

$$N = \langle \omega(\xi), y \rangle + \langle \Omega(x; \xi), z\bar{z} \rangle.$$

N is assumed to be *regular* on the domain $D(s, r) \times \Pi$ in the following sense: for each $\xi \in \Pi$, $i_\xi N$ is real analytic on $D(s, r)$, and for each $w \in D(s, r)$, $i_w N$ is Lipschitz on Π, as required in assumption C* of the KAM theorem.

The right hand side R is also assumed to be regular on $D(s, r) \times \Pi$ and of the form

$$R = \sum_{2|m| + |n + \bar{n}| \leq 2} R_{mn\bar{n}}(x; \xi) y^m z^n \bar{z}^{\bar{n}} \tag{19.1}$$

in usual multi-index notation, where momentarily m, n and $\bar{n}$ also denote multi-indices. Hence, if we define

$$\deg(y^m z^n \bar{z}^{\bar{n}}) = 2\,|m| + |n + \bar{n}|, \tag{19.2}$$

then R is a polynomial in $y, z, \bar{z}$ of degree 2 with coefficients depending regularly on x and ξ. Moreover, the Hamiltonian vector field associated with R is assumed to define a regular map

$$X_R \colon \mathcal{S}_{p,\mathbb{C}} \to \mathcal{S}_{q,\mathbb{C}}, \qquad p - q \leq \delta.$$

Note that $X_H = (\partial_y H, -\partial_x H, \mathrm{i}\partial_{\bar{z}} H, -\mathrm{i}\partial_z H)^T$ is the vector field of a Hamiltonian H in the complex notation.

The *mean value* of a function u on $\mathbb{T}^n$ is defined as usual by

$$[u] = \frac{1}{(2\pi)^n} \int_{\mathbb{T}^n} u(x)\,\mathrm{d}x.$$

The part of R in *generalized normal form* is defined as

$$\langle R \rangle = [R_{000}] + \sum_{|m|=1} [R_{m00}] y^m + \sum_{|n|=1} R_{0n\bar{n}}(x; \xi) z^n \bar{z}^n.$$

In the sequel we omit the term $[R_{000}]$, since it only depends on ξ and does not affect the dynamics. Note that in $\langle R \rangle$ the coefficients $R_{0n\bar{n}}$ are *not* averaged over $\mathbb{T}^n$.

Given N and R, we now seek a solution $\hat{N}$ and F of the linearized equation of the same form as N and R, respectively. To this end, write $\Omega = \bar{\Omega} + \tilde{\Omega}$, where

$$\bar{\Omega} = [\Omega], \qquad \tilde{\Omega} = \Omega - [\Omega],$$

are the x-independent and x-dependent parts of the exterior frequencies Ω, respectively. We will impose small divisor and growth conditions on $\bar{\Omega}$ and a smallness condition on $\tilde{\Omega}$. To formulate these we define

$$\langle k\rangle = \max(1, |k|), \qquad [l]_\delta = \max\Big(1, \Big|\sum_{j\geq 1} j l_j\Big| \cdot \sum_{j\geq 1} j^\delta \left|l_j\right|\Big)$$

for $k \in \mathbb{Z}^n$ and $l \in \mathbb{Z}^\infty$. For example,

$$[l]_\delta = \begin{cases} j^{1+\delta}, & l = e_j, \\ |i-j|(i^\delta + j^\delta), & l = e_i - e_j \neq 0 \end{cases} .$$

Moreover, for a function u analytic on $D(s)$ we set

$$\|u\|_{s,\tau} = \sum_{k\in\mathbb{Z}^n} \left|\hat{u}_k\right| |k|^\tau \mathrm{e}^{|k|s},$$

as in assumption V above. Observe that by standard estimates for Fourier coefficients,

$$\|u\|_{s,\tau} \leq \frac{c}{\sigma^{\tau+n}} |u|^{\sup}_{D(s+\sigma)} \tag{19.3}$$

for $\sigma > 0$ with a constant c depending only on n.

Lemma 19.1. *Suppose that uniformly on* Π,

$$\begin{aligned} |\langle k,\omega(\xi)\rangle + \langle l,\bar{\Omega}(\xi)\rangle| &\geq \alpha\frac{[l]_\delta}{\langle k\rangle^\tau}, && k\neq 0,\ |l|\leq 2 \\ |\langle l,\bar{\Omega}(\xi)\rangle| &\geq m\,[l]_{d-1}, && 0<|l|\leq 2, \\ j^{-\delta}\left\|\tilde{\Omega}_j\right\|_{s,\tau} &\leq \alpha\gamma_0, && j\geq 1, \end{aligned} \tag{19.4}$$

with constants $\tau \geq n$, $d > 1$, $0 < \gamma_0 \leq 1$, $m > 0$, *and a parameter* $0 < \alpha \leq m$. *If* γ_0 *is sufficiently small, then the linearized equation* $\{F,N\} + \hat{N} = R$ *has a solution* F, $\hat{N}$, *which is unique with the normalization* $\langle F\rangle = 0$, $\langle\hat{N}\rangle = \hat{N}$, *is regular on* $D(s,r)\times\Pi$ *in the above sense, and satisfies for* $0 < \sigma < s$ *the estimates*

$$\begin{aligned} \|X_F\|^{\sup}_{r,p;D(s-\sigma,r)\times\Pi} &\leq \frac{B_\sigma}{\alpha}\|X_R\|^{\sup}_{r,q;D(s,r)\times\Pi}, \\ \|X_F\|^{\mathrm{lip}}_{r,p;D(s-\sigma,r)\times\Pi} &\leq \frac{B_\sigma}{\alpha}\left(\|X_R\|^{\mathrm{lip}}_{r,q;D(s,r)\times\Pi} + \frac{M}{\alpha}\|X_R\|^{\sup}_{r,q;D(s,r)\times\Pi}\right), \end{aligned}$$

and

$$\begin{aligned} \|X_{\hat{N}}\|^{\sup}_{r,q;D(s-\sigma,r)\times\Pi} &\leq B_\sigma\|X_R\|^{\sup}_{r,q;D(s,r)\times\Pi}, \\ \|X_{\hat{N}}\|^{\mathrm{lip}}_{r,q;D(s-\sigma,r)\times\Pi} &\leq B_\sigma\left(\|X_R\|^{\mathrm{lip}}_{r,q;D(s,r)\times\Pi} + \frac{M}{\alpha}\|X_R\|^{\sup}_{r,q;D(s,r)\times\Pi}\right). \end{aligned}$$

Here, $M = |\omega|_{\Pi}^{\mathrm{lip}} + |\Omega|_{-\delta;D(s)\times\Pi}^{\mathrm{lip}}$, *and*

$$B_\sigma = \frac{a}{\sigma^b} \exp\left(\left(\frac{c}{\sigma}\right)^{1/\beta}\right), \qquad \frac{1}{\beta} = \frac{\delta}{d-1-\delta}, \tag{19.5}$$

with constants a, b, c *depending only on* n *and* τ.

The crucial, somewhat hidden feature of the first two of these estimates is the 'p' on their left hand sides and the 'q' on their right hand sides. That is, the solution X_F is bounded in the stronger norm $\|\cdot\|_{r,p}$ rather than $\|\cdot\|_{r,q}$.

We also note that B_σ is increasing fast as σ tends to zero, but not too fast for a subsequent iteration of Newton type to work. This kind of estimates also occur if one uses the more general approximation functions of Rüssmann [19, 116, 117] rather then the usual diophantine conditions – see for example [108].

Proof. Decompose $R = R^0 + R^1 + R^2$, where R^j comprises all terms in the expansion of R with $|n+\bar{n}| = j$. Decompose similarly F, N and $\hat{N}$, where necessarily $N^1 = 0$ and $\hat{N}^1 = 0$ by normalization. Comparing coefficients the linearized equation decomposes into

$$\begin{aligned} \{F^0, N^0\} + \hat{N}^0 &= R^0, \\ \{F^1, N\} \qquad &= R^1, \\ \{F^2, N\} + \hat{N}^2 &= R^2 - \{F^0, N^2\}. \end{aligned} \tag{19.6}$$

We will see that with the chosen normalization and the diophantine conditions these equations determine $\hat{N}^0$, F^0, F^1 and then $\hat{N}^2$, F^2 uniquely.

The first equation is independent of $z, \bar{z}$ and amounts to the classical, finite-dimensional pde

$$\partial_\omega F^0 + \hat{N}^0 = R^0, \qquad \partial_\omega = \sum_{1\le i\le n} \omega_i \partial_{x_i}.$$

This leads to $\hat{N}^0 = [R^0]$ and $\partial_\omega F^0 = R^0 - [R^0]$ with $[F^0] = 0$. Their estimates are standard and of the same form – indeed much better – than the ones for F^1, F^2 and $\hat{N}^2$ obtained below. For later reference we record that

$$\begin{aligned} \|X_{F^0}\|_{r,p;D(s-2\sigma,r)}^{\mathrm{sup}} &\le \frac{B_\sigma}{\alpha} \|X_R\|_{r,q;D(s,r)}^{\mathrm{sup}}, \\ \|X_{F^0}\|_{r,p;D(s-4\sigma,r)}^{\mathrm{lip}} &\le \frac{B_\sigma}{\alpha} \left(\|X_R\|_{r,q;D(s,r)}^{\mathrm{lip}} + \frac{M}{\alpha} \|X_R\|_{r,q;D(s,r)}^{\mathrm{sup}} \right), \end{aligned}$$

where $M = |\omega|_{\Pi}^{\mathrm{lip}} + |\Omega|_{-\delta;D(s)\times\Pi}^{\mathrm{lip}}$ and B_σ as in (19.5). We drop the factor Π from the domains during this proof for brevity, since it stays the same throughout. Note that X_{F^0} does not have any $z, \bar{z}$-component, so $\|X_{F^0}\|_{r,p}$ does not depend on p.

Consider the second equation in (19.6). Writing

$$R^1 = R^{10} + R^{01} = \langle \mathcal{R}^{10}, z\rangle + \langle \mathcal{R}^{01}, \bar{z}\rangle$$

and similarly F^1 it decomposes into

$$\{F^{ij}, N\} = R^{ij}, \qquad i + j = 1,$$

and it suffices to study each equation individually. We have $\mathcal{R}^{10} = R_z\big|_{z,\bar{z}=0}$ and thus

$$\frac{1}{r}\left\|\mathcal{R}^{10}\right\|_{q;D(s)}^{\sup} \le \|X_R\|_{r,q;D(s,r)}^{\sup},$$

where $D(s) = \{|\mathrm{Im}\, x| < s\}$. Writing $R^{10} = \langle\mathcal{R}^{10}, z\rangle = \sum_{j\ge 1} R_j(x;\xi) z_j$, and similarly F^{10}, the equation $\{F^{10}, N\} = R^{10}$ further decomposes into

$$\mathrm{i}\partial_\omega F_j + \Omega_j F_j = \mathrm{i}\partial_\omega F_j + \bar{\Omega}_j F_j + \tilde{\Omega}_j F_j = \mathrm{i} R_j, \qquad j \ge 1.$$

Here we did not indicate the dependence on x and ξ. Recall that $\tilde{\Omega}_j = \tilde{\Omega}_j(x;\xi)$. To obtain a solution of these equations with useful estimates we want to apply Kuksin's Lemma of section 23. The assumptions of this lemma are now verified.

By the diophantine condition in (19.4) we have uniformly on Π

$$|\langle k, \omega\rangle| \ge \frac{\alpha}{\langle k\rangle^\tau}, \qquad k \ne 0,$$

$$\left|\langle k, \omega\rangle + \bar{\Omega}_j\right| \ge \frac{\alpha j^{1+\delta}}{\langle k\rangle^\tau}, \qquad k \in \mathbb{Z}^n, \quad j \ge 1.$$

So assumption U of Kuksin's lemma is satisfied with $\lambda = \bar{\Omega}_j$ and $l_j = j^{1+\delta}$ in place of l. Moreover, $[\tilde{\Omega}] = 0$, and by the smallness condition in (19.4) we have

$$\left\|\tilde{\Omega}_j\right\|_{s,\tau} \le \alpha\gamma_0 j^\delta, \qquad j \ge 1,$$

uniformly on Π. So assumption V of Kuksin's lemma is satisfied with $\gamma_j = \gamma_0 j^\delta$ in place of γ. Finally, by the growth condition in (19.4), we have $\left|\bar{\Omega}_j\right| \ge m j^d$ for $j \ge 1$, and one verifies that

$$\left|\bar{\Omega}_j\right| \ge |\omega|_\Pi^{\sup} \gamma_j^{1+\beta}, \qquad j \ge 1,$$

with $\beta = (d - 1 - \delta)/\delta$, if γ_0 in the definition of γ_j is sufficiently small. An explicit and stronger condition on γ_0 is given in (19.8) below.

Thus, Kuksin's lemma and its corollaries apply, and so the unique solution F_j satisfies the estimate

$$\left|F_j\right|_{D(s-2\sigma)}^{\sup} \le \frac{B_\sigma}{\alpha l_j}\left|R_j\right|_{D(s-\sigma)}^{\sup}, \qquad j \ge 1.$$

Here and in the following, B_σ stands for a function of σ of the form (19.5) with various constants a, b, c, which only depend on n and τ, but not on j. Since $l_j = j^{1+\delta}$ and $p - q \le \delta$ this and Lemma M.2 imply

$$\left\|\mathcal{F}^{10}\right\|_{p;D(s-2\sigma)}^{\sup} \le \frac{B_\sigma}{\alpha}\left\|\mathcal{R}^{10}\right\|_{q;D(s)}^{\sup} \le \frac{B_\sigma}{\alpha} r\, \|X_R\|_{r,q;D(s,r)}^{\sup}.$$

The same estimate holds for $\mathcal{F}^{01}$. Multiplying $\mathcal{F}^{10}$ with z and $\mathcal{F}^{01}$ with $\bar{z}$ and using $p \geq 0$ this gives

$$\frac{1}{r^2}\left|F^1\right|^{\sup}_{D(s-2\sigma,r)} \leq \frac{B_\sigma}{\alpha}\left\|X_R\right\|^{\sup}_{r,q;D(s,r)},$$

and finally with Cauchy's estimate

$$\left\|X_{F^1}\right\|^{\sup}_{r,p;D(s-3\sigma)} \leq \frac{B_\sigma}{\alpha}\left\|X_R\right\|^{\sup}_{r,q;D(s,r)}.$$

To obtain Lipschitz estimates we study first the differences $\Delta F_j = i_\xi F_j - i_\zeta F_j$ for $\xi, \zeta \in \Pi$. We obtain $\mathrm{i}\Delta R_j = \mathrm{i}\partial_\omega \Delta F_j + \Omega_j \Delta F_j + \mathrm{i}\partial_{\Delta\omega} F_j + F_j \Delta\Omega_j$ from (19), hence

$$\mathrm{i}\partial_\omega \Delta F_j + \Omega_j \Delta F_j = \mathrm{i}\Delta R_j - \mathrm{i}\partial_{\Delta\omega} F_j - F_j \Delta\Omega_j,$$

with $\Omega_j = \bar{\Omega}_j + \tilde{\Omega}_j$. The right hand side is known, so ΔF_j uniquely solves the same kind of equation as F_j. So we obtain

$$\begin{aligned}\left|\Delta F_j\right|^{\sup}_{D(s-3\sigma)} &\leq \frac{B_\sigma}{\alpha l_j}\left(\left|\Delta R_j\right|^{\sup}_{D(s-\sigma)} + \frac{1}{\sigma}\left|F_j\right|^{\sup}_{D(s-2\sigma)}\left(|\Delta\omega| + \left|\Delta\Omega_j\right|^{\sup}_{D(s)}\right)\right) \\ &\leq \frac{B_\sigma}{\alpha l_j}\left|\Delta R_j\right|^{\sup}_{D(s-\sigma)} + \frac{B_\sigma^2}{\alpha^2 l_j^2}\left|R_j\right|^{\sup}_{D(s-\sigma)}\left(|\Delta\omega| + \left|\Delta\Omega_j\right|^{\sup}_{D(s)}\right).\end{aligned}$$

Multiplying by $l_j = j^{1+\delta}$ and going to the vector norms for $\mathcal{F}^{10}$ with the help of Lemma M.2 using $p - q \leq \delta$, we obtain

$$\begin{aligned}\left\|\Delta\mathcal{F}^{10}\right\|^{\sup}_{p;D(s-3\sigma)} \leq\ & \frac{B_\sigma}{\alpha}\left\|\Delta\mathcal{R}^{10}\right\|^{\sup}_{q;D(s)} \\ &+ \frac{B_\sigma^2}{\alpha^2}\left\|\mathcal{R}^{10}\right\|^{\sup}_{q;D(s)}\left(|\Delta\omega| + |\Delta\Omega|^{\sup}_{-\delta;D(s)}\right).\end{aligned}$$

Dividing by $|\xi - \zeta| \neq 0$ and taking the supremum over Π,

$$\left\|\mathcal{F}^{10}\right\|^{\mathrm{lip}}_{p;D(s-3\sigma)} \leq \frac{B_\sigma^2}{\alpha}\left(\left\|\mathcal{R}^{10}\right\|^{\mathrm{lip}}_{q;D(s)} + \frac{M}{\alpha}\left\|\mathcal{R}^{10}\right\|^{\sup}_{q;D(s)}\right).$$

The same estimate applies to $\mathcal{F}^{01}$. So for the vector field of F^1 we finally get

$$\left\|X_{F^1}\right\|^{\mathrm{lip}}_{r,p;D(s-4\sigma)} \leq \frac{B_\sigma^2}{\alpha}\left(\left\|X_R\right\|^{\mathrm{lip}}_{r,q;D(s,r)} + \frac{M}{\alpha}\left\|X_R\right\|^{\sup}_{r,q;D(s,r)}\right).$$

This concludes the discussion of F^1.

Consider now the third equation in (19.6). Writing $R^2 = R^{20} + R^{11} + R^{02}$ and similarly F^2 and N^2 this equation decomposes into

$$\left\{F^{ij}, N\right\} = R^{ij} - \left\langle R^{ij}\right\rangle, \qquad \hat{N}^{11} = \left\langle R^{11}\right\rangle - \left\{F^0, N^2\right\},$$

while $\hat{N}^{ij} = 0$ for $i \neq j$.

Consider the equation for F^{11}, which is slightly more complicated than the ones for F^{20} and F^{02}. Writing $R^{11} = \langle \mathcal{R}^{11} z, \bar{z} \rangle$, we have $\mathcal{R}^{11} = R_{z\bar{z}}|_{z,\bar{z}=0}$. Thus $\mathcal{R}^{11}$ is the Jacobian of R_z with respect to $\bar{z}$ at $\bar{z} = 0$. By Cauchy's inequality we have an estimate in the induced operator norm:

$$\|\mathcal{R}^{11}\|^{\sup}_{q,p;D(s)} \leq \frac{1}{r} \|R_z\|^{\sup}_{q;D(s,r)} \leq \|X_R\|^{\sup}_{r,q;D(s,r)},$$

where $\|\cdot\|_{q,p}$ denotes the operator norm induced by $\|\cdot\|_p$ and $\|\cdot\|_q$ in the source and target spaces, respectively.

Now write more explicitly

$$R^{11} = \sum_{i,j\geq 1} R_{ij}(x;\xi) z_i \bar{z}_j,$$

and similarly F^{11}. The equation $\{F^{ij}, N\} = R^{ij} - \langle R^{ij} \rangle$ decomposes into

$$\mathrm{i}\partial_\omega F_{ij} + (\Omega_i - \Omega_j) F_{ij} = \mathrm{i} R_{ij}, \qquad i \neq j,$$

and $\mathrm{i}\partial_\omega F_{jj} = 0$. The normalization $\langle F^{11} \rangle = 0$ enforces $F_{jj} = 0$, so it suffices to consider the first equation. Letting $\Omega_{ij} = \Omega_i - \Omega_j = \bar{\Omega}_{ij} + \tilde{\Omega}_{ij}$ with $\bar{\Omega}_{ij} = [\Omega_{ij}]$ it becomes

$$\mathrm{i}\partial_\omega F_{ij} + \bar{\Omega}_{ij} F_{ij} + \tilde{\Omega}_{ij} F_{ij} = \mathrm{i} R_{ij}, \qquad i \neq j,$$

to which we apply Kuksin's lemma. To this end we now verify assumptions U, V and W.

Again, by the diophantine condition in (19.4) we have

$$|\langle k,\omega \rangle| \geq \frac{\alpha}{\langle k \rangle^\tau}, \qquad \left|\langle k,\omega \rangle + \bar{\Omega}_{ij}\right| \geq \frac{\alpha l_{ij}}{\langle k \rangle^\tau}, \tag{19.7}$$

for $k \neq 0$ uniformly on Π with $l_{ij} = |i-j|(i^\delta + j^\delta)$. So assumption U is satisfied with $\lambda = \bar{\Omega}_{ij}$ and l_{ij} in place of l. Moreover, we have $[\tilde{\Omega}_{ij}] = 0$, and by the smallness condition in (19.4),

$$\left\|\tilde{\Omega}_{ij}\right\|_{s,\tau} \leq \|\tilde{\Omega}_i\|_{s,\tau} + \left\|\tilde{\Omega}_j\right\|_{s,\tau} \leq \alpha\gamma_0 (i^\delta + j^\delta).$$

So assumption V is satisfied with $\gamma_{ij} = \gamma_0(i^\delta + j^\delta)$ in place of γ. Finally, we have $\left|\bar{\Omega}_{ij}\right| \geq m\,|i-j|(i^{d-1} + j^{d-1})$ by the growth condition in (19.4), and one finds

$$\left|\bar{\Omega}_{ij}\right| \geq |\omega|^{\sup}_{\Pi} \gamma_{ij}^{1+\beta},$$

if for example

$$\gamma_0 \leq c_\beta \frac{m}{|\omega|^{\sup}_{\Pi}}, \qquad \beta = \frac{d-1-\delta}{\delta}, \tag{19.8}$$

with some constant c_β depending only on β.

So Kuksin's lemma and its corollaries apply, and we obtain

$$\left|F_{ij}\right|^{\sup}_{D(s-2\sigma)} \leq \frac{B_\sigma}{\alpha l_{ij}} \left|R_{ij}\right|^{\sup}_{D(s-\sigma)},$$

or

$$(i^\delta + j^\delta)\left|F_{ij}\right|^{\sup}_{D(s-2\sigma)} \leq \frac{B_\sigma}{\alpha}\frac{1}{|i-j|} \left|R_{ij}\right|^{\sup}_{D(s-\sigma)}, \qquad i \neq j.$$

With Lemma M.3 this yields

$$\begin{aligned}\left\|\mathcal{F}^{11}\right\|^{\sup}_{p,p;D(s-2\sigma)}, &\left\|\mathcal{F}^{11}\right\|^{\sup}_{q,q;D(s-2\sigma)} \\ &\leq \frac{B_\sigma}{\alpha}\left\|\mathcal{R}^{11}\right\|^{\sup}_{q,p;D(s)} \leq \frac{B_\sigma}{\alpha}\left\|X_R\right\|^{\sup}_{r,q;D(s,r)}. \end{aligned} \tag{19.9}$$

The same, and even better estimates hold for $\mathcal{F}^{20}$ and $\mathcal{F}^{02}$. Multiplying with $z, \bar{z}$ we then get

$$\frac{1}{r^2}\left|F^2\right|^{\sup}_{D(s-2\sigma,r)} \leq \frac{B_\sigma}{\alpha}\left\|X_R\right\|^{\sup}_{r,q;D(s,r)},$$

and finally with Cauchy's estimate

$$\left\|X_{F^2}\right\|^{\sup}_{r,p;D(s-3\sigma,r)} \leq \frac{B_\sigma}{\alpha}\left\|X_R\right\|^{\sup}_{r,q;D(s,r)},$$

in complete analogy to the estimate for X_{F^1} above.

The estimate for the Lipschitz semi-norm of X_{F^2} is obtained by the same arguments as the one for X_{F^1}, and the result is analogous. We therefore omit it.

For the contribution to the normal form we obtain

$$\hat{N}^2 = \hat{N}^{11} = \langle\hat{\Omega}(x;\xi), z\bar{z}\rangle = \sum_{j\geq 1}\hat{\Omega}_j(x;\xi)z_j\bar{z}_j$$

with $\hat{\Omega}_j(x;\xi) = R_{jj} + \{\tilde{\Omega}_j, F^0\} = R_{jj} + \langle\partial_x\tilde{\Omega}_j, \partial_y F^0\rangle$. We find that

$$\left|R_{jj}\right|^{\sup}_{D(s-2\sigma)} \leq j^{p-q}\left\|X_R\right\|^{\sup}_{r,q;D(s,r)} \leq j^\delta\left\|X_R\right\|^{\sup}_{r,q;D(s,r)}$$

and

$$\begin{aligned}\left|\langle\partial_x\tilde{\Omega}_j, \partial_y F^0\rangle\right| &\leq \left|\partial_x\Omega_j\right|_{D(s-2\sigma)}\left|\partial_y F^0\right|_{D(s-2\sigma)} \\ &\leq \left\|\Omega_j\right\|_{s,\tau}\left\|X_{F^0}\right\|^{\sup}_{r,p;D(s-2\sigma,r)} \\ &\leq j^\delta\gamma_0 B_\sigma\left\|X_R\right\|^{\sup}_{r,q;D(s,r)}. \end{aligned}$$

Hence, $j^{-\delta}\left\|\hat{\Omega}_j\right\|_{s-4\sigma,\tau} \leq B_\sigma\left\|X_R\right\|^{\sup}_{r,q;D(s,r)}$, using (19.3). Together with the estimate for $\hat{N}^0 = [R^0]$ this implies

$$\left\|X_{\hat{N}}\right\|^{\sup}_{r,q;D(s-4\sigma,r)} \leq B_\sigma\left\|X_R\right\|^{\sup}_{r,q;D(s,r)}.$$

As to the Lipschitz estimate we have

$$\Delta\hat{\Omega}_j = \Delta R_{jj} + \langle \partial_x \Delta\tilde{\Omega}_j, \partial_y F^0 \rangle + \langle \partial_x \tilde{\Omega}_j, \partial_y \Delta F^0 \rangle.$$

This gives

$$\begin{aligned}
\left|\Delta\hat{\Omega}_j\right|^{\text{sup}}_{D(s-4\sigma)} &\le \left|\Delta R_{jj}\right|^{\text{sup}}_{D(s-4\sigma)} + \frac{1}{\sigma}\left|\Delta\Omega_j\right|^{\text{sup}}_{D(s-3\sigma)}\left|\partial_y F^0\right|^{\text{sup}}_{D(s-4\sigma)} \\
&\quad + \frac{1}{\sigma}\left|\Omega_j\right|^{\text{sup}}_{D(s-3\sigma)}\left|\Delta\partial_y F^0\right|^{\text{sup}}_{D(s-4\sigma)} \\
&\le j^\delta \left\|\Delta X_R\right\|^{\text{sup}}_{r,q;D(s,r)} + j^\delta \left|\Delta\Omega\right|^{\text{sup}}_{-\delta;D(s)} \frac{B_\sigma}{\alpha}\left\|X_R\right\|^{\text{sup}}_{r,q;D(s,r)} \\
&\quad + j^\delta \frac{\alpha\gamma_0}{\sigma}\left\|\Delta X_{F^0}\right\|^{\text{sup}}_{r,p;D(s-4\sigma,r)}.
\end{aligned}$$

Dividing by j^δ and $|\xi - \zeta| \neq 0$, taking the supremum over Π and using the estimate of $\|X_{F^0}\|^{\text{lip}}$ above this gives

$$\begin{aligned}
|\hat{\Omega}|^{\text{lip}}_{-\delta;\Pi} &\le (1 + \gamma_0 B_\sigma)\left\|X_R\right\|^{\text{lip}}_{r,q;D(s,r)} \\
&\quad + (B_\sigma + \gamma_0 B_\sigma)\frac{M}{\alpha}\left\|X_R\right\|^{\text{sup}}_{r,q;D(s,r)} \\
&\le B_\sigma \left(\left\|X_R\right\|^{\text{lip}}_{r,q;D(s,r)} + \frac{M}{\alpha}\left\|X_R\right\|^{\text{sup}}_{r,q;D(s,r)} \right).
\end{aligned}$$

From these estimates the ones for $X_{\hat{N}}$ are readily derived.

The final estimates of the lemma are obtained by replacing σ by $\sigma/4$ throughout the proof. This affects only the constants a, b, c in the expression of B_σ and completes the proof. □

For later reference the estimates of Lemma 19.1 may be condensed as follows. For $\lambda \ge 0$, define

$$\|X\|^\lambda_r = \|X\|^{\text{sup}}_r + \lambda \|X\|^{\text{lip}}_r.$$

The symbol 'λ' in $\|X\|^\lambda_r$ will always be used in this role and never has the meaning of exponentiation.

Lemma 19.2. *The estimates of Lemma 19.1 imply that*

$$\begin{aligned}
\|X_F\|^\lambda_{r,p;D(s-\sigma,r)\times\Pi} &\le \frac{B_\sigma}{\alpha}\|X_R\|^\lambda_{r,q;D(s,r)\times\Pi}, \\
\|X_{\hat{N}}\|^\lambda_{r,q;D(s-\sigma,r)\times\Pi} &\le B_\sigma\|X_R\|^\lambda_{r,q;D(s,r)\times\Pi},
\end{aligned}$$

for $0 < \sigma < s$ and $0 \le \lambda \le \alpha/M$ with another B_σ of the same form as in Lemma 19.1.

The preceding lemma also gives us an estimate of $\|DX_F\|^\lambda_{r,p,p;D(s-2\sigma,r/2)\times\Pi}$ with the help of Cauchy's estimate. However, for the estimate (20.4) below we also need an estimate in terms of the operator norm $\|\cdot\|_{r,q,q}$.

Lemma 19.3. *Under the assumptions of Lemma 19.1,*

$$\|DX_F\|^{\lambda}_{r,p,p;D(s-\sigma,r)\times\Pi},\ \|DX_F\|^{\lambda}_{r,q,q;D(s-\sigma,r)\times\Pi} \le \frac{B_\sigma}{\alpha}\|X_R\|^{\lambda}_{r,q;D(s,r)\times\Pi}.$$

Proof. It remains to consider the estimate of $\|DX_F\|^{\lambda}_{r,q,q}$. We have

$$DX_F = \begin{pmatrix} F_{yx} & 0 & 0 & 0 \\ -F_{xx} & -F_{xy} & -F_{xz} & -F_{x\bar{z}} \\ \mathrm{i}F_{\bar{z}x} & 0 & \mathrm{i}F_{\bar{z}z} & \mathrm{i}F_{\bar{z}\bar{z}} \\ -\mathrm{i}F_{zx} & 0 & -\mathrm{i}F_{zz} & -\mathrm{i}F_{z\bar{z}} \end{pmatrix}.$$

The lower right 2×2-block is independent of z and $\bar{z}$ and defines a bounded operator on ℓ^2_q by the estimate (19.9). Since F_z, $F_{\bar{z}}$ are in ℓ^2_p and $p \ge q \ge 0$, the operators F_{xz}, $F_{x\bar{z}}$ are bounded in ℓ^2_p by Cauchy's estimate, and hence bounded in ℓ^2_q as well.

As to the estimates we note that

$$\|DX_F\|_{r,q,q} = \left\|W_r^{-1}DX_F W_r\right\|_{q,q},$$

where $W_r = \mathrm{diag}(1, r^2, r, r)$ according to the above block decomposition. In other words, the weighted norm $\|DX_F\|_{r,q,q}$ is obtained by first multiplying the entries of DX_F with the corresponding entries in the matrix

$$\begin{pmatrix} 1 & r^2 & r & r \\ r^{-2} & 1 & r^{-1} & r^{-1} \\ r^{-1} & r & 1 & 1 \\ r^{-1} & r & 1 & 1 \end{pmatrix}$$

and then taking the norm $\|\cdot\|_{q,q}$.

In all the cases where an x-derivative is involved, such as F_{xz}, the proper estimate of each block is now obtained from the corresponding estimate of the vector field X_F, such as F_z, Cauchy's estimate with respect to x, and the fact that $\ell^2_p \subset \ell^2_q$. In the other cases we use the estimate (19.9).

Finally, replacing again σ by $\sigma/2$ we obtain the claimed supremum estimates. The Lipschitz estimates are obtained in a similar fashion. □

20 The KAM Step

At the general ν-th step of the iteration scheme we are given a Hamiltonian

$$H_\nu = N_\nu + P_\nu,$$

where

$$N_\nu = \langle \omega_\nu(\xi), y\rangle + \langle \Omega_\nu(x;\xi), z\bar{z}\rangle$$

is a generalized normal form, and P_ν is a small perturbation of it. Both are assumed to be regular on $D(s_\nu, r_\nu) \times \Pi_\nu$ in the sense described on page 152, where Π_ν is some closed bounded subset of $\mathbb{R}^n$. On Π_ν we assume that

$$\begin{aligned} |\langle k, \omega_\nu(\xi)\rangle + \langle l, \bar{\Omega}_\nu(\xi)\rangle| &\geq \alpha_\nu \frac{[l]_\delta}{\langle k\rangle^\tau}, \qquad && k \neq 0, |l| \leq 2 \\ |\langle l, \bar{\Omega}_\nu(\xi)\rangle| &\geq m_\nu \, [l]_{d-1}, && 0 < |l| \leq 2, \\ j^{-\delta} \left\| \tilde{\Omega}_{\nu,j} \right\|_{s_\nu,\tau} &\leq \alpha_\nu \gamma_0, && j \geq 1, \end{aligned} \tag{20.1}$$

where we let $\Omega_\nu = \bar{\Omega}_\nu + \tilde{\Omega}_\nu$ with $\bar{\Omega}_\nu = [\Omega_\nu]$. Moreover, we assume that

$$|\omega_\nu|^{\mathrm{lip}}_{\Pi_\nu} + |\Omega_\nu|^{\mathrm{lip}}_{-\delta; D(s_\nu)\times\Pi_\nu} \leq M_\nu .$$

For the rest of this section we drop the subscript 'ν' from the notation, and write the symbol '+' for '$\nu + 1$' for simplicity. Thus, $P = P_\nu$, $P_+ = P_{\nu+1}$, and so on. Also, we write $u \lessdot v$, if $u \leq cv$ with some constant $c \geq 1$, which only depends on n and τ.

To perform one step of the iteration we assume that

$$\|X_P\|^{\mathrm{sup}}_{r,q; D(s,r)\times\Pi} + \frac{\alpha}{M} \|X_P\|^{\mathrm{lip}}_{r,q; D(s,r)\times\Pi} \leq \frac{\alpha\eta^2}{B^0_\sigma}$$

with some $0 < \eta < 1/16$ and $0 < \sigma < s$, $\sigma \leq 1$, where B^0_σ is a function of σ of the form (19.5) with certain sufficiently large constants a, b, c depending only on n and τ, which could be made explicit. Using the notation of Lemma 19.2 the smallness condition may also be written as

$$\|X_P\|^{\lambda}_{r,q; D(s,r)\times\Pi} \leq \frac{\alpha\eta^2}{B^0_\sigma} \tag{20.2}$$

for $0 \leq \lambda \leq \alpha/M$. – From now on we also drop the factor 'Π' from the notation of domains, since it stays the same throughout this section.

Approximation of P

We approximate P by its Taylor polynomial R of order 2 in $y, z, \bar{z}$ in the sense of (19.2), which is of the form (19.1). This amounts to corresponding approximations of the partial derivatives P_x, P_y, P_z, $P_{\bar{z}}$, which constitute the vector field X_P. Since P is real analytic, R_x, R_y, R_z, $R_{\bar{z}}$ and their remainders are given by certain Cauchy integrals, which we can estimate exactly as in a finite dimensional setting. We therefore obtain

$$\begin{aligned} \|X_R\|^{\lambda}_{r,q; D(s,r)} &\lessdot \|X_P\|^{\lambda}_{r,q; D(s,r)}, \\ \|X_P - X_R\|^{\lambda}_{\eta r,q; D(s,4\eta r)} &\lessdot \eta \, \|X_P\|^{\lambda}_{r,q; D(s,r)}, \end{aligned} \tag{20.3}$$

for $0 \leq \lambda \leq \alpha/M$ and $0 < \eta < 1/16$.

Solution of the Linearized Equation

The hypotheses (20.1) are of exactly the same form as those of Lemma 19.1 for the linearized equation. Hence, if γ_0 is sufficiently small, Lemma 19.1 applies, and we can solve $\{F,N\} + \hat{N} = R$. With Lemmas 19.2 and 19.3 and estimates (20.3) we obtain

$$\begin{aligned} \|X_{\hat{N}}\|^{\lambda}_{r,q;D(s-\sigma,r)} &\leq B_\sigma \, \|X_P\|^{\lambda}_{r,q;D(s,r)}, \\ \|X_F\|^{\lambda}_{r,p;D(s-\sigma,r)} &\leq \frac{B_\sigma}{\alpha} \, \|X_P\|^{\lambda}_{r,q;D(s,r)}, \end{aligned} \tag{20.4}$$

for $0 \leq \lambda \leq \alpha/M$. The bound in the second line also holds for $\|DX_F\|^{\lambda}_{r,p,p;D(s-\sigma,r)}$ and for $\|DX_F\|^{\lambda}_{r,q,q;D(s-\sigma,r)}$. Together with the smallness assumption (20.2) we get

$$\|X_F\|^{\lambda}_{r,p;D(s-\sigma,r)}, \, \|DX_F\|^{\lambda}_{r,q,q;D(s-\sigma,r)} \leq \eta^2 \frac{B_\sigma}{B^0_\sigma}. \tag{20.5}$$

Coordinate Transformation

We now choose B^0_σ in such a way that the preceding estimate in particular gives

$$\|X_F\|^{\sup}_{r,p;D(s-\sigma,r)} \leq \eta^2 \frac{B_\sigma}{B^0_\sigma} \leq \frac{\eta^2 \sigma}{c_0}$$

with some suitable constant $c_0 \geq 1$. Then the flow X^t_F of the vector field X_F exists on $D(s-2\sigma, r/2)$ for $-1 \leq t \leq 1$ and takes this domain into $D(s-\sigma, r)$. Similarly, it takes $D(s-3\sigma, r/4)$ into $D(s-2\sigma, r/2)$. Together with Lemma M.4 we obtain

$$\begin{aligned} \left\|X^t_F - \mathrm{id}\right\|^{\lambda}_{r,p;D(s-2\sigma,r/2)} &\lessdot \|X_F\|^{\lambda}_{r,p;D(s-\sigma,r)}, \\ \left\|DX^t_F - I\right\|^{\lambda}_{r,q,q;D(s-3\sigma,r/4)} &\lessdot \|DX_F\|^{\lambda}_{r,q,q;D(s-\sigma,r)}, \end{aligned} \tag{20.6}$$

for $0 \leq \lambda \leq \alpha/M$. The latter estimate also holds in the $\|\cdot\|_{r,p,p}$-norm, which we do not write down explicitly.

In particular, we notice that X^t_F for $-1 \leq t \leq 1$ is a symplectic transformation, which transforms Hamiltonian vector fields of order $p-q$ according to the transformation rule of appendix K, since DX^t_F is a bounded linear operator both on $T\mathcal{S}_{p,\mathbb{C}}$ and $T\mathcal{S}_{q,\mathbb{C}}$.

The New Hamiltonian

Subjecting the Hamiltonian $H = N + P$ defined on $D(s,r)$ to the symplectic transformation $\Phi = X^t_F\big|_{t=1}$ we obtain the new Hamiltonian $H \circ \Phi = N_+ + P_+$ on $D(s-\sigma, r/2)$, where $N_+ = N + \hat{N}$ and

$$P_+ = (P - R) \circ X^1_F + \int_0^1 \{R(t), F\} \circ X^t_F \, \mathrm{d}t,$$

with $R(t) = (1-t)\hat{N} + tR$. Hence, as X_F^t is symplectic,

$$X_{P_+} = (X_F^1)^*(X_P - X_R) + \int_0^1 (X_F^t)^* \left[X_{R(t)}, X_F\right] \mathrm{d}t.$$

We will show below that for any vector field Y,

$$\left\|(X_F^t)^* Y\right\|_{\eta r, q; D(s-4\sigma, \eta r)}^{\lambda} \lessdot \|Y\|_{\eta r, q; D(s-2\sigma, 4\eta r)}^{\lambda} \tag{20.7}$$

for $0 \le t \le 1$. We already estimated $\|X_P - X_R\|_{\eta r, q}^{\lambda}$ in (20.3), so it remains to consider the commutator $[X_{R(t)}, X_F]$.

First, we have

$$\begin{aligned} \|X_{R(t)}\|_{r,q;D(s-\sigma,r)}^{\lambda} &\le \|X_{\hat{N}}\|_{r,q;D(s-\sigma,r)}^{\lambda} + \|X_R\|_{r,q;D(s-\sigma,r)}^{\lambda} \\ &\le B_\sigma \|X_P\|_{r,q;D(s,r)}^{\lambda} \end{aligned}$$

by (20.3) and (20.4). Moreover, we have the pointwise estimate

$$\begin{aligned} \left\|\left[X_{R(t)}, X_F\right]\right\|_{r,q} &\le \left\|DX_{R(t)} \cdot X_F\right\|_{r,q} + \left\|DX_F \cdot X_{R(t)}\right\|_{r,q} \\ &\le \left\|DX_{R(t)}\right\|_{r,q,p} \|X_F\|_{r,p} + \|DX_F\|_{r,q,q} \left\|X_{R(t)}\right\|_{r,q}. \end{aligned}$$

By the product rule for Lipschitz-norms and Cauchy's estimate we thus obtain

$$\begin{aligned} &\left\|\left[X_{R(t)}, X_F\right]\right\|_{r,q;D(s-2\sigma,r/2)}^{\lambda} \\ &\quad\lessdot \left\|DX_{R(t)}\right\|_{r,q,p;D(s-2\sigma,r/2)}^{\lambda} \|X_F\|_{r,p;D(s-2\sigma,r/2)}^{\lambda} \\ &\qquad + \|DX_F\|_{r,q,q;D(s-2\sigma,r/2)}^{\lambda} \left\|X_{R(t)}\right\|_{r,q;D(s-2\sigma,r/2)}^{\lambda} \\ &\quad\lessdot \frac{B_\sigma}{\alpha} \left(\|X_P\|_{r,q;D(s,r)}^{\lambda}\right)^2 \end{aligned}$$

for $0 \le \lambda \le \alpha/M$. Hence, also

$$\begin{aligned} \left\|\left[X_{R(t)}, X_F\right]\right\|_{\eta r,q;D(s-2\sigma,r/2)}^{\lambda} &\le \frac{1}{\eta^2} \left\|\left[X_{R(t)}, X_F\right]\right\|_{r,q;D(s-2\sigma,r/2)}^{\lambda} \\ &\lessdot \frac{B_\sigma}{\alpha\eta^2} \left(\|X_P\|_{r,q;D(s,r)}^{\lambda}\right)^2. \end{aligned}$$

Together with the estimate of $X_P - X_R$ in (20.3) and with (20.7) we finally arrive at the estimate

$$\left\|X_{P_+}\right\|_{\eta r,q;D(s-4\sigma,\eta r)}^{\lambda} \lessdot \eta \|X_P\|_{r,q;D(s,r)}^{\lambda} + \frac{B_\sigma}{\alpha\eta^2} \left(\|X_P\|_{r,q;D(s,r)}^{\lambda}\right)^2 \tag{20.8}$$

for $0 \le \lambda \le \alpha/M$. This is the bound for the new perturbation.

Proof of Estimate (20.7)

Fix $-1 \le t \le 1$ and set $\Phi = X_F^t$. Consider the pull back $\Phi^* Y = D\Phi^{-1} Y \circ \Phi$. In view of the estimate for $X_F^t - \mathrm{id}$, Φ maps the domain $U = D(s - 4\sigma, \eta r)$ into the domain $V = D(s - 3\sigma, 2\eta r)$. Hence,

$$\left\| \Phi^* Y \right\|_{\eta r, q; U}^{\mathrm{sup}} \le \left\| D\Phi^{-1} \right\|_{\eta r, q, q; V}^{\mathrm{sup}} \left\| Y \right\|_{\eta r, q; V}^{\mathrm{sup}},$$

and by Lemma M.4,

$$\left\| D\Phi^{-1} \right\|_{\eta r, q, q; V}^{\mathrm{sup}} \le 1 + \frac{1}{\eta^2} \left\| DX_F^{-t} - I \right\|_{r, q, q; V}^{\mathrm{sup}} \lessdot 1$$

by (20.5) and (20.6). So we have $\|\Phi^* Y\|_{\eta r, q; U}^{\mathrm{sup}} \lessdot \|Y\|_{\eta r, q; V}^{\mathrm{sup}}$.

For the Lipschitz semi-norm we have to take into account that both Φ and Y depend on the parameters in Π. Therefore,

$$\begin{aligned}
\left\| \Delta \Phi^* Y \right\|_{\eta r, q; U}^{\mathrm{sup}} &\le \left\| \Delta D\Phi^{-1} \right\|_{\eta r, q, q; U}^{\mathrm{sup}} \left\| Y \circ \Phi \right\|_{\eta r, q; U}^{\mathrm{sup}} \\
&\quad + \left\| D\Phi^{-1} \right\|_{\eta r, q, q; U}^{\mathrm{sup}} \left\| \Delta (Y \circ \Phi) \right\|_{\eta r, q; U}^{\mathrm{sup}} \\
&\lessdot \left\| \Delta (D\Phi^{-1} - I) \right\|_{\eta r, q, q; U}^{\mathrm{sup}} \left\| Y \right\|_{\eta r, q; V}^{\mathrm{sup}} \\
&\quad + \left\| \Delta Y \right\|_{\eta r, q; V}^{\mathrm{sup}} + \left\| DY \right\|_{\eta r, q, p; V}^{\mathrm{sup}} \left\| \Delta \Phi \right\|_{\eta r, p; U}^{\mathrm{sup}} \\
&\lessdot \frac{1}{\eta^2} \left\| \Delta (D\Phi^{-1} - I) \right\|_{r, q, q; U}^{\mathrm{sup}} \left\| Y \right\|_{\eta r, q; V}^{\mathrm{sup}} \\
&\quad + \left\| \Delta Y \right\|_{\eta r, q; V}^{\mathrm{sup}} + \frac{1}{\sigma} \left\| Y \right\|_{\eta r, q, p; W}^{\mathrm{sup}} \left\| \Delta \Phi \right\|_{\eta r, p; U}^{\mathrm{sup}}
\end{aligned}$$

with $W = D(s - 2\sigma, 4\eta r)$ by Cauchy's estimate. Dividing by $|\xi - \zeta|$, taking the supremum over Π and using the Lipschitz estimates for X_F^t and DX_F^t from (20.6) we obtain

$$\left\| \Phi^* Y \right\|_{\eta r, q; U}^{\mathrm{lip}} \lessdot \left\| Y \right\|_{\eta r, q; V}^{\mathrm{lip}} + \frac{M}{\alpha} \left\| Y \right\|_{\eta r, q; W}^{\mathrm{sup}}.$$

From these estimates (20.7) follows.

The New Generalized Normal Form

This is

$$N_+ = N + \hat{N} = \langle \omega_+(\xi), y \rangle + \langle \Omega_+(x; \xi), z\bar{z} \rangle$$

with $\omega_+ = \omega + \hat{\omega}$ and $\Omega_+ = \Omega + \hat{\Omega}$. For $\hat{N}$, defined similarly in terms of $\hat{\omega}$ and $\hat{\Omega}$, we have the estimate

$$\|X_{\hat{N}}\|_{r, q; D(s-\sigma, r)}^{\lambda} \le B_\sigma \, \|X_P\|_{r, q; D(s, r)}^{\lambda}$$

for $0 \le \lambda \le \alpha/M$. The weighted norm implies that we have $|\hat{\omega}| \le \|X_{\hat{N}}\|_{r,q}^{\mathrm{sup}}$ and

$\|\hat{\Omega}z\|_q \le r\,\|X_{\hat{N}}\|_{r,q}^{\sup}$ on $D(s,r)$, and consequently $|\hat{\Omega}|_{q-p} \le \|X_{\hat{N}}\|_{r,q}^{\sup}$. The same holds for the Lipschitz semi-norms. Since $p - q \le \delta$ we obtain

$$|\hat{\omega}|_{\Pi}^{\lambda} + |\hat{\Omega}|_{-\delta;\,D(s-\sigma)}^{\lambda} \le B_\sigma\,\|X_P\|_{r,q;\,D(s,r)}^{\lambda} \tag{20.9}$$

for $0 \le \lambda \le \alpha/M$.

In order to control the assumptions of the KAM step for the iteration we notice that the last estimate also implies

$$\sup j^{-\delta}\left\|\hat{\Omega}_j\right\|_{s-2\sigma,\tau} \le B_\sigma\,\|X_P\|_{r,q;\,D(s,r)}^{\sup}$$

with a B_σ with different constants a, b, c using (19.3). Moreover,

$$\begin{aligned} |\langle l,\hat{\Omega}\rangle|_{D(s-\sigma)}^{\sup} &\le |l|_\delta\,|\hat{\Omega}|_{-\delta;\,D(s-\sigma)}^{\sup} \\ &\le [l]_{d-1} B_\sigma\,\|X_P\|_{r,q;\,D(s,r)}^{\lambda} \\ &\le \hat{m}\,[l]_{d-1}, \end{aligned}$$

if $B_\sigma\,\|X_P\|_{r,q;\,D(s,r)}^{\lambda} \le \hat{m}$. Finally, for $k \ne 0$,

$$\begin{aligned} |\langle k,\hat{\omega}\rangle + \langle l,\hat{\Omega}\rangle| &\le |k|\,|\hat{\omega}| + [l]_{d-1}\,|\hat{\Omega}|_{-\delta} \\ &\le |k|\,[l]_{d-1}\cdot B_\sigma\,\|X_P\|_{r,q;\,D(s,r)}^{\lambda} \\ &\le \hat{\alpha}\,\frac{[l]_{d-1}}{\langle k\rangle^{\tau}}, \end{aligned} \tag{20.10}$$

provided

$$B_\sigma\,\|X_P\|_{r,q;\,D(s,r)}^{\lambda} \le \frac{\hat{\alpha}}{\langle k\rangle^{\tau+1}}.$$

In the next section we will make sure that $\hat{m} \ll m$ and $\hat{\alpha} \ll \alpha$, so that conditions (20.1) are essentially preserved.

21 Iteration and Convergence

To iterate the KAM step infinitely often we now choose sequences for all its parameters. The guiding principle is to choose η so as to minimize the bound on the next perturbation, to keep α, m and M essentially constant, and to keep close track of the size ε of X_P.

First we make some heuristic considerations. We define B_σ^0, which is used in the smallness condition (20.2) of the KAM step, to be the largest among all functions B_σ encountered during the KAM step. Further, we define c_0 to be twice the largest of all the constants c that are implicit in the '$\lessdot$'-notation. To make both terms in the bound for $X_{P_{\nu+1}}$ of equal size, namely

$$\left\|X_{P_{\nu+1}}\right\|_{\eta_\nu r_\nu,q}^{\lambda_\nu} \lessdot \eta_\nu\varepsilon_\nu + \frac{B_{\sigma_\nu}^0}{\alpha_\nu\eta_\nu^2}\,\varepsilon_\nu^2,$$

we choose η_ν by

$$\eta_\nu^3 = \frac{B_\nu^0 \varepsilon_\nu}{\alpha_\nu}, \qquad B_\nu^0 = B_{\sigma_\nu}^0.$$

One then obtains

$$\varepsilon_{\nu+1} = \frac{c_0}{2}\left(\eta_\nu \varepsilon_\nu + \frac{B_\nu^0}{\alpha_\nu \eta_\nu^2}\varepsilon_\nu^2\right) = c_0 \eta_\nu \varepsilon_\nu = \left(\frac{B_\nu}{\alpha_\nu}\right)^{\kappa-1} \varepsilon_\nu^\kappa,$$

where $\kappa = 4/3$ and

$$B_\nu = c_0^3 B_\nu^0 = c_0^3 B_{\sigma_\nu}^0. \tag{21.1}$$

Note that B_ν is different from B_σ.

Iteration of the identity for $\varepsilon_{\nu+1}$ then gives

$$\varepsilon_\nu = \left(\varepsilon_0 \prod_{\mu=0}^{\nu-1} \left(\frac{B_\mu}{\alpha_\mu}\right)^{\frac{\kappa-1}{\kappa^{\mu+1}}}\right)^{\kappa^\nu}. \tag{21.2}$$

We will choose B_ν and α_ν so that the product in this expression increases with ν. Further, choosing $\sigma_\nu \sim \sigma_0/\nu^2$ and $\alpha_\nu \geq \alpha_0/2$ the product will converge as $\nu \to \infty$. So the scheme will converge exponentially fast, if we roughly choose

$$\varepsilon_0 < \prod_{\mu=0}^{\infty} \left(\frac{\alpha_\mu}{B_\mu}\right)^{\frac{\kappa-1}{\kappa^{\mu+1}}}.$$

We now give the precise set up. For $\nu \geq 0$ let

$$\alpha_\nu = \frac{\alpha_0}{2}(1+2^{-\nu}), \quad m_\nu = \frac{m_0}{2}(1+2^{-\nu}), \quad M_\nu = M_0(2-2^{-\nu})$$

and

$$\sigma_\nu = \frac{\sigma_0}{1+\nu^2}, \qquad \sigma_0 = \frac{s_0}{20}, \qquad s_{\nu+1} = s_\nu - 4\sigma_\nu,$$

so that $s_0 > s_1 > s_2 > \cdots > s_0/2$. Define B_ν and ε_ν by (21.1) and (21.2), and let

$$\eta_\nu^3 = \frac{B_\nu^0 \varepsilon_\nu}{\alpha_\nu}, \qquad r_{\nu+1} = \eta_\nu r_\nu.$$

This defines the domains $D_\nu = D(s_\nu, r_\nu)$. Finally, we let

$$\lambda_\nu = \frac{\alpha_\nu}{M_\nu}, \qquad K_\nu = K_0^{\kappa^\nu}, \qquad K_0 = \gamma_0^{-\frac{1}{\tau+1}},$$

where $\kappa = 4/3$, and $\gamma_0 \leq 1$ is the small parameter of Lemma 19.1 appearing in (19.4) and (20.1) and required to satisfy (19.8).

Lemma 21.1 (Iterative Lemma). *Suppose that*

$$\varepsilon_0 \le \frac{\alpha_0\gamma_0}{4}\prod_{\mu=0}^{\infty} B_\mu^{-\frac{\kappa-1}{\kappa^{\mu+1}}}, \qquad \alpha_0 \le \frac{m_0}{2}. \tag{21.3}$$

Suppose $H_\nu = N_\nu + P_\nu$ is regular on $D_\nu \times \Pi_\nu$, where N_ν is a generalized normal form with coefficients satisfying $|\omega_\nu|^{\mathrm{lip}}_{\Pi_\nu} + |\Omega_\nu|^{\mathrm{lip}}_{-\delta;D_\nu\times\Pi_\nu} \le M_\nu$ and

$$\begin{aligned} |\langle k,\omega_\nu(\xi)\rangle + \langle l,\bar\Omega_\nu(\xi)\rangle| &\ge \alpha_\nu\frac{[l]_\delta}{\langle k\rangle^\tau}, && k\ne 0,\ |l|\le 2\\ |\langle l,\bar\Omega_\nu(\xi)\rangle| &\ge m_\nu\,[l]_{d-1}, && 0<|l|\le 2,\\ j^{-\delta}\left\|\tilde\Omega_{\nu,j}\right\|_{s_\nu,\tau} &\le (\alpha_0-\alpha_\nu)\gamma_0, && j\ge 1, \end{aligned} \tag{21.4}$$

on Π_ν, and P_ν satisfies

$$\left\|X_{P_\nu}\right\|^{\lambda_\nu}_{r_\nu,q;D_\nu\times\Pi_\nu} \le \varepsilon_\nu.$$

Then there exists a Lipschitz family of real analytic symplectic coordinate transformations $\Phi_{\nu+1}\colon D_{\nu+1}\times\Pi_\nu \to D_\nu$ and a closed subset

$$\Pi_{\nu+1} = \Pi_\nu \smallsetminus \bigcup_{\substack{|k|>K_\nu\\ |l|\le 2}} \mathcal{R}^{\nu+1}_{kl}(\alpha_{\nu+1}) \subset \Pi_\nu,$$

where

$$\mathcal{R}^{\nu+1}_{kl}(\alpha) = \left\{\xi\in\Pi_\nu\colon\ |\langle k,\omega_{\nu+1}(\xi)\rangle + \langle l,\bar\Omega_{\nu+1}(\xi)\rangle| < \alpha\frac{[l]_\delta}{\langle k\rangle^\tau}\right\},$$

such that for $H_{\nu+1} = H_\nu\circ\Phi_{\nu+1} = N_{\nu+1} + P_{\nu+1}$ the same assumptions as above are satisfied with '$\nu+1$' in place of 'ν'.

In comparing assumptions (20.1) and (21.4) note that $\alpha_0 - \alpha_\nu \le \alpha_\nu$.

Proof. In view of the definition of η_ν, namely $\eta_\nu^3 = B_\nu^0\varepsilon_\nu/\alpha_\nu$, the smallness condition of the KAM step, namely $\varepsilon_\nu \le \alpha_\nu\eta_\nu^2/B_\nu^0$, is equivalent to

$$\varepsilon_\nu \le \frac{\alpha_\nu}{B_\nu^0}.$$

To verify this inequality we argue as follows. As B_ν and α_ν^{-1} are increasing with ν,

$$\frac{B_\nu}{\alpha_\nu} = \prod_{\mu=\nu}^{\infty}\left(\frac{B_\nu}{\alpha_\nu}\right)^{\frac{\kappa-1}{\kappa^{\mu-\nu+1}}} \le \left(\prod_{\mu=\nu}^{\infty}\left(\frac{B_\mu}{\alpha_\mu}\right)^{\frac{\kappa-1}{\kappa^{\mu+1}}}\right)^{\kappa^\nu}.$$

By the definition of ε_ν in (21.2), the bound $\alpha_\nu \ge \alpha_0/2$ and the smallness condition on ε_0 in (21.3),

$$\frac{\varepsilon_\nu B_\nu^0}{\alpha_\nu} \le c_0^{-3}\left(\varepsilon_0\prod_{\mu=0}^{\infty}\left(\frac{B_\mu}{\alpha_\mu}\right)^{\frac{\kappa-1}{\kappa^{\mu+1}}}\right)^{\kappa^\nu} \le c_0^{-3}\left(\frac{\gamma_0}{2}\right)^{\kappa^\nu} \le 1.$$

So the smallness condition of the KAM step is satisfied for each $\nu \geq 0$. In particular, if $c_0 \geq 16$, then

$$\frac{\varepsilon_\nu B_\nu^0}{\alpha_\nu} \leq \frac{\gamma_0}{2^{\nu+2}} \tag{21.5}$$

in view of $\gamma_0 \leq 1$ and $\kappa = 4/3$, and also $\eta_\nu < 1/16$.

By the KAM step there exists a transformation $\Phi_{\nu+1}\colon D_{\nu+1} \times \Pi_\nu \to D_\nu$ taking H_ν into $H_{\nu+1} = N_{\nu+1} + P_{\nu+1}$. By (20.8) and (21.2) the new perturbation $P_{\nu+1}$ then satisfies the estimate

$$\begin{aligned} \left\| X_{P_{\nu+1}} \right\|_{r_{\nu+1},q;D_{\nu+1}\times\Pi_\nu}^{\lambda_{\nu+1}} &\leq \frac{c_0}{2}\left(\eta_\nu \varepsilon_\nu + \frac{B_\nu^0}{\alpha_\nu \eta_\nu^2}\varepsilon_\nu^2\right) \\ &= c_0 \left(\frac{B_\nu^0}{\alpha_\nu}\right)^{\kappa-1} \varepsilon_\nu^\kappa = \left(\frac{B_\nu}{\alpha_\nu}\right)^{\kappa-1} \varepsilon_\nu^\kappa = \varepsilon_{\nu+1}. \end{aligned}$$

Consider now the new normal form $N_{\nu+1}$ and its coefficients. By the estimates (20.9) and the ones following it and (21.5) we have, with $\lambda_\nu = \alpha_\mu / M_\nu$,

$$\begin{aligned} |\hat{\omega}_\nu|_{\Pi_\nu}^{\mathrm{lip}} + |\hat{\Omega}_\nu|_{-\delta;D_{\nu+1}\times\Pi_\nu}^{\mathrm{lip}} &\leq \frac{M_\nu}{\alpha_\nu} B_\nu^0 \varepsilon_\nu \leq \frac{M_\nu}{2^{\nu+2}}, \\ j^{-\delta} \left\| \hat{\Omega}_{\nu,j} \right\|_{s_{\nu+1},\tau} &\leq B_\nu^0 \varepsilon_\nu \leq \frac{\alpha_\nu \gamma_0}{2^{\nu+2}} \leq (\alpha_\nu - \alpha_{\nu+1})\gamma_0, \end{aligned}$$

and, as $\alpha_0 \gamma_0 \leq m_0$,

$$\frac{|\langle l, \hat{\Omega}_\nu\rangle|_{D_{\nu+1}\times\Pi_\nu}^{\mathrm{sup}}}{[l]_{d-1}} \leq B_\nu^0 \varepsilon_\nu \leq \frac{\alpha_\nu\gamma_0}{2^{\nu+2}} \leq m_\nu - m_{\nu+1}.$$

From this it follows that all assumptions in Lemma 21.1 except the diophantine conditions in (21.4) also hold for '$\nu + 1$' in place of 'ν'.

As to the diophantine conditions we use that

$$B_\nu^0 \varepsilon_\nu \leq \frac{\alpha_\nu \gamma_0^{\kappa^\nu}}{2^{\nu+2}} \leq \frac{\alpha_\nu - \alpha_{\nu+1}}{K_\nu^{\tau+1}}$$

by the definition of K_ν. Hence, by (20.10),

$$|\langle k, \hat{\omega}_\nu\rangle + \langle l, \hat{\Omega}_\nu\rangle| \leq (\alpha_\nu - \alpha_{\nu+1}) \frac{|k|\,[l]_{d-1}}{K_\nu^{\tau+1}} \leq (\alpha_\nu - \alpha_{\nu+1}) \frac{[l]_{d-1}}{\langle k\rangle^\tau}$$

for $\langle k\rangle \leq K_\nu$ on $D_{\nu+1} \times \Pi_\nu$. Using the diophantine conditions for ω_ν and $\bar{\Omega}_\nu$ on Π_ν we see that they also hold on Π_ν for the new frequencies $\omega_{\nu+1}$ and $\bar{\Omega}_{\nu+1}$ up to $\langle k\rangle \leq K_\nu$. It remains to remove from Π_ν the union of the open resonant zones $\mathcal{R}_{kl}^{\nu+1}(\alpha_{\nu+1})$ for $\langle k\rangle > K_\nu$ and $|l| \leq 2$ to obtain the parameter domain $\Pi_{\nu+1}$ on which all the diophantine conditions are satisfied. □

With (20.4), (20.6) and (20.9) we also obtain the following estimates. We use that $r_{\nu+1} = \eta_\nu r_\nu \le r_\nu/4$ and $\eta_\nu < 1/16$.

Lemma 21.2. *For $\nu \ge 0$,*

$$\|\Phi_{\nu+1} - \mathrm{id}\|^{\lambda_\nu}_{r_\nu,p;D_{\nu+1}\times\Pi_\nu}, \|D\Phi_{\nu+1} - I\|^{\lambda_\nu}_{r_\nu,q,q;D_{\nu+1}\times\Pi_\nu} \le \frac{B^0_\nu}{\alpha_\nu}\varepsilon_\nu,$$

and

$$|\omega_{\nu+1} - \omega_\nu|^{\lambda_\nu}_{\Pi_\nu}, |\Omega_{\nu+1} - \Omega_\nu|^{\lambda_\nu}_{-\delta;D_{\nu+1}\times\Pi_\nu} \le B^0_\nu\varepsilon_\nu.$$

The estimate of $D\Phi_{\nu+1} - I$ holds also with 'p' for 'q'.

We are now in a position to prove the KAM theorem. Suppose its assumptions are satisfied. To apply Lemma 21.1 with $\nu = 0$, set

$$N_0 = N, \quad P_0 = P, \quad s_0 = s, \quad r_0 = r,$$

and similarly $M_0 = M$, $\alpha_0 = \alpha$, $\lambda_0 = \lambda = \alpha/M$ and $m_0 = m$. Define γ in Theorem 18.1 by setting

$$\gamma = \gamma_0\gamma_s, \qquad \gamma_s = \frac{1}{4}\prod_{\mu=0}^{\infty} B_\mu^{-\frac{\kappa-1}{\kappa^{\mu+1}}}, \tag{21.6}$$

where γ_0 is the same parameter as before. Note that γ_s only depends on n, τ and s through the definition of the B_μ, while γ_0 will depend on m, LM and other parameters as made explicit in equation (22.2), in addition to satisfying (19.8).

The smallness condition (21.3) of Lemma 21.1 is then satisfied by the assumption of the KAM theorem:

$$\varepsilon_0 \stackrel{\text{def}}{=} \|X_{P_0}\|^{\lambda_0}_{r_0,q;D_0\times\Pi_0} = \|X_P\|^{\lambda}_{r,q;D(s,r)\times\Pi} \le \alpha\gamma = \alpha_0\gamma_0\gamma_s.$$

The small divisor conditions are satisfied by setting

$$\Pi_0 = \Pi \setminus \bigcup_{\substack{(k,l)\neq(0,0)\\ |l|\le 2}} \mathscr{R}^0_{kl}(\alpha_0),$$

that is, by removing all resonance zones defined in terms of the unperturbed frequencies. The other two conditions in (21.4) follow immediately from assumption A* and $\tilde{\Omega}_0 = 0$.

Hence, Lemma 21.1 applies, and we obtain a decreasing sequence of domains $D_\nu \times \Pi_\nu$ and a sequence of transformations

$$\Phi^\nu = \Phi_1 \circ \cdots \circ \Phi_\nu\colon \; D_\nu \times \Pi_{\nu-1} \to D_0,$$

such that $H \circ \Phi^\nu = H_\nu + P_\nu$ for $\nu \ge 1$. Moreover, the estimates of Lemma 21.2 hold.

To prove the convergence of the Φ^ν we consider the operator norms

$$\|L\|_{r,\tilde{r},p} = \sup_{W \neq 0} \frac{\|LW\|_{r,p}}{\|W\|_{\tilde{r},p}}.$$

These norms satisfy $\|AB\|_{r,\tilde{r},p} \le \|A\|_{r,r,p} \|B\|_{\tilde{r},\tilde{r},p}$ for $r \ge \tilde{r}$ as $\|W\|_{r,p} \le \|W\|_{\tilde{r},p}$. With the mean value theorem we then obtain

$$\left\|\Phi^{\nu+1} - \Phi^\nu\right\|^{\sup}_{r_0,p;D_{\nu+1}} \le \|D\Phi^\nu\|^{\sup}_{r_0,r_\nu,p;D_\nu} \|\Phi_{\nu+1} - \mathrm{id}\|^{\sup}_{r_\nu,p;D_{\nu+1}}$$

and, by the chain rule,

$$\|D\Phi^\nu\|^{\sup}_{r_0,r_\nu,p;D_\nu} \le \prod_{\mu=1}^{\nu} \|D\Phi_\mu\|^{\sup}_{r_\mu,r_\mu,p;D_\mu} \le \prod_{\mu=1}^{\infty}\left(1 + \frac{1}{2^{\mu+2}}\right) \le 2$$

for all $\nu \ge 1$, using the estimates of Lemma 21.2 together with (21.5) and the fact that the r_ν are decreasing. Also,

$$\begin{aligned}\left\|\Phi^{\nu+1} - \Phi^\nu\right\|^{\mathrm{lip}}_{r_0,p;D_{\nu+1}} &\le \|D\Phi^\nu\|^{\mathrm{lip}}_{r_0,r_\nu,p;D_\nu} \|\Phi_{\nu+1} - \mathrm{id}\|^{\sup}_{r_\nu,p;D_{\nu+1}} \\ &\quad + \|D\Phi^\nu\|^{\sup}_{r_0,r_\nu,p;D_\nu} \|\Phi_{\nu+1} - \mathrm{id}\|^{\mathrm{lip}}_{r_\nu,p;D_{\nu+1}},\end{aligned}$$

and again by the chain rule and the preceding estimate of $D\Phi_\nu$,

$$\begin{aligned}\|D\Phi^\nu\|^{\mathrm{lip}}_{r_0,r_\nu,p;D_\nu} &\le \sum_{\mu=1}^{\nu} \|D\Phi_\mu\|^{\mathrm{lip}}_{r_\mu,r_\mu,p;D_\mu} \prod_{\rho\neq\mu} \|D\Phi_\rho\|^{\sup}_{r_\rho,r_\rho,p;D_\rho} \\ &\le 2\sum_{\mu=1}^{\nu} \|D\Phi_\mu - I\|^{\mathrm{lip}}_{r_\mu,r_\mu,p;D_\mu} \\ &\le 2\sum_{\mu=1}^{\infty} \frac{M_\mu}{\alpha_\mu}\frac{\varepsilon_\mu B^0_\mu}{\alpha_\mu} \le \frac{M_0}{\alpha_0}.\end{aligned}$$

It follows that

$$\left\|\Phi^{\nu+1} - \Phi^\nu\right\|^{\lambda_0}_{r_0,p;D_{\nu+1}\times\Pi_\nu} \lessdot \|\Phi_{\nu+1} - \mathrm{id}\|^{\lambda_\nu}_{r_\nu,p;D_{\nu+1}\times\Pi_\nu}.$$

This shows that the Φ^ν converge uniformly on

$$\bigcap_{\nu\ge 0} D_\nu \times \Pi_\nu = D_* \times \Pi_\alpha,$$

where $\Pi_\alpha = \bigcap_{\nu\ge 0} \Pi_\nu$ and $D_* = D(s_*) \times \{0\} \times \{0\} \times \{0\}$, to a Lipschitz continuous family of real analytic torus embeddings

$$\Phi\colon\ \mathbb{T}^n \times \Pi_\alpha \to \mathcal{S}_p,$$

for which the estimates of Theorem 18.1 hold. Similarly, the frequencies ω_ν converge uniformly on Π_α to a Lipschitz continuous limit ω_*, and the frequencies Ω_ν converge uniformly on $D_* \times \Pi_\alpha$ to a regular limit Ω_*, with estimates as in Theorem 18.1. The embedded tori are invariant under the perturbed Hamiltonian flow, and the flow on them is linear, because

$$\begin{aligned}
&\left\| X_H \circ \Phi^\nu - D\Phi^\nu \cdot X_{N_\nu} \right\|^{\sup}_{r_0,q;D_\nu \times \Pi_\alpha} \\
&\quad \leq \| D\Phi^\nu \|^{\sup}_{r_0,r_\nu,q;D_\nu \times \Pi_\alpha} \left\| (\Phi^\nu)^* X_H - X_{N_\nu} \right\|^{\sup}_{r_\nu,q;D_\nu \times \Pi_\alpha} \\
&\quad \lessdot \left\| X_{P_\nu} \right\|^{\sup}_{r_\nu,q;D_\nu \times \Pi_\alpha}.
\end{aligned}$$

Hence in the limit, $X_H \circ \Phi = D\Phi \cdot X_{N_*}$ on D_* for each $\xi \in \Pi_\alpha$, where N_* is the generalized normal form with coefficients ω_* and Ω_*.

It remains to prove the claims about the set $\Pi \setminus \Pi_\alpha$. This is the subject of the next section.

22 The Excluded Set of Parameters

During the iterative construction we obtain a decreasing sequence of closed sets $\Pi_0 \supset \Pi_1 \supset \ldots$ such that $\Pi_\alpha = \bigcap_{\nu \geq 0} \Pi_\nu$ and

$$\Pi \setminus \Pi_\alpha = \bigcup_{\nu \geq 0} \bigcup_{|k| > K_{\nu-1}, |l| \leq 2} \mathcal{R}^\nu_{kl}(\alpha_\nu),$$

where $K_{-1} = 0$, $K_\nu = \gamma_0^{\kappa^\nu/\tau+1}$ as defined in the iteration, and

$$\mathcal{R}^\nu_{kl}(\alpha_\nu) = \left\{ \xi \in \Pi_{\nu-1} \colon \; |\langle k, \omega_\nu(\xi) \rangle + \langle l, \bar{\Omega}_\nu(\xi) \rangle| < \alpha_\nu \frac{[l]_\delta}{\langle k \rangle^\tau} \right\}$$

with $\Pi_{-1} = \Pi$. Here, ω_ν and $\bar{\Omega}_\nu$ are defined and Lipschitz continuous on $\Pi_{\nu-1}$, and $\omega_0 = \omega$, $\Omega_0 = \Omega$ are the frequencies of the unperturbed system. Recall that $\mathcal{R}^0_{0l} = \emptyset$ for $1 \leq |l| \leq 2$ by assumption A*. Moreover, using a telescoping argument together with the estimates of Lemma 21.2 as well as (21.5) we have

$$|\omega_\nu - \omega|^{\lambda_\nu}_{\Pi_\nu}, \, |\bar{\Omega}_\nu - \Omega|^{\lambda_\nu}_{-\delta;\Pi_\nu} \leq \frac{\alpha_0 \gamma_0}{4}$$

for all $\nu \geq 0$.

We write $\Pi \setminus \Pi_\alpha = \Xi^1_\alpha \cup \Xi^2_\alpha$, where

$$\Xi^1_\alpha = \bigcup_{0 < |k| \leq K_0, |l| \leq 2} \mathcal{R}^0_{kl}(\alpha_0),$$

$$\Xi^2_\alpha = \bigcup_{\nu \geq 0} \bigcup_{|k| > \max(K_0, K_{\nu-1}), |l| \leq 2} \mathcal{R}^\nu_{kl}(\alpha_\nu).$$

The set Ξ^1_α is defined in terms of the original frequencies and thus known *a priori*. Since $|k| \le K_0$, it sort of describes the 'coarse structure' of $\Pi \smallsetminus \Pi_\alpha$. The set Ξ^2_α depends on the perturbation and is only known *a posteriori*. Since here $|k| > K_0$, it describes the 'fine structure' of $\Pi \smallsetminus \Pi_\alpha$.

We first note that for each k there are at most finitely many nonempty resonance zones $\mathcal{R}^\nu_{kl}(\alpha_\nu)$.

Lemma 22.1. *If $\mathcal{R}^\nu_{kl}(\alpha_\nu) \neq \emptyset$, then*

$$[l]_{d-1} \le \theta\, |k|,$$

with $\theta = 4(1 + |\omega|^{\sup}_\Pi)/m$ and $m = m_0$.

Proof. If there exists $\xi \in \mathcal{R}^\nu_{kl}(\alpha_\nu)$, then (21.4) implies that

$$\begin{aligned} |\langle k, \omega_\nu(\xi)\rangle| &\ge |\langle l, \bar{\Omega}_\nu(\xi)\rangle| - \alpha_\nu \frac{[l]_\delta}{\langle k\rangle^\tau} \\ &\ge m_\nu\, [l]_{d-1} - \alpha_\nu\, [l]_\delta \\ &\ge \frac{m}{4}\, [l]_{d-1}, \end{aligned}$$

since $[l]_\delta \le [l]_{d-1}$ for $\delta < d-1$ and $\alpha_\nu \le m_\nu/2$, $m_\nu \ge m/2$ by construction. Hence,

$$\frac{m}{4}\, [l]_{d-1} \le |k|\, |\omega_\nu(\xi)| \le |k|(1 + |\omega|^{\sup}_\Pi). \qquad \square$$

We now prove that the measure of the 'coarse structure' tends to zero as α tends to zero.

Proposition 22.2. $\operatorname{meas}(\Xi^1_\alpha) \to 0$ *as* $\alpha \to 0$.

Proof. By the preceding lemma for $\nu = 0$, for each k there are only finitely many l for which $\mathcal{R}^0_{kl}(\alpha_\nu)$ is not empty. Moreover, $|k| \le K_0$. Hence, Ξ^1_α is a *finite* union of resonance zones. For each of its members we know that $\operatorname{meas}(\mathcal{R}^0_{kl}(\alpha)) \to 0$ as $\alpha \to 0$ by assumption B* for $l \neq 0$, and by elementary volume estimates for $l = 0$. This proves the proposition. $\square$

In the remainder of this section we estimate the measure of Ξ^2_α.

Proposition 22.3. *If γ_0 is sufficiently small and $\tau \ge n + 1 + 2/(d-1)$, then*

$$\operatorname{meas}\left(\Xi^2_\alpha\right) = O(\alpha).$$

The implicit constants are made explicit below. The two proposition together imply that $\operatorname{meas}(\Pi \smallsetminus \Pi_\alpha) \to 0$ as $\alpha \to 0$.

As a preparatory step we extend the frequencies ω_ν and Ω_ν to Lipschitz maps defined on all of Π. Indeed, by Lemma M.5 each component of $\omega_\nu - \omega$ and of $\Omega_\nu - \Omega$ has a Lipschitz continuous extension from Π_ν to Π which preserves minimum, maximum and Lipschitz semi-norm. Since we use the sup-norm for ω, and a weighted sup-norm for Ω, doing this for each component we obtain extensions $\breve{\omega}_\nu \colon \Pi \to \mathbb{R}^n$ of ω_ν and $\breve{\Omega}_\nu \colon \Pi \to \ell^\infty_{-\delta}$ of Ω_ν with $|\breve{\omega}_\nu - \omega|^\lambda_\Pi = |\omega_\nu - \omega|^\lambda_{\Pi_\nu}$ and $|\breve{\Omega}_\nu - \Omega|^\lambda_{-\delta;\Pi} = |\Omega_\nu - \Omega|^\lambda_{-\delta;\Pi_\nu}$. It follows that $\mathcal{R}^\nu_{kl}(\alpha_\nu)$ is contained in

$$\breve{\mathcal{R}}^\nu_{kl}(\alpha_\nu) = \left\{ \xi \in \Pi \colon \left| \langle k, \breve{\omega}_\nu(\xi) \rangle + \langle l, \breve{\Omega}_\nu(\xi) \rangle \right| < \alpha_\nu \frac{[l]_\delta}{\langle k \rangle^\tau} \right\}.$$

Moreover, we need not distinguish between the different values of ν in $\breve{\omega}_\nu$ and $\breve{\Omega}_\nu$. In the sequel we only use the fact that $\breve{\omega}_\nu$ and $\breve{\Omega}_\nu$ are Lipschitz maps ω' and Ω' on Π which satisfy the estimates

$$\left|\omega' - \omega\right|^\lambda_\Pi, \left|\Omega' - \Omega\right|^\lambda_{-\delta;\Pi} \leq \frac{\alpha_0 \gamma_0}{4} \leq \frac{\alpha}{4LM} \tag{22.1}$$

for $0 \leq \lambda \leq \alpha/M$, if we assume that $\gamma_0 \leq 1/LM$. Here we use for the first time the bound

$$\left|\omega^{-1}\right|^{\mathrm{lip}}_{\omega(\Pi)} \leq L < \infty$$

introduced in section 18 right before Theorem 18.1.

Henceforth we consider functions ω', Ω' which satisfy these estimates and may also depend on k and l, and define

$$\mathcal{R}'_{kl}(\alpha) = \left\{ \xi \in \Pi \colon \left|\langle k, \omega'(\xi) \rangle + \langle l, \Omega'(\xi) \rangle\right| < \alpha \frac{[l]_\delta}{\langle k \rangle^\tau} \right\}$$

and

$$\Xi'_\alpha = \bigcup_{\nu \geq 1} \bigcup_{|k| > K_{\nu-1}, |l| \leq 2} \mathcal{R}'_{kl}(\alpha).$$

It suffices to show that $\operatorname{meas}(\Xi'_\alpha) = O(\alpha)$, which will prove Proposition 22.3.

Lemma 22.4. *If $|k| \geq 8LM\,|l|_\delta$ and $\tau > n$, then*

$$\operatorname{meas}\left(\mathcal{R}'_{kl}(\alpha)\right) \leq g \frac{\alpha}{\langle k \rangle^\tau},$$

with $g = 40\theta L^n M^{n-1} \rho^{n-1}$, where $\rho = \operatorname{diam} \Pi$ and θ is defined in Lemma 22.1.

Proof. We introduce the unperturbed frequencies $\zeta = \omega(\xi)$ as parameters over the domain $Z = \omega(\Pi)$ and consider the resonance zones $\dot{\mathcal{R}}_{kl} = \omega(\mathcal{R}'_{kl})$ in Z. Writing $\dot{\omega}$ and $\dot{\Omega}$ for the frequencies ω' and Ω' as functions of ζ, we then have by (22.1)

$$|\dot{\omega} - \mathrm{id}|^{\mathrm{lip}}_Z \leq \frac{1}{4}, \qquad |\dot{\Omega}|^{\mathrm{lip}}_{-\delta;Z} \leq 2LM,$$

using $\lambda = \alpha/M$ and $LM \geq 1$.

Now consider $\dot{\mathcal{R}}_{kl}(\alpha)$, and let $\phi(\zeta) = \langle k, \dot{\omega}(\zeta)\rangle + \langle l, \dot{\Omega}(\zeta)\rangle$. Choose a vector $v \in \{-1, 1\}^n$ such that $\langle k, v\rangle = |k|$, and write $\zeta = \zeta(r) = rv + w$ with $r \in \mathbb{R}$ and $w \perp v$. As a function of r, we then have, for $t > s$ and $\zeta(t), \zeta(s) \in Z$,

$$\begin{aligned}\langle k, \dot{\omega}(\zeta)\rangle|_s^t &= \langle k, \zeta\rangle|_s^t + \langle k, \dot{\omega}(\zeta) - \zeta\rangle|_s^t \\ &\geq |k|\,|t-s| - \frac{1}{4}|k|\,|t-s| = \frac{3}{4}|k|\,|t-s|\end{aligned}$$

and

$$\begin{aligned}|\langle l, \dot{\Omega}(\zeta)\rangle||_s^t &\leq 2\,|l|_\delta\,|\dot{\Omega}|_{-\delta}^{\mathrm{lip}}\,|t-s| \\ &\leq 4LM\,|l|_\delta\,|t-s| \leq \frac{1}{2}|k|\,|t-s|,\end{aligned}$$

where for the last inequality we used the assumption of the lemma. Hence we have $\phi(rv+w)|_s^t \geq \frac{1}{4}|k|\,|t-s|$ uniformly in w. It follows that for each w the point set

$$\{r : rv + w \in Z, |\phi(rv+w)| < \delta\}$$

is contained in the interval $|r - r_0(w)| < 4\delta/|k|$, where r_0 can be chosen to depend miserably on w. Hence, with Fubini's theorem and $\delta = \alpha\,[l]_\delta/\langle k\rangle^\tau$,

$$\operatorname{meas}(\dot{\mathcal{R}}_{kl}(\alpha)) \leq 4(\operatorname{diam} Z)^{n-1}\alpha\,\frac{[l]_\delta}{\langle k\rangle^{\tau+1}}.$$

Going back to the original parameter domain Π by the inverse frequency map ω^{-1} and noting that $\operatorname{diam} Z \leq M \operatorname{diam}\Pi$, $[l]_\delta \leq [l]_{d-1} \leq \theta\,|k|$ by Lemma 22.1, and $\operatorname{meas}(\mathcal{R}'_{kl}) \leq L^n \operatorname{meas}(\dot{\mathcal{R}}_{kl})$, the final estimate follows. □

For the next lemma let $\sigma = \min(\delta, d-1-\delta) > 0$ and

$$K_* = 8LM \cdot L_*^{\delta/\sigma}, \qquad L_* = 16LM\theta.$$

Lemma 22.5. *The measure estimate of Lemma 22.4 holds for any resonance zone $\mathcal{R}'_{kl}(\alpha)$ with $|k| \geq K_*$ and $|l| \leq 2$.*

Proof. Let $|k| \geq K_*$. If $l = 0$, then Lemma 22.4 applies. So assume that $l \neq 0$. If $\mathcal{R}'_{kl}(\alpha)$ is empty, there is nothing to do. Otherwise, we also want to apply Lemma 22.4 and thus verify its assumption. By Lemma 22.1 and our choice of σ,

$$\theta\,|k| \geq [l]_{d-1} \geq \frac{1}{2}[l]_{d-1-\delta}\,|l|_\delta \geq \frac{1}{2}[l]_\sigma\,|l|_\delta.$$

If $[l]_\sigma \geq L_* = 16LM\theta$, then division by θ gives $|k| \geq 8LM\,|l|_\delta$, and Lemma 22.4 applies. On the other hand, if $[l]_\sigma \leq L_*$, then $|l|_\delta \leq L_*^{\delta/\sigma}$, hence again

$$8LM\,|l|_\delta \leq 8LM \cdot L_*^{\delta/\sigma} = K_* \leq |k|.$$

So Lemma 22.4 applies also in this case. □

Proof of Proposition 22.3. Choose γ_0 sufficiently small so that

$$K_0 = \gamma_0^{-\frac{1}{\tau+1}} \geq K_*, \tag{22.2}$$

and thus $K_\nu \geq K_0 \geq K_*$ for all $\nu \geq 0$. Then the estimate of Lemma 22.4 applies to *all* resonance zones in the union of Ξ'_α. For a fixed k, it suffices to consider l with $[l]_{d-1} \leq \theta\,|k|$ according to Lemma 22.1. Taking into account that $|l|_{d-1} \leq 2\,[l]_{d-1}$ we get

$$\operatorname{card}\left\{l: \ |l| \leq 2, [l]_{d-1} \leq \theta\,|k|\right\} \lessdot \theta^s\,|k|^s, \qquad s = \frac{2}{d-1}.$$

Hence, by Lemmas 22.4 and 22.5,

$$\operatorname{meas}\left(\bigcup_l \mathcal{R}'_{kl}(\alpha)\right) \lessdot \alpha g \theta^s \frac{1}{|k|^{\tau-s}}.$$

If we choose $\tau \geq n+1+s$, then $\sum_{|k|>K_\nu} |k|^{s-\tau} \lessdot K_\nu^{-1}$ and hence

$$\operatorname{meas}\left(\bigcup_{|k|>K_\nu,|l|\leq 2} \mathcal{R}'_{kl}(\alpha)\right) \lessdot \alpha g \theta^s \cdot \frac{1}{K_\nu}.$$

The sum of the latter inequality over all ν converges, and we obtain the estimate of Proposition 22.3. □

VI

Kuksin's Lemma

23 Kuksin's Lemma

We consider the following first order partial differential equation coming up in the proof of the classical KAM theorem:

$$-\mathrm{i}\partial_\omega u + \lambda u + b(x)u = f, \qquad x \in \mathbb{T}^n, \tag{23.1}$$

for functions on the torus $\mathbb{T}^n = \mathbb{R}^n/2\pi\mathbb{Z}^n$, where

$$\partial_\omega = \sum_{\nu=1}^{n} \omega_\nu \partial_{x_\nu}, \qquad \omega = (\omega_1, \dots, \omega_n) \in \mathbb{R}^n.$$

We make the following assumptions, where $|k| = |k_1| + \cdots + |k_n|$ for $k \in \mathbb{Z}^n$.

Assumption U. The frequency vector ω is *diophantine*: there are constants $\alpha > 0$, $\tau > n$ and $l > 0$ such that

$$|\langle k, \omega\rangle + \lambda| \geq \frac{\alpha l}{|k|^\tau},$$
$$|\langle k, \omega\rangle| \geq \frac{\alpha}{|k|^\tau},$$

for all $0 \neq k \in \mathbb{Z}^n$. Also, $|\lambda| \geq \alpha l$.

Assumption V. The function b is analytic on some complex strip

$$D(s) = \{x \colon |\mathrm{Im}\, x| < s\} \subset \mathbb{C}^n$$

around $\mathbb{T}^n$ with mean value zero: $[b] = \int_{\mathbb{T}^n} b(x)\,\mathrm{d}x = 0$. Moreover,

$$\|b\|_{s,\tau} \overset{\text{def}}{=} \sum_{k \in \mathbb{Z}^n} |\hat{b}_k|\, |k|^\tau \mathrm{e}^{|k|s} \leq \gamma\alpha$$

for $b = \sum_k \hat{b}_k \mathrm{e}^{\mathrm{i}\langle k, x\rangle}$ with some $\gamma > 0$ and the same τ as in assumption U. We also assume that $s \leq 1$ for simplicity.

Assumption W. The function f is analytic on the same complex strip $D(s)$ and bounded in a weighted norm:

$$\|f\|_s \stackrel{\text{def}}{=} \sum_k |\hat{f}_k| e^{|k|s} < \infty.$$

Lemma 23.1 (Kuksin [74, 73]). *Under assumptions U, V and W, equation* (23.1) *has a unique solution $u = Lf$ that is analytic on $D(s)$ and satisfies*

$$\|u\|_{s-\sigma} \leq \frac{c e^{2\gamma}}{\alpha l \sigma^\tau} \|f\|_s, \qquad 0 < \sigma \leq s. \tag{23.2}$$

Moreover, if

$$\frac{\gamma}{\lambda} \leq \frac{\sigma}{5\,|\omega|},$$

with $|\omega| = \max |\omega_i|$, then also

$$\|u\|_{s-\sigma} \leq \frac{c}{\alpha l \sigma^{\tau+n+3}} \|f\|_s, \qquad 0 < \sigma \leq s. \tag{23.3}$$

The constant c can be chosen so that it only depends on τ in the first case, and only on τ and n in the second case.

The second estimate – which is the difficult one – is not needed if there is a uniform bound on γ. But if a whole family of equations with no such bound is considered, then it gives a uniform bound for $\|u\|_{s-\sigma}$ provided λ is substantially larger than γ.

Corollary 23.2. *If, in addition to the assumptions of Kuksin's lemma,*

$$\lambda \geq |\omega| \gamma^{1+\beta}$$

with some $\beta > 0$, then $u = Lf$ satisfies the estimate

$$\|u\|_{s-\sigma} \leq \frac{c e^{2(5/\sigma)^{1/\beta}}}{\alpha l \sigma^{\tau+n+3}} \|f\|_s, \qquad 0 < \sigma \leq s,$$

where c is a constant which depends only on n and τ.

Proof. Indeed, with this assumption we have $\lambda/\gamma \geq |\omega|\gamma^\beta$. Hence, the second estimate (23.3) applies, if $\gamma^\beta \geq 5/\sigma$. Otherwise, the first estimate (23.2) applies, with $\gamma \leq (5/\sigma)^{1/\beta}$. Combining these two estimates we obtain the estimate of the Corollary. □

For later reference we note that an analogous estimate holds also in terms of the sup-norm $|u|_s = \sup_{x \in D(s)} |u(x)|$. This follows from standard estimates for the Fourier coefficients of analytic functions, compare (19.3).

Corollary 23.3. *Under the same assumptions as in Corollary 23.2,*

$$|u|_{s-\sigma} \le \frac{c\mathrm{e}^{2(6/\sigma)^{1/\beta}}}{\alpha l \sigma^{\tau+2n+3}} |f|_s, \qquad 0 < \sigma \le s,$$

where c is another constant which depends only on n and τ.

The proof of the lemma has two parts. First, the solution $u = Lf$ is constructed by converting (23.1) via an integrating factor into an equation with constant coefficients. This also provides the first estimate (23.2). Second, for γ/λ sufficiently small, one considers a sequence of equations (23.1) where the frequency vector ω is approximated by rational ones and the variable coefficient b by trigonometric polynomials. Each of these equations has a unique periodic solution, which can be represented by an oscillatory integral with a complex valued phase function. A contour deformation is then constructed by a contraction argument, and using Cauchy's theorem, these oscillatory integrals are estimated uniformly. Taking limits one obtains the second estimate (23.3).

This scheme was first described by Kuksin [74]. For the convenience of the reader we present here our own version of it.

Before giving the details we observe that we may divide equation (23.1) by α and thus replace ω, λ, b and f by $\alpha^{-1}\omega$, $\alpha^{-1}\lambda$, $\alpha^{-1}b$ and $\alpha^{-1}f$, respectively. In the assumptions U, V and W, the constant α is then replaced by 1. Henceforth, we can assume the normalization $\alpha = 1$ for the rest of our considerations.

The proof of the lemma now takes N steps, where $N = 7$.

1. Existence and Uniqueness

To obtain an integrating factor B in (23.1), solve

$$\partial_\omega B = b, \qquad [B] = 0.$$

Then

$$\begin{aligned} \|B\|_s = \sum_{k\neq 0} |\hat{B}_k| \mathrm{e}^{|k|s} &= \sum_{k\neq 0} \frac{|\hat{b}_k|}{|\langle k,\omega\rangle|} \mathrm{e}^{|k|s} \\ &\le \sum_{k\neq 0} |\hat{b}_k|\, |k|^\tau \mathrm{e}^{|k|s} \\ &\le \|b\|_{s,\tau} \le \gamma. \end{aligned}$$

Hence, B is analytic in $D(s)$. Now set $u = \mathrm{e}^{-\mathrm{i}B} v$, $f = \mathrm{e}^{-\mathrm{i}B} g$ to obtain

$$-\mathrm{i}\partial_\omega v + \lambda v = g.$$

This equation has a unique solution v by comparing Fourier coefficients:

$$(\langle k,\omega\rangle + \lambda)\hat{v}_k = \hat{g}_k, \qquad k \in \mathbb{Z}^n.$$

We get

$$
\begin{aligned}
\|v\|_{s-\sigma} &= \sum_k \frac{|\hat{g}_k|}{|\langle k,\omega\rangle + \lambda|} \mathrm{e}^{|k|(s-\sigma)} \\
&\le \sum_k \frac{1}{l} |\hat{g}_k|(1+|k|^\tau)\mathrm{e}^{|k|(s-\sigma)} \\
&\le \frac{1}{l} \|g\|_s \sup_{t\ge 0} (1+t^\tau)\mathrm{e}^{-\sigma t} \\
&\le \frac{1}{l} \|g\|_s \frac{c}{\sigma^\tau}
\end{aligned}
$$

with some constant c depending only on τ.

Going back we obtain the unique analytic solution $u = \mathrm{e}^{-\mathrm{i}B} v$ of (23.1) with the estimates

$$
\begin{aligned}
\|u\|_{s-\sigma} &\le \mathrm{e}^{\|B\|_s} \|v\|_{s-\sigma} \le \mathrm{e}^{\gamma} \|v\|_{s-\sigma}, \\
\|g\|_s &\le \mathrm{e}^{\|B\|_s} \|f\|_s \quad \le \mathrm{e}^{\gamma} \|f\|_s,
\end{aligned}
$$

where we made use of the fact that the norm is multiplicative: $\|uv\|_s \le \|u\|_s \|v\|_s$. Hence, combining the inequalities above, we obtain

$$
\|u\|_{s-\sigma} \le \frac{c}{l} \|f\|_s \frac{\mathrm{e}^{2\gamma}}{\sigma^\tau}
$$

as stated in (23.2).

2. Approximation

To obtain estimate (23.3) we now assume that

$$
\frac{\gamma}{\lambda} \le \frac{\sigma}{5|\omega|},
$$

and approximate ω by rational frequency vectors ω_ν. The following is proven at the end of this section. Recall that $\alpha = 1$.

Lemma 23.4 (Approximation Lemma). *There exist a sequence of frequency vectors*

$$
\omega_\nu = \frac{2\pi}{T_\nu} \cdot m_\nu
$$

with $m_\nu \in \mathbb{Z}^n$, $T_\nu \to \infty$, and a sequence of constants $K_\nu \to \infty$ such that for all ν,

(a) $$|\omega - \omega_\nu| \le \frac{2\pi}{T_\nu},$$

(b) $$|\langle k,\omega_\nu\rangle| \ge \frac{1}{2|k|^\tau}, \qquad 0 < |k| < K_\nu,$$

(c) $$|\langle k,\omega_\nu + \lambda\rangle| \ge \frac{l}{2|k|^{\tau+n+2}}, \qquad |k| \ne 0.$$

Consequently, also

(d) $$\frac{\gamma}{\lambda} \leq \frac{\sigma}{4\,|\omega_\nu|}$$

for all large ν.

3. The Periodic Problem

We now consider the approximate, periodic problem

$$-\mathrm{i}\partial_{\omega_\nu} u + \lambda u + b_\nu(x) u = f,$$

where

$$\omega_\nu = \frac{2\pi}{T_\nu} \cdot m_\nu, \qquad b_\nu = \sum_{0<|k|<K_\nu} \hat{b}_k \mathrm{e}^{\mathrm{i}\langle k, x\rangle}.$$

Fixing ν we drop it from the notation in the following, which hopefully does not lead to confusion.

We may solve again $\partial_\omega B = b$ with $[B] = 0$. Using (b) of the Approximation Lemma we get, as before,

$$\|B\|_s \leq 2\,\|b\|_{s,\tau} \leq 2\gamma,$$

and with $u = \mathrm{e}^{-\mathrm{i}B} v$, $f = \mathrm{e}^{-\mathrm{i}B} g$, we obtain the equation $-\mathrm{i}\partial_\omega v + \lambda v = g$. Its solution has the integral representation

$$v(x) = \eta_T \int_0^T \mathrm{e}^{\mathrm{i}\lambda t} g(x + t\omega)\,\mathrm{d}t, \qquad \eta_T = \frac{\mathrm{i}}{\mathrm{e}^{\mathrm{i}\lambda T} - 1}.$$

To obtain this identity, write $w(t) = v(x + t\omega)$ and $h(t) = g(x + t\omega)$. Both functions are periodic in t with period T. Substituting into the preceding equation we find

$$\frac{\mathrm{d}}{\mathrm{d}t} w(t) + \mathrm{i}\lambda w(t) = \mathrm{i}h(t).$$

With an integrating factor $\mathrm{e}^{\mathrm{i}\lambda t}$ and a subsequent integration over $[0, T]$ this leads to the given integral representation.

From this and the substitutions $u = \mathrm{e}^{\mathrm{i}B} v$ and $f = \mathrm{e}^{\mathrm{i}B} g$ results the representation of u as an oscillatory integral:

$$\begin{aligned} u(x) &= \eta_T \int_0^T \mathrm{e}^{\mathrm{i}\lambda t + \mathrm{i}B(x+t\omega) - \mathrm{i}B(x)} f(x + t\omega)\,\mathrm{d}t \\ &= \eta_T \int_0^T \mathrm{e}^{\mathrm{i}\lambda(t + \delta[B(x+t\omega) - B(x)])} f(x + t\omega)\,\mathrm{d}t \end{aligned} \tag{23.4}$$

with

$$\delta = \frac{1}{\lambda}.$$

Our aim is to obtain a *uniform* estimate for $x \in D(s - 2\sigma)$. In the following we therefore *fix* x with $|\mathrm{Im}\, x| < s - 2\sigma$ and derive such an estimate for $|u(x)|$.

4. Stationary Phase Transformation

To estimate the last integral we deform the path of integration into a curve Γ in $\mathbb{C}$ with the same endpoints. To this end we want to find a change of the integration variable $t = \phi(r)$ with $\phi(0) = 0$ and $\phi(T) = T$ such that

$$\phi(r) + \delta\big(B(x + \phi(r)\omega) - B(x)\big) = r. \tag{23.5}$$

We then obtain

$$u(x) = \eta_T \int_0^T \mathrm{e}^{\mathrm{i}\lambda r} f(x + \phi(r)\omega)\phi'(r)\, \mathrm{d}r,$$

which we will use to estimate u.

To find ϕ we lift the problem to the torus $\mathbb{T}^n$ and write

$$\phi(r) = r + \psi(r\omega), \qquad \psi\colon \mathbb{T}^n \to \mathbb{C}.$$

Also, let $r\omega = \xi \in \mathbb{T}^n$. From (23.5) we obtain the equation

$$\psi(\xi) + \delta\big(B(x + \xi + \psi(\xi)\omega) - B(x)\big) = 0 \tag{23.6}$$

which we solve by a contraction argument.

Consider the space $\mathcal{A}$ of functions $\psi\colon D(\sigma) \to \mathbb{C}$ with

(i) ψ analytic and 2π-periodic on $D(\sigma)$,

(ii) $\psi(0) = 0$,

(iii) $|\psi|_\sigma = \sup_{\xi \in D(\sigma)} |\psi(\xi)| \le 4\gamma\delta$.

Define a map $\mathcal{T}$ on $\mathcal{A}$ by

$$\mathcal{T}(\psi)(\xi) = -\delta\left(B(x + \xi + \psi(\xi)\omega) - B(x)\right).$$

To show that $\mathcal{T}$ is well defined we have to verify that the argument of B is in the strip $D(s)$. To this end we observe that

$$\gamma\delta = \frac{\gamma}{\lambda} \le \frac{\sigma}{4\,|\omega|}$$

by (d) in the Approximation Lemma. Then

$$|\mathrm{Im}(\psi(\xi)\omega)| \le |\psi|_\sigma\, |\omega| \le 4\gamma\delta\,|\omega| \le \sigma,$$

hence $|\mathrm{Im}(x + \xi + \psi(\xi)\omega)| < s - 2\sigma + \sigma + \sigma = s$. It follows that $B(x + \xi + \psi(\xi)\omega)$ and thus $\mathcal{T}(\psi)$ are well defined.

$\mathcal{T}$ maps $\mathcal{A}$ into $\mathcal{A}$: (i) and (ii) clearly hold, and, as $|B|_\sigma \le \|B\|_s \le 2\gamma$,

$$|\mathcal{T}(\psi)|_\sigma \le 2\delta\,|B|_\sigma \le 4\gamma\delta.$$

Further, $\mathcal{T}$ is a contraction:

$$\begin{aligned}\mathcal{T}(\psi)(\xi) - \mathcal{T}(\chi)(\xi) &= \delta\big(B(x+\xi+\chi\omega) - B(x+\xi+\psi\omega)\big)\\ &= \delta\int_0^1 \langle dB(x+\xi(t)), (\psi-\chi)\omega\rangle\,\mathrm{d}t\\ &= \delta\int_0^1 b(x+\xi(t))(\psi-\chi)\,\mathrm{d}t\end{aligned}$$

with $\xi(t) = \xi + ((1-t)\chi + t\psi)\omega$. Hence, using assumption V and the fact that $|\mathrm{Im}(x+\xi(t))| < s$ for $0 \le t \le 1$,

$$|\mathcal{T}(\psi) - \mathcal{T}(\chi)|_\sigma \le \delta\,\|b\|_{s,\tau}\,|\psi-\chi|_\sigma \le \gamma\delta\,|\psi-\chi|_\sigma \le \frac{1}{2}\,|\psi-\chi|_\sigma,$$

since $\gamma\delta \le \sigma/4\,|\omega| \le 1/4\,|\omega|$, and $|\omega| \ge 1/2$ for all approximating frequencies.

Thus there exists a unique analytic solution $\psi \in \mathcal{A}$ of equation (23.6). Consequently, ϕ with $\phi(r) = r + \psi(r\omega)$ solves equation (23.5). In particular, $\phi - \mathrm{id}$ has period T, since $\omega = 2\pi m/T$. Moreover, differentiating the fixed point equation (23.6) with respect to r one gets

$$\partial_\omega\psi = -\delta\,\langle dB, \omega + (\partial_\omega\psi)\omega\rangle = -\delta b(1+\partial_\omega\psi)$$

and, again using assumption V and $\gamma\delta \le 1/2$,

$$|\partial_\omega\psi|_\sigma \le \gamma\delta(1+|\partial_\omega\psi|_\sigma) \le \frac{1}{2}(1+|\partial_\omega\psi|_\sigma),$$

hence $|\partial_\omega\psi|_\sigma \le 1$.

5. Transformation and Estimate

We observe that ϕ maps the interval $[0,T]$ into a curve Γ in $D(\sigma)$ with the *same* endpoints as $[0,T]$. By Cauchy's theorem, the integral in (23.4) over $[0,T]$ is the same as over Γ. Thus, substituting $t = \phi(r) = r + \psi(r\omega)$ in the integral (23.4) we obtain

$$u(x) = \eta_T\int_0^T \mathrm{e}^{\mathrm{i}\lambda r} f(x+\phi(r)\omega)\phi'(r)\,\mathrm{d}r = \eta_T\int_0^T \mathrm{e}^{\mathrm{i}\lambda r} h(x;\xi)\big|_{\xi=r\omega}\mathrm{d}r$$

with

$$h(x;\xi) = f(x+\xi+\psi(\xi)\omega)(1+\partial_\omega\psi(\xi))$$

analytic on $D(\sigma)$, and

$$|h(x;\,\cdot)|_\sigma \le 2\,|f|_s.$$

The above integral for $u(x)$ is now easily evaluated. Since h is periodic in ξ,

$$h(x;\xi) = \sum_k \hat{h}_k(x) \mathrm{e}^{\mathrm{i}\langle k,\xi\rangle}, \qquad |\hat{h}_k| \le |h|_\sigma \mathrm{e}^{-|k|\sigma}.$$

Moreover

$$\int_0^T \mathrm{e}^{\mathrm{i}\lambda r} \mathrm{e}^{\mathrm{i}\langle k,\xi\rangle}\big|_{\xi=r\omega} \mathrm{d}r = \int_0^T \mathrm{e}^{\mathrm{i}(\lambda+\langle k,\omega\rangle)r} \mathrm{d}r = -\mathrm{i}\frac{\mathrm{e}^{\mathrm{i}\lambda T}-1}{\langle k,\omega\rangle+\lambda} = \frac{\eta_T^{-1}}{\langle k,\omega\rangle+\lambda}.$$

Hence,

$$u(x) = \sum_k \frac{\hat{h}_k(x)}{\langle k,\omega\rangle+\lambda}.$$

With (c) of the Approximation Lemma, one obtains for every $x \in D(s-2\sigma)$ the estimate

$$|u(x)| \le \frac{2}{l}\sum_{k\neq 0} |k|^{\tau+n+2} \mathrm{e}^{-|k|\sigma}\, |h|_\sigma + \frac{1}{\lambda}\,|h|_\sigma \le \frac{1}{l}\,|f|_s \cdot \frac{c}{\sigma^{\tau+n+3}},$$

where c depends only on n and τ. We used that $\lambda \ge l$ by assumption U for $\alpha = 1$.

6. Taking Limits

Summarizing the preceding arguments we obtain the following result. For every frequency vector

$$\omega_\nu = \frac{2\pi}{T_\nu}\cdot m_\nu, \qquad m_\nu \in \mathbb{Z}^n,$$

satisfying (b), (c) and (d) the equation $-\mathrm{i}\partial_{\omega_\nu} u + \lambda u + b_\nu u = f$ has a unique analytic solution u_ν satisfying

$$|u_\nu|_{s-2\sigma} \le \frac{1}{l}\,|f|_s \cdot \frac{c}{\sigma^{\tau+n+3}}, \qquad 0 < \sigma \le s.$$

Letting $\nu \to \infty$ and taking (a) of the Approximation Lemma into account, we obtain approximating sequences $\omega_\nu \to \omega$, $b_\nu \to b$. The corresponding u_ν are uniformly bounded. Hence we can choose a convergent subsequence converging to some solution u of $-\mathrm{i}\partial_\omega u + \lambda u + bu = f$ with

$$|u|_{s-2\sigma} \le \frac{1}{l}\,|f|_s \cdot \frac{c}{\sigma^{\tau+n+3}}.$$

From this the second estimate (23.3) of the lemma follows. Incidentally, since the solution u is unique, actually the whole sequence u_ν converges to u.

7. Proof of the Approximation Lemma

For any t-interval I of the form

$$I = [T, T + \Delta], \qquad T \geq \Delta = \frac{2\pi}{|\omega|}$$

there is an integer vector $m \in \mathbb{Z}^n$ such that $|t\omega - 2\pi m| \leq 2\pi$ for $t \in I$. Hence,

$$|\omega - \omega_t| \leq \frac{2\pi}{T}, \qquad \omega_t = \frac{2\pi}{t} \cdot m,$$

for each $t \in I$. This gives (a).

To obtain (b) we estimate, for arbitrary $t \in I$ and using assumption U,

$$|\langle k, \omega_t \rangle| \geq |\langle k, \omega \rangle| - |\langle k, (\omega - \omega_t) \rangle| \geq \frac{1}{|k|^\tau} - |k| \frac{2\pi}{T} \geq \frac{1}{2\,|k|^\tau}$$

for $4\pi\,|k|^{\tau+1} \leq T$, that is, for

$$|k| \leq K_T \stackrel{\text{def}}{=} \left(\frac{T}{4\pi} \right)^{1/(\tau+1)}.$$

This gives (b).

For (c) we estimate similarly, again using assumption U,

$$|\langle k, \omega_t \rangle + \lambda| \geq |\langle k, \omega \rangle + \lambda| - |\langle k, (\omega - \omega_t) \rangle| \geq \frac{l}{|k|^\tau} - |k| \frac{2\pi}{T} \geq \frac{l}{2\,|k|^\tau}$$

for $t \in I$ and $4\pi\,|k|^{\tau+1} \leq lT$. So it remains to consider k with

$$4\pi\,|k|^{\tau+1} \geq lT. \tag{23.7}$$

Assume that for some $t \in I$,

$$|\langle k, \omega_t \rangle + \lambda| \leq \frac{l}{2},$$

otherwise there is nothing to do. In particular, we have $|\langle k, \omega_t \rangle| \geq l/2$, since $\lambda \geq l$. As long as this holds, we have for $\varphi(t) := \langle k, \omega_t \rangle + \lambda$ the estimate

$$\left|\varphi'(t)\right| = \frac{1}{t}\,|\langle k, \omega_t \rangle| \geq \frac{l}{2t} \geq \frac{l}{4T},$$

as $t \leq T + \Delta \leq 2T$. Hence the measure of the subset of the t-interval I where

$$|\langle k, \omega_t \rangle + \lambda| \leq \frac{l}{2\,|k|^{\tau+n+2}} \tag{23.8}$$

can be estimated by

$$\frac{4T}{l} \frac{l}{2\,|k|^{\tau+n+2}} = \frac{2T}{|k|^{\tau+n+2}}.$$

Summing over all $k \in \mathbb{Z}^n$ with (23.7) the total measure of the subset of t-values in I satisfying (23.8) can be estimated by

$$c(n) \cdot \frac{T}{K_0^{\tau+2}},$$

where $c(n)$ depends only on n, and K_0 is the lower bound of $|k|$ in (23.7). That is,

$$K_0 = c(l)T^{1/(\tau+1)}.$$

Hence the measure is bounded by

$$c(n,l)T^{-1/(\tau+1)} < \frac{2\pi}{|\omega|} = |I|,$$

if T is sufficiently large. Note that the dependence of the constant on l is irrelevant here, since we want to make T large anyhow, after l is given. Hence for each large T we can find $t \in [T, T+\Delta]$ so that (23.8) does *not* hold. That is, for such t, (c) holds. This proves the approximation lemma.

VII

Background Material

A Analyticity

We discuss the notion of an analytic map between two complex Banach spaces and prove two frequently used characterizations of such maps. Much of this material is taken from appendix A in [112].

Let E and F be complex Banach spaces with norms $|\cdot|$ and $\|\cdot\|$, respectively, and let $U \subset E$ be open. A map

$$f\colon\ U \to F$$

is *analytic on* U, if it is continuously differentiable on U. This is the straightforward generalization of the notion of an analytic function of one complex variable.

It is convenient to introduce another notion of analyticity. The map $f: U \to F$ is *weakly analytic on* U, if for each $u \in U$, $h \in E$ and $L \in F^*$, the function

$$z \mapsto Lf(u + zh)$$

is analytic in some neighbourhood of the origin in $\mathbb{C}$ in the usual sense. The *radius of weak analyticity of* f *at* u is the supremum of all $r \geq 0$ such that the above function is defined and analytic in the disc $|z| < 1$ for all $L \in F^*$ and $h \in E$ with $|h| < r$.

It is easy to see that the radius r of weak analyticity at u is equal to the distance ρ of u to the boundary of U. Clearly, $r \leq \rho$ by definition. On the other hand, if L and h are given with $|h| < \rho$, then the function $z \mapsto Lf(u + zh)$ is well defined on the disc $|z| < 1$ and analytic in some neighbourhood of each point in it, since f is weakly analytic on all of U. Consequently, this function is analytic on $|z| < 1$, and so also $r \geq \rho$. It follows that $r = \rho$.

The notion of a weakly analytic map is weaker than that of an analytic map. For instance, every globally defined, but unbounded linear operator is weakly analytic, but not analytic. Remarkably, a weakly analytic map is analytic, if in addition it is locally bounded.

Before we get to this result, we state two basic lemmata.

Lemma A.1 (Cauchy's Formula). *Suppose f is weakly analytic and continuous on U. Then, for every $u \in U$ and $h \in E$,*

$$f(u+zh) = \frac{1}{2\pi \mathrm{i}} \int_{|\zeta|=\rho} \frac{f(u+\zeta h)}{\zeta - z} \, \mathrm{d}\zeta$$

for $|z| < \rho < r/|h|$, where r is the radius of weak analyticity of f at u.

Proof. Fix $u \in U$, and let $r > 0$ be the radius of weak analyticity of f at u. Then the open ball of radius r around u is contained in U. For every $h \in E$, the integral

$$\frac{1}{2\pi \mathrm{i}} \int_{|\zeta|=\rho} \frac{f(u+\zeta h)}{\zeta - z} \, \mathrm{d}\zeta, \qquad |z| < \rho < r/|h|,$$

is well defined, since f is continuous and $|\zeta h| < r$, $|\zeta - z| > 0$ for $|\zeta| = \rho$. Then, for every $L \in F^*$,

$$\frac{1}{2\pi \mathrm{i}} L \int_{|\zeta|=\rho} \frac{f(u+\zeta h)}{\zeta - z} \, \mathrm{d}\zeta = \frac{1}{2\pi \mathrm{i}} \int_{|\zeta|=\rho} \frac{Lf(u+\zeta h)}{\zeta - z} \, \mathrm{d}\zeta = Lf(u+zh)$$

by the usual Cauchy formula. Since this holds for all L, the statement follows. □

Lemma A.2 (Cauchy's Estimate). *Let f be an analytic map from the open ball of radius r around u in E into F, such that $\|f\| \le M$ on this ball. Then*

$$\|d_u f\| = \max_{h \neq 0} \frac{\|d_u f(h)\|}{|h|} \le \frac{M}{r}.$$

Proof. Let $h \neq 0$ in E. Then $\phi(z) = f(u+zh)$ is an analytic map from the complex disc $|z| < r/|h|$ in $\mathbb{C}$ into F that is uniformly bounded by M. Hence

$$\|d_0 \phi\| = \|d_u f(h)\| \le \frac{M}{r} \cdot |h|$$

by the usual Cauchy inequality. The above statement follows, since $h \neq 0$ was arbitrary. □

The statement of the lemma is particularly transparent, when f is a complex valued function. Then $d_u f$ is an element in the dual space E^* to E, and the induced operator norm is the norm $|\cdot|_{E^*}$ dual to $|\cdot|_E$. So, for instance, if f is bounded in absolute value by M on the balls

$$|u|_\infty, \quad |u|_2, \quad |u|_1 \quad < \quad r,$$

in standard spaces, then

$$|d_0 f|_1, \quad |d_0 f|_2, \quad |d_0 f|_\infty \quad \le \quad \frac{M}{r},$$

respectively, in both finite and infinite dimensional settings.

Next we have a simple result about the analyticity of maps into sequence spaces with the sup-norm. Let $U \subset E$ be open, and $\ell^\infty(F)$ the space of all sequences in a complex Banach space F endowed with the sup-norm $|\cdot|_\infty$.

Theorem A.3. *Suppose*

$$f\colon\ U \to \ell^\infty(F), \qquad u \mapsto f(u) = (f_n(u))_{n\geq 1},$$

is locally bounded, and each coordinate function f_n is analytic on U with values in a complex Banach space F. Then f is analytic as a map from U into $\ell^\infty(F)$.

Proof. We verify directly that f is continuously differentiable at each point $u \in U$, with $d_u f = (d_u f_n)_{n\geq 1}$. Indeed,

$$|f(u+h) - f(u) - d_u f(h)|_\infty = \sup_{n\geq 1} |f_n(u+h) - f_n(u) - d_u f_n(h)|_F.$$

Since f is bounded in a fixed ball around u, each coordinate function f_n is bounded by the same constant M on this ball. Cauchy's estimate applied to f_n then gives

$$|f_n(u+h) - f_n(u) - d_u f_n(h)|_F \leq M\,|h|_E^2$$

for all small $|h|_E$, uniformly in n. Hence,

$$|f(u+h) - f(u) - d_u f(h)|_\infty \leq M\,|h|_E^2.$$

The continuity of $d_u f$ follows by the same token. □

We now turn to the basic characterization of analytic maps between complex Banach spaces. An infinitely often differentiable function f is said to be represented by its Taylor series near a point u, if

$$f(u+h) = \sum_{n\geq 0} \frac{1}{n!} d_u^n f(h, \ldots, h),$$

for all sufficiently small h, with the series converging absolutely and uniformly.

Theorem A.4. *Let $f\colon U \to F$ be a map from an open subset U of a complex Banach space E into a complex Banach space F. Then the following three statements are equivalent.*

(1) *f is weakly analytic and locally bounded on U.*

(2) *f is analytic on U.*

(3) *f is infinitely often differentiable on U, and is represented by its Taylor series in a neighbourhood of each point in U.*

Proof. (1) ⇒ (2) Suppose f is weakly analytic and locally bounded. We first show that f is continuous.

Fix $u \in U$ and choose $r > 0$ so small that

$$\sup_{|h| \le r} \|f(u+h)\| = M < \infty.$$

As $Lf(u+zh)$ is continuous, we obtain, by the usual Cauchy formula,

$$Lf(u+zh) - Lf(u) = \frac{z}{2\pi \mathrm{i}} \int_{|\zeta|=1} \frac{Lf(u+\zeta h)}{(\zeta - z)\zeta} \, \mathrm{d}\zeta$$

for $|z| < 1$ and $|h| < r$, and for any $L \in F^*$. Hence, for $|z| < \frac{1}{2}$,

$$\left| \frac{Lf(u+zh) - Lf(u)}{z} \right| \le 2M \, \|L\|,$$

where $\|L\|$ denotes the operator norm of L. This estimate holds for all $L \in F^*$ uniformly for $|z| < \frac{1}{2}$ and $|h| < r$. Consequently,

$$\left\| \frac{f(u+zh) - f(u)}{z} \right\| \le 2M$$

for $|z| < \frac{1}{2}$ and $|h| < r$. From this, the continuity of f follows.

Now, f being weakly analytic and continuous, Cauchy's formula applies, and

$$f(u+zh) = \frac{1}{2\pi \mathrm{i}} \int_{|\zeta|=1} \frac{f(u+\zeta h)}{\zeta - z} \, \mathrm{d}\zeta$$

for $|z| < 1$ and $|h| < r$. It follows that f has a directional derivative in every direction h, namely

$$\delta_u(h) = \lim_{z \to 0} \frac{f(u+zh) - f(u)}{z} = \frac{1}{2\pi \mathrm{i}} \int_{|\zeta|=1} \frac{f(u+\zeta h)}{\zeta^2} \, \mathrm{d}\zeta.$$

In fact, this limit is uniform in $|v - u| < r/2$ and $|h| < r/2$, since

$$\left\| \frac{f(v+zh) - f(v)}{z} - \delta_v(h) \right\| = \left\| \frac{z}{2\pi \mathrm{i}} \int_{|\zeta|=1} \frac{f(v+\zeta h)}{\zeta^2(\zeta - z)} \, \mathrm{d}\zeta \right\| \le 2M \, |z|$$

for $|z| < \frac{1}{2}$. It follows from this that f is continuously differentiable, hence analytic on U.

(2) ⇒ (3) Suppose f is analytic on U. As before, fix $u \in U$ and $r > 0$ such that $\sup_{|h| \le r} \|f(u+h)\| = M < \infty$. For $h \in E$ and $n \ge 0$, define

$$P_n(h) = \frac{n!}{2\pi \mathrm{i}} \int_{|\zeta|=\rho} \frac{f(u+\zeta h)}{\zeta^{n+1}} \, \mathrm{d}\zeta,$$

where $\rho > 0$ is chosen sufficiently small. The integral is independent of ρ as long as $\rho \le r/|h|$, since f is analytic. For instance, $P_0(h) = f(u)$ and $P_1(h) = d_u f(h)$. We show that $P_n(h)$ is the n-th directional derivative of f in the direction h.

First of all, Cauchy's formula and the expansion

$$\frac{1}{\zeta - 1} = \sum_{n=0}^{m} \frac{1}{\zeta^{n+1}} + \frac{1}{\zeta^{m+1}(\zeta - 1)}$$

give

$$f(u+h) - \sum_{n=0}^{m} \frac{1}{n!} P_n(h) = \frac{1}{2\pi \mathrm{i}} \int_{|\zeta|=\rho} \frac{f(u+\zeta h)}{\zeta^{m+1}(\zeta - 1)} \, \mathrm{d}\zeta$$

for $|h| < r$. Choosing $\rho = r/|h|$ for $h \neq 0$, the norm of the right hand side is bounded by

$$M \left(\frac{|h|}{r} \right)^m \frac{|h|}{r - |h|} = \frac{M}{r^m (r - |h|)} |h|^{m+1}.$$

Consequently,

$$f(u+h) = \sum_{n=0}^{\infty} \frac{1}{n!} P_n(h)$$

for $|h| < r$. Moreover, the sum converges uniformly in every ball $|h| \leq \rho < r$.

We now show that each P_n is a homogeneous polynomial of degree n in h. That is, there exists a bounded symmetric n-linear map A_n such that $P_n = \hat{A}_n$, the polynomial associated to A_n by evaluating it on the diagonal.

Consider the F-valued map A_n defined by

$$\begin{aligned} &A_n(h_1, \ldots, h_n) \\ &\quad = \left(\frac{1}{2\pi \mathrm{i}} \right)^n \int_{|\zeta_1|=\varepsilon} \cdots \int_{|\zeta_n|=\varepsilon} \frac{f(u + \zeta_1 h_1 + \cdots + \zeta_n h_n)}{\zeta_1^2 \ldots \zeta_n^2} \, \mathrm{d}\zeta_1 \ldots \mathrm{d}\zeta_n, \end{aligned}$$

where $\varepsilon > 0$ is sufficiently small, say $\varepsilon < \min_{1 \leq i \leq n} r/|h_i|$. For every $L \in F^*$, the map $(z_1, \ldots, z_n) \mapsto Lf(u + z_1 h_1 + \cdots + z_n h_n)$ is analytic in a neighbourhood of the origin in $\mathbb{C}^n$. Hence, by the usual Cauchy formula for n complex variables,

$$LA_n(h_1, \ldots, h_n) = \frac{\mathrm{d}}{\mathrm{d}z_1} \cdots \frac{\mathrm{d}}{\mathrm{d}z_n} Lf(u + z_1 h_1 + \cdots + z_n h_n) \Big|_{z_1, \ldots, z_n = 0}.$$

It follows that A_n is linear and symmetric in all arguments. A_n is also bounded by a straightforward estimate. Finally, using Cauchy's formula again,

$$LA_n(h, \ldots, h) = \left(\frac{\mathrm{d}}{\mathrm{d}z} \right)^n Lf(u + zh) \Big|_{z=0} = LP_n(h)$$

for all L. Therefore, $A_n(h, \ldots, h) = \hat{A}_n(h) = P_n(h)$, as we wanted to show.

Thus, on the ball of radius r around u, the map f is represented by a power series, which converges uniformly on every smaller ball around u. It is a basic fact that such a map is infinitely often differentiable. In particular, $d_u^n f = A_n$ for all $n \geq 0$.

(3) ⇒ (1) This is trivial. □

A special case of the preceding theorem arises for maps into a Hilbert space.

Theorem A.5. *Let $f: U \to H$ be a map from an open subset U of a complex Banach space into a Hilbert space with orthonormal basis $(e_n)_{n\geq 1}$. Then f is analytic on U if and only if f is locally bounded, and each "coordinate function"*

$$f_n = \langle f, e_n \rangle : \quad U \to \mathbb{C}$$

is analytic on U. Moreover, the derivative of f is given by

$$df(h) = \sum_{n\geq 1} df_n(h) e_n.$$

Proof. Let $L \in H^*$. By the Riesz representation theorem, there is a unique element ℓ in H such that $L\phi = \langle \phi, \ell \rangle$ for all ϕ in H. Write $\ell = \sum_{n\geq 1} \lambda_n e_n$, and set

$$\ell_m = \sum_{n=1}^{m} \lambda_n e_n, \qquad m \geq 1.$$

Then L is the operator norm limit of the functionals L_m defined by $L_m\phi = \langle \phi, \ell_m \rangle$. That is, as $m \to \infty$,

$$\sup_{\|\phi\|\leq 1} \|(L - L_m)(\phi)\| \to 0.$$

Now, given x in U, choose $r > 0$ so that f is bounded on the ball of radius r around x. Fix h in the complex Banach space containing U with $\|h\| < r$. On $|z| < 1$, the functions

$$z \mapsto L_m f(x + zh) = \sum_{n=1}^{m} \lambda_n f_n(x + zh), \qquad m \geq 1$$

are analytic by hypotheses and tend uniformly to the function $z \mapsto Lf(x+zh)$, since f is bounded. Hence that function is also analytic on $|z| < 1$. This shows that f is weakly analytic and locally bounded. By Theorem A.4, the function f is analytic.

Conversely, if f is analytic, then f is locally bounded, and each coordinate function f_n is analytic.

Finally, if f is analytic, then $d_x f(h)$ exists and is an element of H, hence can be expanded with respect to the orthonormal basis $(e_n)_{n\geq 1}$. Its n-th coefficient is

$$\langle d_x f(h), e_n \rangle = d_x \langle f, e_n \rangle (h) = d_x f_n(h)$$

by the chain rule, since $\langle \cdot , e_n \rangle$ is a linear function. Thus,

$$d_x f(h) = \sum_{n\geq 1} d_x f_n(h) e_n$$

as was to be proven. □

The next theorem may be considered a generalization of Theorem A.4. We say that a subset $V \subset U$ of an open set U in a complex Banach space is an *analytic subvariety*, if locally it can be represented as the zero set of an analytic function taking values in $\mathbb{C}^n$ for some $n \geq 1$.

Theorem A.6. *Let $V_1, \ldots, V_m$ be analytic subvarieties of an open subset U in a complex Banach space E. If f is a complex-valued function on U, which is*

(i) *analytic on $U \smallsetminus (V_1 \cup \cdots \cup V_m)$,*

(ii) *continuous on U, and,*

(iii) *when restricted to each of the V_i, weakly analytic on V_i,*

then f is analytic on U.

Proof. We are going to show that not only the restriction of f, but f itself is weakly analytic in every point in $V_1 \cup \cdots \cup V_m$. Since f is also locally bounded by continuity, f is then analytic on U by Theorem A.4.

Let D be a one-dimensional complex disc around an arbitrary point in U. Locally, we can write

$$V_i = \{q \in U : \varphi_i(q) = 0\}, \qquad 1 \leq i \leq m,$$

with analytic, vector-valued functions φ_i. When restricted to D, each function φ_i either vanishes identically or has only a finite number of zeroes in D, possibly after shrinking D a bit.

If at least one φ_i vanishes identically, then D is contained in some subvariety V_i, and f is analytic on $D \subset V_i$ by assumption (iii). Otherwise, none of the functions φ_i vanishes identically on D, and consequently

$$D \cap (V_1 \cup \cdots \cup V_m)$$

is a finite set. Outside this set in D, f is analytic by hypotheses (i), and on all of D, f is continuous by hypotheses (ii). It follows that these singularities are removable, and that f is analytic on all of D.

Since the disc D was arbitrary, it follows that f is weakly analytic. Hence f is analytic by Theorem A.4. □

Finally, we introduce the notion of a real analytic map. Let E, F be real Banach spaces, let $E_{\mathbb{C}}$, $F_{\mathbb{C}}$ be their complexifications, and let $U \subset E$ be open. A map

$$f\colon \; U \to F$$

is *real analytic* on U, if for each point in U there is a neighbourhood $V \subset E_{\mathbb{C}}$ and an analytic map

$$g\colon \; V \to F_{\mathbb{C}},$$

such that

$$f = g \quad \text{on} \quad U \cap V.$$

It follows that a real analytic map can be extended into a Taylor series with real coefficients in a ball at each point. The converse is also true.

B Spectra

In this appendix we collect some basic facts about the spectra of Schrödinger operators on a finite interval. The main purpose is to fix notions and notations. Only in a few cases do we provide proofs, otherwise we refer for example to the references [80, 82, 84] and [112].

Fundamental Solution

We consider the differential equation

$$-y'' + qy = \lambda y \tag{B.1}$$

on the compact interval [0, 1] depending on a potential $q \in L^2_{\mathbb{C}} = L^2_{\mathbb{C}}([0,1])$ and a complex parameter $\lambda \in \mathbb{C}$.

By definition, a function y is a *solution* of this equation, if it is continuously differentiable, y' is absolutely continuous, and the equation holds almost everywhere for y''. One *fundamental solution* is given by the particular solutions y_1 and y_2 satisfying the initial conditions

$$\begin{aligned} y_1(0,\lambda,q) &= 1, & y_2(0,\lambda,q) &= 0,\\ y_1'(0,\lambda,q) &= 0, & y_2'(0,\lambda,q) &= 1. \end{aligned}$$

Any other solution of (B.1) is a linear combination of y_1 and y_2 with coefficients determined by its initial values.

The associated *Floquet matrix* is the 2×2-matrix

$$F(\lambda,q) = \begin{pmatrix} m_1 & m_2 \\ m_1' & m_2' \end{pmatrix}(\lambda,q) = \begin{pmatrix} y_1 & y_2 \\ y_1' & y_2' \end{pmatrix}(1,\lambda,q),$$

where we introduce the convenient notation

$$m_i = y_i|_{x=1}, \qquad m_i' = y_i'|_{x=1}$$

for the values of this fundamental solution at $x = 1$. The Floquet matrix describes the shift of initial data at $x = 0$ to initial data at $x = 1$, since

$$\begin{pmatrix} y(1) \\ y'(1) \end{pmatrix} = F(\lambda,q)\begin{pmatrix} y(0) \\ y'(0) \end{pmatrix}$$

for any solution y. Its determinant is 1 in view of the Wronskian identity

$$W(y_1,y_2) \overset{\text{def}}{=} y_1 y_2' - y_1' y_2 \equiv 1.$$

Its trace,

$$\Delta \overset{\text{def}}{=} \operatorname{tr} F = m_1 + m_2',$$

is called the *discriminant* of q and is fundamental in discussing its periodic spectrum defined below.

We note the following asymptotic behavior of the m-functions [112, chapter 1].

Proposition B.1.

$$m_1(\lambda) = \cos\sqrt{\lambda} + O\left(\frac{c_\lambda}{|\lambda|^{1/2}}\right),$$
$$m_2(\lambda) = \frac{\sin\sqrt{\lambda}}{\sqrt{\lambda}} + O\left(\frac{c_\lambda}{|\lambda|}\right),$$

and

$$m_1'(\lambda) = -\sqrt{\lambda}\sin\sqrt{\lambda} + O(c_\lambda),$$
$$m_2'(\lambda) = \cos\sqrt{\lambda} + O\left(\frac{c_\lambda}{|\lambda|^{1/2}}\right),$$

locally uniformly on $\mathbb{C} \times L^2_{\mathbb{C}}$ *with* $c_\lambda = \mathrm{e}^{|\mathrm{Im}\sqrt{\lambda}|}$.

From this proposition one immediately obtains a first asymptotic estimate for the Δ-function as well. But we also need the following refined estimate, which can be found in [84, Section 1.4] – see also Theorem C.3.

Proposition B.2. *For* $q \in H^2_{0,\mathbb{C}}$,

$$\Delta(\lambda) = 2\cos\sqrt{\lambda} + \frac{\|q\|^2}{4}\frac{\sin\sqrt{\lambda}}{\lambda^{3/2}} + O\left(\frac{c_\lambda}{|\lambda|^2}\right)$$

locally uniformly on $\mathbb{C} \times H^2_{0,\mathbb{C}}$ *with* $c_\lambda = \mathrm{e}^{|\mathrm{Im}\sqrt{\lambda}|}$.

As functions of λ and q, m_1 and m_2 as well as their x-derivatives are *compact* on $\mathbb{C} \times L^2_{\mathbb{C}}$, that is, continuous with respect to the weak topology. Moreover, they are also real analytic functions of λ and q, and their respective derivatives are denoted $\dot{m}_i$ and dm_i. The latter has a representation in terms of a unique gradient, denoted

$$\partial m_i = \frac{\partial m_i}{\partial q(x)},$$

such that

$$dm_i(v) = \int_0^1 \partial m_i(x) v(x)\,\mathrm{d}x$$

for all $v \in L^2_{\mathbb{C}}$. The same notation is used for all other gradients with respect to q. It is a general phenomenon in this context that such gradients are represented by *products* of solutions of the underlying equation.

Proposition B.3. *For* $i = 1, 2$,

$$\partial m_i = (m_2 y_1 - m_1 y_2) y_i,$$
$$\partial m_i' = (m_2' y_1 - m_1' y_2) y_i.$$

Consequently, $\partial\Delta = m_2 y_1^2 + (m_2' - m_1) y_1 y_2 - m_1' y_2^2$. *The latter gradient also admits the representation*

$$\partial\Delta = m_2(\lambda, T_t q), \qquad T_t q = q(\cdot + t),$$

where q *is understood to be periodically extended beyond* $[0, 1]$.

Proof. We only prove the last assertion. First of all,

$$y_2(x, \lambda, T_t q) = y_2(x+t, \lambda, q) y_1(t, \lambda, q) - y_1(x+t, \lambda, q) y_2(t, \lambda, q),$$

since both sides are solutions of the equation $-y'' + T_t q y = \lambda y$ with the same initial data at $x = 0$. Hence,

$$\begin{aligned} y_2(1, \lambda, T_t q) &= y_2(1+t) y_1(t) - y_1(1+t) y_2(t) \\ &= \left[y_2(1) y_1(t) + y_2'(1)) y_2(t) \right] y_1(t) \\ &\qquad - \left[y_1(1) y_1(t) + y_1'(1) y_2(t) \right] y_2(t) \\ &= m_2 y_1^2(t) + (m_2' - m_1) y_1(t) y_2(t) - m_1' y_2^2(t) \\ &= \frac{\partial\Delta}{\partial q(t)}, \end{aligned}$$

which proves the claim. □

Dirichlet Spectrum

From now on we consider potentials q in the real space $L^2 = L^2_{\mathbb{R}}([0, 1])$. But by analytic continuation, the following results extend to potentials in a sufficiently small complex neighbourhood of L^2 in $L^2_{\mathbb{C}}$.

The spectrum of the differential operator $-\mathrm{d}^2/\mathrm{d}x^2 + q$ with Dirichlet boundary conditions is called the *Dirichlet spectrum of* q. It consists of those complex numbers λ, for which the equation $-y'' + qy = \lambda y$ admits a nontrivial solution vanishing at both endpoints of $[0, 1]$.

Clearly, λ is a Dirichlet eigenvalue of q if and only if $y_2(1, \lambda, q) = 0$. So the Dirichlet spectrum of q is precisely the zero set of the entire function m_2. It turns out that the *Dirichlet eigenvalues* form an unbounded sequence of real numbers

$$\mu_1(q) < \mu_2(q) < \mu_3(q) < \dots,$$

where each eigenvalue is a simple root of m_2, as $\dot{m}_2$ does not vanish there in view of the next proposition below. Hence its algebraic multiplicity is one and thus equal to its geometric multiplicity, which is the dimension of the associated eigenspace. With μ_n we can thus associate a unique normalized *Dirichlet eigenfunction* g_n by requiring that $\|g_n\| = 1$ and $g_n'(0) > 0$. Clearly,

$$g_n = \left. \frac{y_2}{\|y_2\|} \right|_{\mu_n},$$

and for the norm of y_2 we have the following result.

Proposition B.4. *At any Dirichlet eigenvalue,* $m_1 m_2' = 1$ *and*

$$\|y_2\|^2 = \frac{\dot{m}_2}{m_1} = \dot{m}_2 m_2' > 0.$$

Proof. Since $m_2 = 0$ at a Dirichlet eigenvalue, we have $m_1 m_2' = 1$ by the Wronskian identity. In view of $m_2(\lambda + \varepsilon, q) = m_2(\lambda, q - \varepsilon)$ and Proposition B.3,

$$\dot{m}_2 = -\int_0^1 \partial m_2 \,\mathrm{d}x = \int_0^1 \left(m_1 y_2^2 - m_2 y_1 y_2\right) \mathrm{d}x = m_1 \|y_2\|^2,$$

which gives the second identity. □

Next we recall the asymptotic behavior of Dirichlet eigenvalues and eigenfunctions and the product formula for m_2 from [112, Theorems 2.4 and 2.5]. Let $[q] = \int_0^1 q(x)\,\mathrm{d}x$ denote the mean value of q.

Proposition B.5. *For* $q \in L^2$, *one has* $\mu_n = n^2\pi^2 + [q] + \ell^2(n)$ *and*

$$g_n = \sqrt{2}\sin \pi n x + O(1/n),$$
$$g_n' = \sqrt{2}\pi n \cos \pi n x + O(1).$$

These estimates hold uniformly on bounded subsets of $[0, 1] \times L^2$.

Proposition B.6. *For* $q \in L^2$,

$$m_2(\lambda) = \prod_{n \geq 1} \frac{\mu_n - \lambda}{n^2\pi^2}.$$

Considered as a function of q in L^2, each Dirichlet eigenvalue is compact and real analytic. Its gradient $\partial\mu_n$ is obtained by differentiating $m_2(\mu_n(q), q) = 0$ with respect to q and using Propositions B.3 and B.4 – see [112, chapter 2].

Proposition B.7. *Each Dirichlet eigenvalue* μ_n *is a compact, real analytic function on* L^2 *with gradient*

$$\partial\mu_n = -\left.\frac{\partial m_2}{\dot{m}_2}\right|_{\mu_n} = g_n^2.$$

Periodic Spectrum

Identify a function $q \in L^2_{\mathbb{C}}$ with its periodic extension beyond $[0, 1]$. The spectrum of the operator $-\mathrm{d}^2/\mathrm{d}x^2 + q$ with periodic boundary conditions on the interval $[0, 2]$ is called the *periodic spectrum of* q. It consists of those complex numbers λ, for which the equation $-y'' + qy = \lambda y$ admits a nontrivial solution with period 2.

As q is extended with period 1, a solution has period 2 if and only if it is either periodic or anti-periodic over $[0, 1]$. Therefore, λ is a periodic eigenvalue of q iff the Floquet matrix F of q either has an eigenvalue 1 or an eigenvalue -1, which in turn

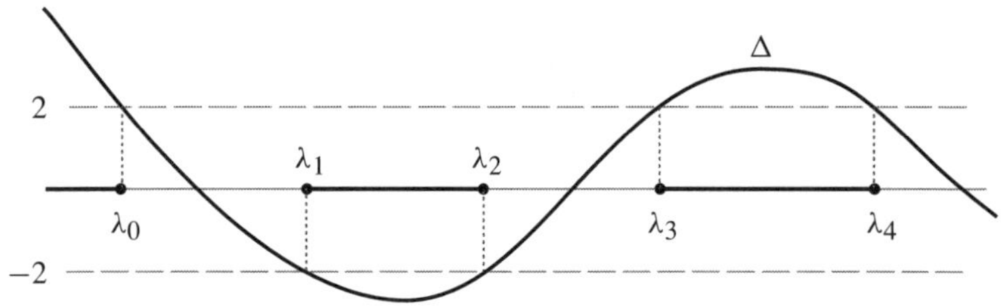

Figure 6 A generic Δ-function

is equivalent to its discriminant Δ being 2 or -2, respectively. Consequently, the periodic spectrum of q is precisely the union of the zero sets of the entire functions $\Delta - 2$ and $\Delta + 2$, or equivalently, the zero set of the entire function $\Delta^2 - 4$.

To locate this zero set in the real case $q \in L^2$, note that $m_1 m_2' = 1$ at any Dirichlet eigenvalue, and $\operatorname{sgn} m_2'(\mu_n) = (-1)^n$, thus

$$\Delta(\mu_n) = \frac{1}{m_2'(\mu_n)} + m_2'(\mu_n) \begin{cases} \geq 2, & n \text{ even}, \\ \leq -2, & n \text{ odd}. \end{cases}$$

Moreover, $\dot{\Delta}$ has exactly n roots below $(n + \frac{1}{2})^2\pi^2$ for all large n by a Counting Lemma completely analogous to [112, Lemma 2.2]. Taking into account its asymptotic behavior for $\lambda \to \pm\infty$, the function Δ therefore must have a shape as depicted in figure 6.

The periodic spectrum of a real potential q therefore consists of an unbounded sequence of *periodic eigenvalues*

$$\lambda_0(q) < \lambda_1(q) \leq \lambda_2(q) < \lambda_3(q) \leq \lambda_4(q) < \dots ,$$

such that $\lambda_{2n-1} \leq \mu_n \leq \lambda_{2n}$ for any $n \geq 1$. Equality may occur in this sequence in every place with a '$\leq$'-sign, and this case occurs precisely if *all* solutions of the corresponding differential equation are 2-periodic.

Thus, also in the periodic case the geometric and algebraic multiplicities of an eigenvalue λ_m coincide, and could be either one or two. If λ_m is simple, then one can associate with it a unique normalized *periodic eigenfunction* f_m by requiring that $\|f_m\| = 1$ and $f_m(0) > 0$, or $f_m(0) = 0$ and $f_m'(0) > 0$.

The results of Propositions B.8–B.13 are formulated for real potentials. By analytic continuation they extend to potentials in a sufficiently small complex neighbourhood of L^2 in $L^2_{\mathbb{C}}$, which can be chosen independently of n.

Proposition B.8. *For $q \in L^2$,*

$$\lambda_{2n-1}, \lambda_{2n} = n^2\pi^2 + [q] + \ell^2(n),$$

Consequently, $\gamma_n = \lambda_{2n} - \lambda_{2n-1} = \ell^2(n)$. These estimates hold uniformly on bounded subsets of L^2.

Proof. These estimates follow by noting that, for example,

$$\lambda_{2n}(q) = \mu_n(T_t q), \quad T_t q = q(\cdot + t)$$

for some $0 \leq t \leq 1$, and applying Proposition B.5. □

We also note the following improved estimate for potentials in $\mathcal{H}_0^N$, which is used in section 11. See [61, 84], and in particular [51] for an elementary proof.

Proposition B.9. *For $q \in \mathcal{H}_0^N$ with $N \geq 1$,*

$$\sum_{n \geq 1} n^{2N} \left(|\gamma_n|^2 + |\mu_n - \tau_n|^2 \right) = O(1)$$

locally uniformly on $\mathcal{H}_0^N$, as well as a small complex neighbourhood of it.

The next product formulas complements the product formula for m_2.

Proposition B.10. *For $q \in L^2$,*

$$\Delta^2(\lambda) - 4 = 4(\lambda_0 - \lambda) \prod_{n \geq 1} \frac{(\lambda_{2n} - \lambda)(\lambda_{2n-1} - \lambda)}{n^4 \pi^4}.$$

At any Dirichlet eigenvalue μ_n,

$$\Delta^2(\mu_n) - 4 = (m_1 + m_2')^2 - 4 m_1 m_2' = (m_1 - m_2')^2$$

by the Wronskian Identity. Hence, one can define a unique, *real analytic* root of this function by

$$\sqrt[*]{\Delta^2(\mu_n) - 4} = m_1(\mu_n) - m_2'(\mu_n),$$

compare equation (6.2).

Like the Dirichlet eigenvalues, each periodic eigenvalue is a compact function of q in L^2, and the proof is completely analogous to the one of [112, Theorem 2.3]. Unlike the former, however, the latter is a real analytic function of the potential only when it is simple.

Proposition B.11. *Each periodic eigenvalue λ_n is a compact function on L^2, which is real analytic on the open domain where it is simple, with gradient*

$$\partial \lambda_n = - \left. \frac{\partial \Delta}{\dot{\Delta}} \right|_{\lambda_n} = f_n^2.$$

Moreover, $\gamma_n^2 = (\lambda_{2n} - \lambda_{2n-1})^2$ and $\tau_n = (\lambda_{2n} + \lambda_{2n-1})/2$ are real analytic on all of L^2.

Proof. We prove the statements concerning τ_n. By the product representation of $\Delta^2(\lambda) - 4$ above and the residue theorem,

$$\tau_n = \frac{1}{2\pi \mathrm{i}} \int_{\Gamma_n} \lambda \frac{\Delta(\lambda)\dot{\Delta}(\lambda)}{\Delta^2(\lambda) - 4} \, \mathrm{d}\lambda,$$

with some circuit Γ_n encircling counterclockwise precisely λ_{2n} and λ_{2n-1}. Since Γ_n can be kept fixed locally, this shows that τ_n is a real analytic function of q. □

Proposition B.12.

$$\tau_n = \mu_n + \langle \cos 2\pi n x, q \rangle + O\left(\frac{1}{n}\right)$$

locally uniformly on L^2.

Proof. Write

$$\tau_n - \mu_n = (\tau_n - \mu_n)(tq)\Big|_0^1 = \int_0^1 \langle (\partial \tau_n - \partial \mu_n)(tq), q \rangle \, \mathrm{d}t.$$

We only need to consider the gradients where the n-th gap is open. In this case, the two normalized periodic eigenfunctions can be written as $\sqrt{2}\sin(x - x_n) + O(1/n)$ and $\sqrt{2}\cos(x - x_n) + O(1/n)$. With Propositions B.5, B.7 and B.11 we get

$$\partial \tau_n = 1 + O\left(\frac{1}{n}\right), \qquad \partial \mu_n = 1 - \cos 2\pi n x + O\left(\frac{1}{n}\right),$$

which gives the result. □

Consider now $\dot{\Delta}$. By the Counting Lemma mentioned above, $\dot{\Delta}$ has a *unique* root in each interval $[\lambda_{2n-1}, \lambda_{2n}]$, denoted $\dot{\lambda}_n$, and no other roots. These roots determine the function $\dot{\Delta}$ completely.

Proposition B.13. *For $q \in L^2$,*

$$\dot{\Delta}(\lambda) = -\prod_{n \geq 1} \frac{\dot{\lambda}_n - \lambda}{n^2 \pi^2}$$

and

$$\dot{\lambda}_n - \tau_n = O\left(\frac{\gamma_n^2 \log n}{n}\right)$$

locally uniformly on L^2 for all large n. Also, $\dot{\lambda}_n - \tau_n = O(\gamma_n^2)$ for any fixed $n \geq 1$.

Remark. In fact, we can prove the stronger estimate

$$\dot{\lambda}_n - \tau_n = O(\gamma_n^2 / n)$$

using Proposition D.9 and the remark preceding it.

Proof. We only prove the estimate for $\dot{\lambda}_n$. Fix q in L^2. By Proposition B.10,

$$\Delta^2(\lambda) - 4 = \frac{(\lambda_{2n} - \lambda)(\lambda - \lambda_{2n-1})}{n^2\pi^2} \chi_n(\lambda) \tag{B.2}$$

with

$$\chi_n(\lambda) = 4\frac{\lambda - \lambda_0}{n^2\pi^2} \prod_{m \neq n} \frac{(\lambda_{2m} - \lambda)(\lambda_{2m-1} - \lambda)}{m^4\pi^4}.$$

By the asymptotics of the periodic eigenvalues and Lemma L.1 we have

$$\chi_n(\lambda) = 1 + O\left(\frac{\log n}{n}\right)$$

in a neighbourhood of size $O(1)$ around the interval $[\lambda_{2n-1}, \lambda_{2n}]$, uniformly in a neighbourhood of q. Consequently,

$$\dot{\chi}_n(\lambda) = O\left(\frac{\log n}{n}\right)$$

by Cauchy's estimate on a neighbourhood of similar size.

Now, $\dot{\lambda}_n$ may be characterized as the unique zero of $(\Delta^2)^{\cdot}$ near τ_n. Differentiating equation (B.2) with respect to λ and multiplying with $n^2\pi^2$, we may thus characterize $\dot{\lambda}_n$ as the unique solution of

$$0 = 2(\lambda - \tau_n)\chi_n + (\lambda_{2n} - \lambda)(\lambda - \lambda_{2n-1})\dot{\chi}_n$$

near τ_n. With $4(\lambda_{2n} - \lambda)(\lambda - \lambda_{2n-1}) = \gamma_n^2 - 4(\lambda - \tau_n)^2$, this is equivalent to $\dot{\lambda}_n$ being the unique root of

$$8(\lambda - \tau_n)\chi_n - 4(\lambda - \tau_n)^2\dot{\chi}_n + \gamma_n^2\dot{\chi}_n = 0 \tag{B.3}$$

near τ_n. As q is real, this immediately gives the claimed estimate.

The following argument also applies to q in a complex neighbourhood of L^2. Consider the last term in the last equation, $g_n = \gamma_n^2\dot{\chi}_n$, as a small perturbation of the other terms, $f_n = 8(\lambda - \tau_n)\chi_n - 4(\lambda - \tau_n)^2\dot{\chi}_n$. Clearly, $f_n(\tau_n) = 0$. Moreover, in a neighbourhood U_n of size $O(1)$ around τ_n, the λ-derivative of f_n is

$$\dot{f}_n = 8\chi_n - 4(\lambda - \tau_n)^2\ddot{\chi}_n \sim 8$$

for large n, while $g_n = \gamma_n^2 O(\log n/n)$, locally uniformly in q. It thus follows from the inverse function theorem that for all large n, U_n contains a unique solution $\dot{\lambda}_n$ of equation (B.3), and that $\dot{\lambda}_n - \tau_n = O(g_n)$. This proves the result for large n.

The same argument applies for any individual n when $\gamma_n \to 0$, since then g_n is much smaller than $\dot{f}_n$, and the implicit function theorem applies as well. So also the last statement is proven. □

Isospectral Sets

Let $L_0^2 = \{ q \in L^2 : [q] = 0 \}$, and consider the set

$$\mathrm{Iso}(q) = \{ p \in L_0^2 : \mathrm{spec}(p) = \mathrm{spec}(q) \}$$

of all potentials in L_0^2 with the same periodic spectrum as q. To give a topological description of this set we introduce the following notation. For a real interval $[a, b]$, define

$$[\![a, b]\!] = \begin{cases} \{(a,0), (b,0)\} \cup (a,b) \times \{-1, 1\}, & a < b, \\ \{(a,0)\}, & a = b. \end{cases}$$

Endow this subset of $\mathbb{R}^2$ with the coarsest topology which makes the projections onto the two factors continuous. In this way, $[\![a, b]\!]$ is homeomorphic to a circle with center $(a + b)/2$ and radius $(a - b)/2$, with $(a, 0)$ and $(b, 0)$ being the points of intersection of this circle with the interval $[a, b]$. For the following result see [89].

Theorem B.14. *For any $q \in L_0^2$, the map*

$$\begin{aligned} \mu \times \sigma : \quad \mathrm{Iso}(q) &\to \prod_{n \geq 1} [\![\lambda_{2n-1}, \lambda_{2n}]\!] \\ p &\mapsto (\mu_n(p), \sigma_n(p))_{n \geq 1}, \end{aligned}$$

where

$$\sigma_n(p) = \mathrm{sign} \sqrt[*]{\Delta^2(\mu_n(p), q) - 4)}$$

is a homeomorphism, when the right hand side is endowed with the product topology. Thus, the set Iso(q) *is homeomorphic to a torus, whose dimension equals the number of non-collapsed spectral gaps* $[\lambda_{2n-1}, \lambda_{2n}]$. *In particular, every isospectral set is compact.*

The proof of Theorem B.14 uses the following coordinates, briefly mentioned in section 6, see [43, 89]. For $q \in L_0^2$ and $n \geq 1$ let

$$\hat{\mu}_n(q) = \mu_n(q) - n^2\pi^2, \qquad \kappa_n(q) = \log\,(-1)^n m_2'(\mu_n(q)).$$

It is not difficult to verify that $(\kappa_n)_{n \geq 1}$ belongs to ℓ_1^2, while, by Proposition B.5, $(\hat{\mu}_n)_{n \geq 1}$ belongs to ℓ^2.

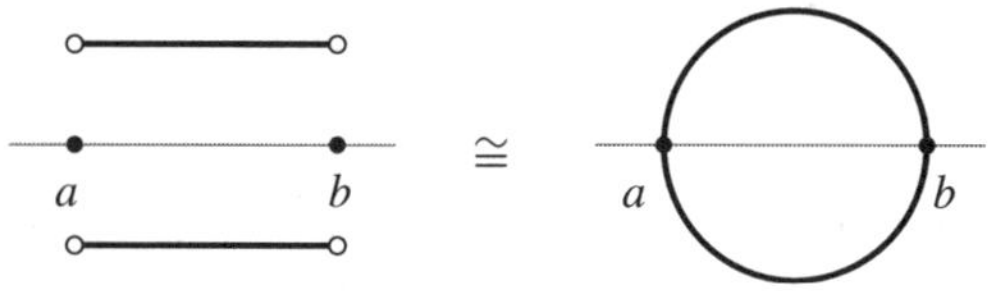

Figure 7 The set $[\![a, b]\!]$

Theorem B.15. *The map*

$$\begin{aligned}\mu\times\kappa\colon\ L_0^2 &\to \ell^2\times\ell_1^2\\ q &\mapsto \big(\hat\mu_n(q),\kappa_n(q)\big)_{n\ge1}\end{aligned}$$

is a real analytic embedding.

For a complete proof of this theorem see [112].

Proof of Theorem B.14. Fix $q\in L_0^2$ and its associated 'circles' $[\![\lambda_{2n-1},\lambda_{2n}]\!]$ for $n\ge1$. The periodic spectrum is the same for each p in $\operatorname{Iso}(q)$, so also their Δ-functions are the same. Hence, for any $n\ge1$ and $p\in\operatorname{Iso}(q)$,

$$\begin{gathered}\lambda_{2n-1}\le\mu_n(p)\le\lambda_{2n},\\ \sqrt[*]{\Delta^2(\mu_n(p))-4}\ne0\quad\text{iff}\quad\lambda_{2n-1}<\mu_n(p)<\lambda_{2n}.\end{gathered}$$

Therefore, the map $\mu\times\sigma$ is well defined on $\operatorname{Iso}(q)$ and takes values in the target space as stated.

To show that this map is onto, pick $(\bar\mu_n,\bar\sigma_n)\in[\![\lambda_{2n-1},\lambda_{2n}]\!]$ arbitrarily for each $n\ge1$. Then define

$$\bar\kappa_n=\log\frac{(-1)^n}{2}\left(\Delta(\bar\mu_n,q)-\bar\sigma_n\sqrt[+]{\Delta^2(\bar\mu_n,q)-4}\right).$$

By the asymptotic behavior of $\bar\mu_n$ and Δ one verifies that $(\bar\kappa_n)_{n\ge1}$ belongs to ℓ_1^2. So by Theorem B.15 there exists a unique potential $p\in L_0^2$ with

$$(\mu\times\kappa)(p)=(\bar\mu_n,\bar\kappa_n)_{n\ge1}.$$

We claim that $p\in\operatorname{Iso}(q)$ with $(\mu\times\sigma)(p)=(\bar\mu_n,\bar\sigma_n)_{n\ge1}$.

To prove this, write

$$m_n=m_2'(\bar\mu_n)=(-1)^n\mathrm{e}^{\bar\kappa_n}$$

for $n\ge1$. We have $m_1(\mu_n)=1/m_2'(\mu_n)$ at any Dirichlet eigenvalue, hence

$$\begin{aligned}\Delta(\bar\mu_n,p)&=\frac{1}{m_n}+m_n,\\ \sqrt[*]{\Delta^2(\bar\mu_n,p)-4}&=\frac{1}{m_n}-m_n,\end{aligned}\tag{B.4}$$

for $n\ge1$. On the other hand, by the definition of $\bar\kappa_n$ we have

$$m_n=\frac12\left(\Delta(\bar\mu_n,q)-\bar\sigma_n\sqrt[+]{\Delta^2(\bar\mu_n,q)-4}\right).$$

From this one directly calculates

$$\frac{1}{m_n} + m_n = \Delta(\bar{\mu}_n, q),$$
$$\frac{1}{m_n} - m_n = \bar{\sigma}_n \sqrt[+]{\Delta^2(\bar{\mu}_n, q) - 4}.$$

Comparing this with equation (B.4) one sees that the Δ-functions of p and q agree at the points $\bar{\mu}_n$, $n \geq 1$. Since the Δ-function is uniquely determined by its values at $\bar{\mu}_n$ due to its interpolation property, we conclude that p and q have the same Δ-function and hence the same periodic spectrum, so p belongs to $\mathrm{Iso}(q)$. Furthermore, $\sigma_n(p) = \bar{\sigma}_n$ for all $n \geq 1$. So

$$(\mu \times \sigma)(p) = (\bar{\mu}_n, \bar{\sigma}_n)_{n \geq 1}.$$

Thus, $\mu \times \sigma$ maps $\mathrm{Iso}(q)$ onto $\prod_{n \geq 1} [\![\lambda_{2n-1}, \lambda_{2n}]\!]$. By the same token, this map is also one-to-one, since $\mu \times \kappa$ is an isomorphism. It is continuous with respect to the product topology, since all data are continuous functions of q.

To prove the continuity of the inverse map, we show that $\mathrm{Iso}(q)$ is compact. By the asymptotics of the Δ-function given in Proposition B.2 the L^2-norm of q is *fixed*, once the periodic spectrum is fixed. Thus, given any sequence in $\mathrm{Iso}(q)$, we can extract a weakly convergent subsequence, since it is bounded. This subsequence converges even strongly, since also the L^2-norms converge. Clearly, the limit function also belongs to $\mathrm{Iso}(q)$ by the continuity of the periodic eigenvalues. Thus, $\mathrm{Iso}(q)$ is compact. □

Density of Finite Gap Potentials

The following result was first proven by Marčhenko and Ostrowski using inverse spectral theory – see [84] and later [49]. The proof given here is more elementary and was initiated by [27].

Theorem B.16. *Finite gap potentials are dense in L_0^2.*

To prove this theorem we introduce for potentials in L_0^2 the complex quantities

$$\alpha_n = \tau_n - \mu_n + \mathrm{i}\, 2\pi n \kappa_n, \qquad n \geq 1.$$

Lemma B.17. *There exists a complex neighbourhood W of L_0^2 so that each α_n is a complex analytic function on W with asymptotics*

$$\alpha_n = \langle \mathrm{e}^{2\pi \mathrm{i} n x}, q \rangle + O\left(\frac{1}{n}\right)$$

locally uniformly on W.

Proof. There exists a complex neighbourhood W of L_0^2 so that μ_n, κ_n and τ_n are real analytic functions on W for all $n \geq 1$. So each α_n is a complex analytic function on W as well. Moreover,

$$\begin{aligned} \tau_n - \mu_n &= \langle \cos 2\pi nx, q \rangle + O(1/n), \\ 2\pi n \kappa_n &= \langle \sin 2\pi nx, q \rangle + O(1/n), \end{aligned}$$

locally uniformly on W by Lemma B.12 and [112, p. 59]. Hence,

$$\alpha_n = \langle \cos 2\pi nx, q \rangle + \mathrm{i} \langle \sin 2\pi nx, q \rangle + O(1/n),$$

which proves the asymptotics. □

Lemma B.18. *For q in L_0^2 and any $n \geq 1$, $\gamma_n(q) = 0$ iff $\alpha_n(q) = 0$.*

Proof. Fix q and n. If $\gamma_n = 0$, then $\mu_n = \tau_n$, and the n-th Dirichlet eigenfunction g_n is also a periodic or anti-periodic eigenfunction. But then

$$\left| y_2'(1, \mu_n) \right| = 1,$$

whence also $\kappa_n = 0$, and thus $\alpha_n = 0$. Conversely, if $\alpha_n = 0$, then $\kappa_n = 0$ implies that g_n is a periodic or anti-periodic eigenfunction, hence μ_n is also a periodic eigenvalue. Since in addition $\mu_n = \tau_n$, the corresponding gap must be collapsed. □

Consider now the map

$$A\colon \; L_0^2 \to \ell_{\mathbb{C}}^2, \quad q \mapsto (\alpha_n(q))_{n\geq 1}.$$

By Lemma B.17 and Theorem A.5, this map is complex analytic. By Lemma B.18, q in L_0^2 is a finite gap potential, if and only if all but finitely many coordinates of $A(q)$ vanish.

To prove Theorem B.16, however, we instead consider the map

$$G = A \circ \Phi\colon \; \ell_{\mathbb{C}}^2 \to \ell_{\mathbb{C}}^2,$$

where

$$\Phi\colon \; \ell_{\mathbb{C}}^2 \to L_0^2, \qquad (\xi_n)_{n\geq 1} \mapsto \operatorname{Re} \sum_{n\geq 1} \xi_n \mathrm{e}^{2\pi \mathrm{i} nx}$$

is the inverse of the restriction of the discrete Fourier transform to L_0^2. Since Φ is a linear isomorphism, it suffices to prove the following statement.

Proposition B.19. *For ξ in a dense subset of $\ell_{\mathbb{C}}^2$, all but finitely many coordinates of $G(\xi)$ vanish.*

Proof. In view of Lemma B.17, G is a *real analytic* map, when considered as a map $(\operatorname{Re}\xi, \operatorname{Im}\xi) \mapsto (\operatorname{Re} G(\xi), \operatorname{Im} G(\xi))$. It is of the form $I + K$, where K maps $\ell^2_{\mathbb{C}}$ into the smaller spaces

$$\ell^2_{\beta,\mathbb{C}} = \left\{ \xi \in \ell^2_{\mathbb{C}} \colon \sum_{n\geq 1} n^{2\beta} \, |\xi_n|^2 < \infty \right\}, \qquad 0 < \beta < \frac{1}{2}.$$

It follows with Cauchy's inequality that on some ball around any given point in $\ell^2_{\mathbb{C}}$, the Jacobian $\mathrm{d}K$ is uniformly bounded as a linear map $\ell^2_{\mathbb{C}} \to \ell^2_{\beta,\mathbb{C}}$. Consequently,

$$\|T_N \mathrm{d}K\| \leq \frac{1}{2}$$

on the same ball for all sufficiently large N in the operator norm on $\ell^2_{\mathbb{C}}$, where T_N denotes the projection onto all *except* the first N coordinates in $\ell^2_{\mathbb{C}}$.

Now fix ξ^{o} in $\ell^2_{\mathbb{C}}$, and let $\varepsilon > 0$ be so small that the preceding estimate holds on the 4ε-ball B around ξ^{o} for all sufficiently large N. We may then fix N so large that also

$$\left\| T_N G(\xi^{\mathrm{o}}) \right\| < \varepsilon.$$

Writing $\xi = \xi_N + \zeta_N$ with $\zeta_N = T_N \xi$ we then have

$$T_N G(\xi) = T_N G(\xi_N + \zeta_N) = \zeta_N + T_N K(\xi_N + \zeta_N)$$

with

$$\left\| d_{\zeta_N} T_N K \right\| \leq \frac{1}{2}$$

uniformly on B. The map

$$\zeta_N \mapsto \zeta_N + T_N K(\xi^{\mathrm{o}}_N + \zeta_N)$$

is thus a local diffeomorphism, and by the inverse function theorem the image of the ball $\|\zeta_N\| < 4\varepsilon$ under this map covers a ball of radius 2ε around $T_N G(\xi^{\mathrm{o}})$. Consequently, in view of $\|T_N G(\xi^{\mathrm{o}})\| < \varepsilon$, there exists $\tilde{\xi} = \xi^{\mathrm{o}}_N + \tilde{\zeta}_N$ with

$$\left\| \tilde{\xi} - \xi^{\mathrm{o}} \right\| = \left\| \tilde{\zeta}_N - \zeta^{\mathrm{o}}_N \right\| < 4\varepsilon$$

such that $T_N G(\tilde{\xi}) = 0$. Since $\varepsilon > 0$ can be chosen arbitrarily small, this proves the claim. □

Remark 1. The proof incidentally shows that for a finite gap potential any finite number of Fourier coefficients can be prescribed arbitrarily.

Remark 2. The preceding proof can be extended to show that finite gap potentials are also dense in any space $\mathcal{H}^N_0$ with $N \geq 1$. Essentially, this requires to appropriately improve the asymptotic estimates of Lemma B.17.

C KdV Hierarchy

There is no generally established notion of *the* KdV hierarchy, and the definitions found in the literature are typically connected with specific ways of constructing it – see [95] and subsequently [31, 35, 41, 78, 83, 87, 104], among others. In this book we define the *KdV hierarchy* as a sequence of Hamiltonian equations

$$u_t = \frac{\mathrm{d}}{\mathrm{d}x}\frac{\partial H^n}{\partial u}, \qquad n \geq 0,$$

with Hamiltonians

$$H^n = \int_{S^1} p_n(u, u_x, \ldots)\,\mathrm{d}x. \tag{C.1}$$

Each p_n is a polynomial in u and its derivatives up to order n, such that these Hamiltonians are in the *KdV algebra*, the Poisson algebra of all Hamiltonians in involution with the KdV actions.

One way to obtain such a sequence is as follows. Consider the Floquet matrix

$$F(\lambda) = \begin{pmatrix} m_1 & m_2 \\ m_1' & m_2' \end{pmatrix}(1,\lambda) = \begin{pmatrix} y_1 & y_2 \\ y_1' & y_2' \end{pmatrix}(1,\lambda)$$

associated with the equation $-y'' + qy = \lambda y$, and its discriminant $\Delta(\lambda) = \operatorname{tr} F(\lambda)$. Since $\det F(\lambda) = 1$ and $\Delta(\lambda) > 2$ for $\lambda < \lambda_0(q)$, where $\lambda_0(q)$ denotes the zero-th periodic eigenvalue of q, the two eigenvalues of $F(\lambda)$ are real, positive, and distinct for $\lambda \to -\infty$. Moreover, exactly one eigenvalue, denoted $w(\lambda)$, is greater than 1. For this eigenvalue one computes

$$\log w(\lambda) = \log \frac{\Delta(\lambda) + \sqrt{\Delta^2(\lambda) - 4}}{2} = \operatorname{arcosh} \frac{\Delta(\lambda)}{2}. \tag{C.2}$$

It turns out that for $\lambda \to -\infty$, this quantity admits an expansion of the form

$$\log w(\lambda) \sim \sqrt{-\lambda} - \sum_{n\geq 0} \frac{1}{4^{n+1}} \frac{G_n}{\sqrt{-\lambda}^{\,2n+3}},$$

whose coefficients G_n are of the form (C.1). As they only depend on Δ and thus on the periodic spectrum of q, these functions belong to the KdV algebra. Indeed, they define a KdV hierarchy, and $H^n = (-1)^n G_n$ is referred to as the n-th KdV Hamiltonian.

Calculation

First note that $w(\lambda)$ for $\lambda < \lambda_0(q)$ is the Floquet multiplier of *any* expanding solution of

$$-y'' + qy = \lambda y \tag{C.3}$$

over the period 1. As these solutions have no roots, we begin by writing them in a special form – see [84]. From now on we assume q to be a smooth, real-valued, 1-periodic function on the real line with mean-value zero.

Lemma C.1. *Let*

$$g = \exp\left(\mu x + \int_0^x \sigma(r, \mu)\,\mathrm{d}r\right).$$

Then g is a solution of (C.3) *with $\lambda = -\mu^2$ and μ sufficiently large, iff σ satisfies the Riccati equation*

$$\sigma' + 2\mu\sigma + \sigma^2 = q. \tag{C.4}$$

In particular, g is a Floquet solution of (C.3) *iff σ is 1-periodic. Its Floquet multiplier is then*

$$w(\lambda) = \exp\left(\mu + \int_0^1 \sigma(r, \mu)\,\mathrm{d}r\right).$$

Proof. Writing g as above, we have

$$g' = (\mu + \sigma)g, \qquad g'' = (\mu + \sigma)^2 g + \sigma' g.$$

Hence, with $\lambda = -\mu^2$,

$$-g'' + (q - \lambda)g = (-2\mu\sigma - \sigma^2 - \sigma' + q)g,$$

from which the first claim follows.

To be a Floquet solution the data of g at 0 and 1 must be proportional, or

$$\frac{g'(0)}{g(0)} = \frac{g'(1)}{g(1)},$$

since g never vanishes. With $g' = (\mu + \sigma)g$ this amounts to $\mu + \sigma(0) = \mu + \sigma(1)$, which gives the second claim. The multiplier of g is then

$$\frac{g(1)}{g(0)} = g(1) = \exp\left(\mu + \int_0^1 \sigma(r, \mu)\,\mathrm{d}r\right). \qquad \square$$

The solution σ of the Riccati equation (C.4) depends on μ and q, and we show that it admits an expansion at $\mu = \infty$. To keep things simple, we consider formal expansions.

Lemma C.2. *Let q be smooth and 1-periodic. Making for $\mu \to \infty$ the formal ansatz*

$$\sigma \sim \sum_{n \geq 0} \frac{s_n}{(-2\mu)^n},$$

σ is a formal solution of (C.4) *iff $s_0 = 0$, $s_1 = -q$ and*

$$s_{n+1} = s_n' + \sum_{0 \leq m \leq n} s_{n-m} s_m, \qquad n \geq 1.$$

Moreover, all s_n are 1-periodic, too.

Proof. Inserting the ansatz into (C.4) and writing $\tilde{\mu} = -2\mu$ we obtain

$$\begin{aligned} q &= \sigma' + 2\mu\sigma + \sigma^2 \\ &= \sum_{n\geq 0} \frac{s_n' - \tilde{\mu} s_n}{\tilde{\mu}^n} + \sum_{n\geq 0} \sum_{0\leq m\leq n} \frac{s_{n-m}s_m}{\tilde{\mu}^n} \\ &= \sum_{n\geq 0} \frac{1}{\tilde{\mu}^n} \Big(s_n' + \sum_{0\leq m\leq n} s_{n-m}s_m \Big) - \sum_{n\geq 0} \frac{s_{n+1}}{\tilde{\mu}^n} - \tilde{\mu} s_0. \end{aligned}$$

Comparing terms of order $-1, 0, 1, \ldots$ in $1/\tilde{\mu}$ gives the first result. As the s_n are polynomial expressions in q and its derivatives, they are 1-periodic. □

By a straightforward calculation,

$$\begin{aligned} s_1 &= -q, \\ s_3 &= -q'' + q^2, \\ s_5 &= -q^{(4)} + 6qq'' + 5q'^2 - 2q^3, \\ s_7 &= -q^{(6)} + 10qq^{(4)} + 28q'q''' \\ &\qquad + 19q''^2 - 30q^2q'' - 50qq'^2 + 5q^4, \end{aligned}$$

while the even coefficients $s_2, s_4, \ldots$ are all exact with 1-periodic primitives. Letting

$$S_n = \int_0^1 s_n(x)\,\mathrm{d}x,$$

the terms $S_2, S_4, \ldots$ thus vanish, while

$$\begin{aligned} S_1 &= -\int_0^1 q\,\mathrm{d}x = 0 \\ S_3 &= \int_0^1 q^2\,\mathrm{d}x, \\ S_5 &= -\int_0^1 (q_x^2 + 2q^3)\,\mathrm{d}x, \\ S_7 &= \int_0^1 (q_{xx}^2 + 10qq_x^2 + 5q^4)\,\mathrm{d}x, \end{aligned}$$

by partial integration.

In this book the S_n define the KdV hierarchy of Hamiltonians through

$$H^n = \frac{(-1)^n}{2} S_{2n+3}.$$

In particular,

$$H^0 = \frac{1}{2}\int_0^1 q^2 \,\mathrm{d}x,$$
$$H^1 = \frac{1}{2}\int_0^1 (q_x^2 + 2q^3) \,\mathrm{d}x,$$
$$H^2 = \frac{1}{2}\int_0^1 (q_{xx}^2 + 10qq_x^2 + 5q^4) \,\mathrm{d}x.$$

Theorem C.3. *For a smooth, 1-periodic potential q with mean value zero,*

$$\operatorname{arcosh} \frac{\Delta(\lambda)}{2} \sim \sqrt{-\lambda} - \sum_{n \geq 0} \frac{(-1)^n}{4^{n+1}} \frac{H^n}{\sqrt{-\lambda}^{2n+3}},$$

for $\lambda \to -\infty$, where each H^n is an integral over $[0, 1]$ *of a polynomial expression in q and its derivatives up to order n given by $H^n = (-1)^n S_{2n+3}/2$.*

Proof. Combining equation (C.2) with the results of the preceding two lemmas we obtain, with $\lambda = -\mu^2$,

$$\operatorname{arcosh} \frac{\Delta(\lambda)}{2} = \log w(\lambda) = \mu + \sum_{n \geq 0} \frac{S_n}{(-2\mu)^n}.$$

Re-indexing the series and setting $H^n = (-1)^n S_{2n+3}/2$ gives the result. □

VIII

Psi-Functions and Frequencies

D Construction of the Psi-Functions

In this appendix we prove the following theorem stated in section 8. In the form presented it is due to [6], but the proof given here is much simpler, and the normalizing constants are explicitly computed. See also [90] for prior results. – For notations we refer to sections 6 and 7.

Theorem D.1. *There exists a complex neighbourhood W of L_0^2 such that for each q in W there exist entire functions ψ_n, $n \geq 1$, satisfying*

$$\frac{1}{2\pi} \int_{\Gamma_m} \frac{\psi_n(\lambda)}{\sqrt[c]{\Delta^2(\lambda) - 4}} \, \mathrm{d}\lambda = \delta_{mn}$$

for all $m \geq 1$. These functions depend analytically on λ and q and admit a product representation

$$\psi_n(\lambda) = \frac{2}{\pi n} \prod_{m \neq n} \frac{\sigma_m^n - \lambda}{m^2 \pi^2},$$

whose complex coefficients σ_m^n depend real analytically on q and satisfy

$$\left| \sigma_m^n - \tau_m \right| \leq C \frac{|\gamma_m|^2}{m}$$

for all m, locally uniformly on W and uniformly in n.

We prove this theorem with the help of the implicit function theorem. To this end we reformulate the statement in terms of a functional equation.

In the following, it is convenient to denote σ_m^n as $\bar{\sigma}_m^n$, and to use the former symbol for general ℓ^2-sequences. Moreover,

$$\bar{\sigma}_m = m^2 \pi^2 + \sigma_m$$

throughout this appendix.

For $\sigma = (\sigma_m)_{m\ge 1}$ in ℓ^2 and $n \ge 1$ define an entire function $\phi_n(\sigma)$ by

$$\phi_n(\sigma, \lambda) = \prod_{m\ne n} \frac{\bar{\sigma}_m - \lambda}{m^2\pi^2}.$$

For q in L_0^2 and $m \ge 1$ define a linear functional $A_m(q)$ on the space of entire functions by

$$A_m(q)\phi = \frac{1}{2\pi}\int_{\Gamma_m} \frac{\phi(\lambda)}{\sqrt[c]{\Delta^2(\lambda, q) - 4}}\,\mathrm{d}\lambda.$$

Locally, one can choose the contours Γ_m to be independent of q, and one can choose them arbitrarily close to the real interval

$$G_m(q) = [\lambda_{2m-1}(q), \lambda_{2m}(q)],$$

so that A_m is actually well defined on the space of real analytic functions on the real line.

For each $n \ge 1$ we then consider on $\ell^2 \times L_0^2$ the functional equation

$$F^n(\sigma, q) = 0,$$

where $F^n = (F_m^n)_{m\ge 1}$ with

$$F_m^n(\sigma, q) = \begin{cases} A_m^n(q)\phi_n(\sigma), & m \ne n, \\ \bar{\sigma}_n - \tau_n(q), & m = n, \end{cases} \tag{D.1}$$

and, for $m \ne n$,

$$A_m^n = w_m^n A_m, \qquad w_m^n = 2\pi m \frac{n^2 - m^2}{n^2}.$$

In fact, each function F_m^n is defined and real analytic on some complex neighbourhood U of $\ell^2 \times L_0^2$, which is independent of n and m.

We show that under some mild provisions there exists a unique solution $\sigma^n(q)$ of $F^n(\sigma, q) = 0$, which is real analytic in q and extends to some complex neighbourhood of L_0^2 independently of n. We then verify that

$$\bar{\sigma}_m^n = \tau_m + O(\gamma_m^2/m),$$

and that this solution satisfies

$$A_n(q)\phi_n(\sigma^n(q)) = \frac{\pi n}{2}.$$

Thus the functions

$$\psi_n = \frac{2}{\pi n}\phi_n(\sigma^n)$$

will have the required properties.

Real Solutions

Before constructing real solutions we first establish the proper setting of the functionals F^n.

Lemma D.2. *For each $n \geq 1$, equation* (D.1) *defines a map*

$$F^n: \ \ell^2 \times L_0^2 \to \ell^2$$
$$(\sigma, q) \mapsto F^n(\sigma, q),$$

which is real analytic and extends analytically to the complex neighbourhood U of $\ell^2 \times L_0^2$ introduced above. Moreover, this neighbourhood U can be chosen so that all F^n are locally uniformly bounded on it.

Proof. Fix n, and consider F_m^n for $m \neq n$. By the definition of ϕ_n and the product formula for $\Delta^2 - 4$ in Proposition B.10,

$$\frac{\phi_n(\sigma, \lambda)}{\sqrt[c]{\Delta^2(\lambda) - 4}} = \frac{\bar{\sigma}_m - \lambda}{\sqrt[s]{(\lambda_{2m} - \lambda)(\lambda - \lambda_{2m-1})}} \zeta_m^n(\lambda) \tag{D.2}$$

for λ near Γ_m with

$$\zeta_m^n(\lambda) = \frac{(-1)^{m+1}}{2 \sqrt[+]{\lambda - \lambda_0}} \frac{n^2\pi^2}{\bar{\sigma}_n - \lambda} \prod_{l \neq m} \frac{\bar{\sigma}_l - \lambda}{\sqrt[+]{(\lambda_{2l} - \lambda)(\lambda_{2l-1} - \lambda)}}. \tag{D.3}$$

The absolute value of the infinite product is $1 + O(\log m/m)$ by Lemma L.2 uniformly on bounded subsets of $\ell^2 \times L_0^2$, since $\lambda = m^2\pi^2 + O(1)$ near Γ_m. On the other hand, locally uniformly around any point in $\ell^2 \times L_0^2$ we may choose $\delta > 0$ and the contours Γ_m in such a way that

$$\inf_{m \neq n} \min_{\lambda \in \Gamma_m} |\bar{\sigma}_n - \lambda| \geq \delta, \qquad \max_{\lambda \in \Gamma_m} |\bar{\sigma}_m - \lambda| = O(\rho_m),$$

where

$$\rho_m \overset{\text{def}}{=} |\bar{\sigma}_m - \tau_m| + \gamma_m + \frac{1}{m}. \tag{D.4}$$

Taking into account the definition of the weights w_m^n, we then get

$$w_m^n \zeta_m^n = 1 + O\left(\frac{\log m}{m}\right) \tag{D.5}$$

near Γ_m, and furthermore

$$\frac{1}{2\pi} \int_{\Gamma_m} \frac{\bar{\sigma}_m - \lambda}{\sqrt[s]{(\lambda_{2m} - \lambda)(\lambda - \lambda_{2m-1})}} w_m^n \zeta_m^n \, \mathrm{d}\lambda = O(\rho_m)$$

by Lemma M.1. These estimates hold uniformly in n.

Taking into account the definition of A^n_m we altogether have

$$F^n_m(\sigma) = O(\rho_m), \qquad m \neq n,$$

locally uniformly on $\ell^2 \times L^2_0$. It follows that F^n maps $\ell^2 \times L^2_0$ into ℓ^2.

Exactly the same arguments can be used to show that F^n maps U into $\ell^2_{\mathbb{C}}$ and that the same estimates hold. Again, the last bound depends on σ and q in a locally uniform fashion, but not on n. Therefore, F^n is locally bounded on the complex neighbourhood U uniformly in n.

We already noticed that each function F^n_m is real analytic on U. Analyticity of the entire map F^n then follows with Theorem A.5. □

Next we consider the Jacobian of F^n with respect to σ. At any given point in $\ell^2 \times L^2_0$ this Jacobian is a bounded linear operator

$$Q^n\colon\ \ell^2 \to \ell^2,$$

which is represented by an infinite matrix (Q^n_{mr}) with elements

$$Q^n_{mr} = \frac{\partial F^n_m}{\partial \sigma_r} = \frac{\partial}{\partial \sigma_r} A^n_m \phi_n = A^n_m \frac{\partial \phi_n}{\partial \sigma_r}, \qquad m, r \neq n,$$

while $Q^n_{mn} = Q^n_{nm} = \delta_{mn}$. But first we make a simple observation, which is used several times below.

Lemma D.3. *If ϕ is real analytic on the real line, and $A_m\phi = 0$ for some $m \geq 1$, then ϕ has a root in $[\lambda_{2m-1}(q), \lambda_{2m}(q)]$.*

Proof. By assumption,

$$A_m\phi = \frac{1}{2\pi}\int_{\Gamma_m} \frac{\phi(\lambda)}{\sqrt[c]{\Delta^2(\lambda) - 4}}\,\mathrm{d}\lambda = 0$$

with a contour Γ_m around G_m sufficiently close to the real axis. If $\gamma_m > 0$, then we may shrink the contour to the interval $[\lambda_{2m-1}, \lambda_{2m}]$ to obtain

$$\frac{1}{\pi}\int_{\lambda_{2m-1}}^{\lambda_{2m}} \frac{\phi(\lambda)}{\sqrt{\Delta^2(\lambda) - 4}}\,\mathrm{d}\lambda = 0,$$

which is possible only when ϕ changes sign in this interval. If $\gamma_m = 0$, we may extract the factor $(\lambda - \tau_m)^2$ from the product representation of $\Delta^2(\lambda) - 4$ and note that the contour integral above turns into a Cauchy integral around τ_m, which then gives $\phi(\tau_m) = 0$. □

This simple lemma is the motivation why we look for entire functions ψ_n of the form

$$\psi_n(\lambda) = c_n \prod_{m \neq n} \frac{\sigma^n_m - \lambda}{m^2\pi^2}$$

in the first place. It also shows that we have to look for the zeroes σ_m^n in the interval $G_m(q) = [\lambda_{2m-1}(q), \lambda_{2m}(q)]$. It therefore makes sense to restrict ourselves to the open domain $V \subset \ell^2 \times L_0^2$ characterized by

$$\frac{\lambda_{2k-2} + \lambda_{2k-1}}{2} < \bar{\sigma}_k < \frac{\lambda_{2k} + \lambda_{2k+1}}{2}, \qquad k \geq 1.$$

As a consequence, any solution (σ, q) in V leads to a monotone sequence $\bar{\sigma}_m^n$, which in turn makes σ unique.

Lemma D.4. *On $V \subset \ell^2 \times L_0^2$, the diagonal elements Q_{mm}^n never vanish and satisfy*

$$Q_{mm}^n = 1 + O\left(\frac{\log m}{m}\right)$$

for $m \neq n$, while

$$Q_{mr}^n = O\left(\frac{\rho_m}{|m^2 - r^2|}\right)$$

for $m \neq r$ and $m, r \neq n$, with ρ_m defined in (D.4). *These estimates hold uniformly in n.*

Proof. By the definition of ϕ_n, for $r \neq n$,

$$\frac{\partial \phi_n}{\partial \sigma_r} = \frac{1}{r^2\pi^2} \prod_{l \neq n,r} \frac{\bar{\sigma}_l - \lambda}{l^2\pi^2} = \frac{\phi_n}{\bar{\sigma}_r - \lambda}.$$

Hence, for λ near Γ_m and $m, r \neq n$,

$$\begin{aligned}\frac{\partial \phi_n}{\partial \sigma_r} \frac{1}{\sqrt[c]{\Delta^2(\lambda) - 4}} &= \frac{1}{\bar{\sigma}_r - \lambda} \frac{\phi_n}{\sqrt[c]{\Delta^2(\lambda) - 4}} \\ &= \frac{\bar{\sigma}_m - \lambda}{\bar{\sigma}_r - \lambda} \frac{\zeta_m^n}{\sqrt[s]{(\lambda_{2m} - \lambda)(\lambda - \lambda_{2m-1})}}\end{aligned}$$

with ζ_m^n as in (D.3). Taking into account the weights in the definition of A_m^n we get

$$Q_{mr}^n = \frac{1}{2\pi} \int_{\Gamma_m} \frac{\bar{\sigma}_m - \lambda}{\bar{\sigma}_r - \lambda} \frac{w_m^n \zeta_m^n}{\sqrt[s]{(\lambda_{2m} - \lambda)(\lambda - \lambda_{2m-1})}} \, d\lambda.$$

For the diagonal element

$$Q_{mm}^n = \frac{1}{2\pi} \int_{\Gamma_m} \frac{w_m^n \zeta_m^n}{\sqrt[s]{(\lambda_{2m} - \lambda)(\lambda - \lambda_{2m-1})}} \, d\lambda,$$

the claimed estimate now follows immediately with (D.5) and Lemma M.1. Moreover, Q_{mm}^n does not vanish by Lemma D.3, since ζ_m^n has no root in $[\lambda_{2m-1}, \lambda_{2m}]$.

To estimate the off-diagonal terms Q^n_{mr}, note that

$$\frac{\bar{\sigma}_m - \lambda}{\bar{\sigma}_r - \lambda} = O\left(\frac{\rho_m}{|m^2 - r^2|}\right)$$

for λ near Γ_m with ρ_m given by (D.4), and apply again (D.5) and Lemma M.1.

As Q^n_{mr} depends on n only through the product $w^n_m \zeta^n_m$, and the latter has been estimated in (D.5) uniformly in n, these estimates hold uniformly in n and locally uniformly on V. □

Lemma D.5. *At any point in V the Jacobian Q^n of F^n with respect to σ is of the form*

$$Q^n = D^n + K^n,$$

where $D^n : \ell^2 \to \ell^2$ is an isomorphism in diagonal form and $K^n : \ell^2 \to \ell^2$ is compact.

Proof. Set $D^n = \operatorname{diag}(Q^n_{mm})$, the diagonal of Q^n. By the preceding lemma, we have

$$0 \neq Q^n_{mm} \sim 1,$$

so $D^n : \ell^2 \to \ell^2$ has a bounded inverse. Moreover, $K^n = Q^n - D^n$ is a bounded linear operator on ℓ^2 with vanishing diagonal and elements

$$K^n_{mr} = Q^n_{mr} = O\left(\frac{\rho_m}{|m^2 - r^2|}\right), \qquad m \neq r,$$

again by the preceding lemma. Clearly,

$$\sum_{m,r} \left|K^n_{mr}\right|^2 < \infty,$$

so K^n is Hilbert-Schmidt, hence compact. □

Lemma D.6. *At any given point in $V \subset \ell^2 \times L^2_0$, each Jacobian Q^n for $n \geq 1$ is one-to-one and hence a linear isomorphism $\ell^2 \to \ell^2$.*

Proof. Fix $n \geq 1$. To show that Q^n is one-to-one, suppose that $Q^n h = 0$ for some $h \in \ell^2$. Then clearly $h_n = 0$, since $Q^n_{nr} = \delta_{nr}$ for $r \geq 1$ by definition. For $m \neq n$, we get

$$0 = \sum_r \frac{\partial F^n_m}{\partial \sigma_r} h_r = \sum_r A^n_m \frac{\partial \phi_n}{\partial \sigma_r} h_r = A^n_m \sum_{r \neq n} \frac{h_r}{\bar{\sigma}_r - \lambda} \phi_n.$$

Thus, using straightforward estimates,

$$\psi(\lambda) \overset{\text{def}}{=} \sum_{r \neq n} \frac{h_r}{\bar{\sigma}_r - \lambda} \phi_n = \sum_{r \neq n} \frac{h_r}{r^2 \pi^2} \prod_{l \neq r,n} \frac{\bar{\sigma}_l - \lambda}{l^2 \pi^2}$$

defines a function that is entire and satisfies $A_m \psi = 0$ for all $m \neq n$, hence has a root ξ_m in each interval $[\lambda_{2m-1}, \lambda_{2m}]$ with $m \neq n$ by Lemma D.3. Consequently, letting $\bar{\sigma}_n = \xi_n = \tau_n$,

$$\psi_*(\lambda) \stackrel{\text{def}}{=} \frac{\bar{\sigma}_n - \lambda}{n^2 \pi^2} \psi = \sum_{r \neq n} \frac{h_r}{\bar{\sigma}_r - \lambda} \phi_*, \qquad \phi_* = \prod_{l \geq 1} \frac{\bar{\sigma}_l - \lambda}{l^2 \pi^2},$$

is also an entire function with roots ξ_m, $m \geq 1$.

Evaluating ψ_* on the circles $|\lambda| = R_k = (k + \frac{1}{2})^2 \pi^2$ we find that

$$\psi_*(\lambda) = \left(\sum_{r \neq n} \frac{h_r}{\bar{\sigma}_r - \lambda} \right) \phi_*(\lambda) = O\left(\frac{1}{k}\right) \frac{\sin \sqrt{\lambda}}{\sqrt{\lambda}}$$

by Lemma L.3. On the other hand, by the same lemma we also have

$$\chi_*(\lambda) \stackrel{\text{def}}{=} \prod_{l \geq 1} \frac{\xi_l - \lambda}{l^2 \pi^2} = \frac{\sin \sqrt{\lambda}}{\sqrt{\lambda}} \left(1 + O\left(\frac{\log k}{k}\right)\right)$$

uniformly on the same circles $|\lambda| = R_k$. The quotient ψ_*/χ_* is thus an entire function with

$$\sup_{|\lambda| = R_k} \left| \frac{\psi_*(\lambda)}{\chi_*(\lambda)} \right| = O\left(\frac{1}{k}\right).$$

By Liouville's theorem this is only possible for $\psi_* = 0$, hence $h = 0$ by evaluating ψ_* at the points $\bar{\sigma}_m$, $m \neq n$.

This shows that Q^n is one-to-one. By the preceding lemma and the Fredholm Alternative, Q^n is thus an isomorphism. □

Lemmas D.2 and D.6 allow us to apply the implicit function theorem to any particular solution of $F^n(\sigma, q) = 0$ in the domain V. The upshot is the following result.

Proposition D.7. *For any $n \geq 1$ there exists a unique real analytic map*

$$\sigma^n \colon \; L_0^2 \to \ell^2$$

with graph in V such that $F^n(\sigma^n(q), q) = 0$ everywhere. Indeed, $\bar{\sigma}_m^n(q) \in G_m(q)$ for each $m \geq 1$ at every point q.

Remark. To be precise, uniqueness holds within the class of all such analytic maps with graph in V.

Proof. First we claim that for any solution of $F^n(\sigma, q) = 0$ in V one has

$$\bar{\sigma}_m^n(q) \in G_m(q), \qquad m \geq 1. \tag{D.6}$$

For $m = n$ this is obvious by definition. For any $m \neq n$, the fact that $A_m \phi_n(\sigma) = 0$ and Lemma D.3 imply that ϕ_n has *some* root ξ_m in G_m. But ϕ_n has *exactly* the roots

$\bar{\sigma}_1 < \bar{\sigma}_2 < \ldots$ with $\bar{\sigma}_m \sim m^2\pi^2$ and $m \neq n$, and *no other* roots. Consequently, $\bar{\sigma}_m = \xi_m \in G_m$ for all $m \neq n$, which proves the claim.

Now, by Lemma D.6 and the implicit function theorem, any particular solution of $F^n(\sigma, q) = 0$ in V can be uniquely extended locally such that σ is given as a real analytic function of q. This local solution can be extended by the continuation method along any path from q to any given point in L_0^2, since $\partial F^n/\partial\sigma$ is a linear isomorphism everywhere on V and the compactness property (D.6) must hold for any continuous extension. Since L_0^2 is simply connected, any particular solution of $F^n(\sigma, q) = 0$ in V thus extends uniquely and globally to a real analytic map $\sigma^n : L_0^2 \to \ell^2$ with graph in V satisfying $F^n(\sigma^n(q), q) = 0$ everywhere.

At $q = 0$ one solution is given by

$$\sigma(0) = 0,$$

as one verifies using Cauchy's formula. Since $G_m(0) = \{m^2\pi^2\}$ for all $m \geq 1$, this solution is also unique. Hence there is one and only one such analytic map. □

Complex Extension

Proposition D.8. *All real analytic maps $\sigma^n : L_0^2 \to \ell^2$ of Proposition D.7 extend to a common complex neighbourhood of L_0^2.*

Proof. To verify that the solutions σ^n of Proposition D.7 all extend to a complex neighbourhood of $\ell^2 \times L_0^2$ independent of n we first show that at every real point q, the inverses of the Jacobians

$$Q^n(\sigma^n(q), q) = \frac{\partial F^n}{\partial \sigma}(\sigma^n(q), q)$$

are bounded uniformly in n.

Consider the Jacobian $Q^n = (Q^n_{mr})$ at a point in V. We have, for $m, r \neq n$,

$$Q^n_{mr} = \frac{w^n_m}{2\pi} \int_{\Gamma_m} \frac{\partial \phi_n}{\partial \sigma_r} \frac{\mathrm{d}\lambda}{\sqrt[c]{\Delta^2(\lambda) - 4}}$$

with

$$Q^n_{mm} = 1 + O\left(\frac{\log m}{m}\right), \qquad Q^n_{mr} = O\left(\frac{\rho_m}{|m^2 - r^2|}\right), \quad m \neq r$$

by Lemma D.4 locally uniformly on V. In this identity for Q^n_{mr} we can pass to the limit $n \to \infty$ to obtain

$$Q^n_{mr} \to Q^*_{mr} = m \int_{\Gamma_m} \frac{\partial \phi_*}{\partial \sigma_r} \frac{\mathrm{d}\lambda}{\sqrt[c]{\Delta^2(\lambda) - 4}}, \qquad \phi_* = \prod_{l \geq 1} \frac{\bar{\sigma}_l - \lambda}{l^2\pi^2},$$

where ϕ_* is the limit of

$$\phi_n = \prod_{l \neq n} \frac{\bar{\sigma}_l - \lambda}{l^2\pi^2} = \frac{n^2\pi^2}{\bar{\sigma}_n - \lambda}\phi_*.$$

By Lemma D.4, the Q^*_{mr} satisfy the same asymptotic estimates as Q^n_{mr} and define a bounded operator Q^* on ℓ^2. Moreover, the same estimates imply that $Q^n \to Q^*$ in the ℓ^2-operator norm locally uniformly on V.

The diagonal elements Q^*_{mm} do not vanish by Lemma D.3, since $\partial\phi_*/\partial\sigma_m$ has no root in G_m. Hence, by the same arguments used in the proofs of Lemmas D.5 and D.6, Q^* is boundedly invertible on ℓ^2 at every point in V. As the set

$$\Pi(q) = \prod_{m\geq 1} \left(G_m(q) - m^2\pi^2\right)$$

is compact in ℓ^2, Q^* is indeed uniformly boundedly invertible for σ in $\Pi(q)$ for any fixed q. By continuity, then also $Q^n(\sigma, q)$ is uniformly boundedly invertible for all large n for σ in $\Pi(q)$, and hence for all n.

By Lemma D.2 the maps F^n are analytic and locally uniformly bounded on a common complex neighbourhood of V uniformly in n. Using Cauchy's estimate, the variation δQ of Q^n with respect to σ and q can thus be kept as small as needed by restricting oneself to a sufficiently small complex neighbourhood of $\Pi(q) \times \{q\}$. Using the standard estimate

$$\left\|(Q+\delta Q)^{-1}\right\| \leq 2\left\|Q^{-1}\right\| \qquad \text{if} \quad 2\left\|\delta Q\right\| \leq \left\|Q^{-1}\right\|^{-1}$$

for a perturbation δQ of $Q = Q^n$, this gives us a similar uniform bound on the inverses of the Jacobians on this complex neighbourhood. The result then follows by the implicit function theorem. □

Remark. Continuing the preceding proof one can actually show that

$$\sigma^n(q) \to \dot\lambda(q) = (\dot\lambda_m(q))_{m\geq 1} \quad \text{as} \quad n \to \infty,$$

where $\dot\lambda_m$ denote the roots of $\dot\Delta$. This follows from the fact that

$$A_m \dot\Delta = \frac{1}{2\pi}\int_{\Gamma_m} \frac{\dot\Delta(\lambda)}{\sqrt{\Delta^2(\lambda)-4}}\,\mathrm{d}\lambda = 0$$

for all $m \geq 1$ and a uniqueness argument.

Asymptotics

Proposition D.9. *The components of* $\sigma^n = (\sigma^n_m)$ *satisfy*

$$\left|\bar\sigma^n_m - \tau_m\right| \leq C\frac{\gamma_m^2}{m}$$

for all m*, locally uniformly in a complex neighbourhood of* L^2_0 *and uniformly in* n*.*

Proof. Fix $\sigma^n = (\sigma_m^n)$, and drop the superscript n for this proof. There is nothing to prove for $\bar{\sigma}_n = \tau_n$, so consider $\bar{\sigma}_m$ with $m \neq n$. In view of their construction by the implicit function theorem, we have a first crude estimate

$$\bar{\sigma}_m = \tau_m + O(1), \tag{D.7}$$

which we refine now.

Since σ^n solves $F^n(\sigma, q) = 0$, we have

$$\begin{aligned} 0 &= \frac{1}{2\pi} \int_{\Gamma_m} \frac{\phi_n(\sigma, \lambda)}{\sqrt[c]{\Delta^2(\lambda) - 4}} \, \mathrm{d}\lambda \\ &= \frac{1}{2\pi} \int_{\Gamma_m} \frac{\bar{\sigma}_m - \lambda}{\sqrt[s]{(\lambda_{2m} - \lambda)(\lambda - \lambda_{2m-1})}} \zeta_m(\lambda) \, \mathrm{d}\lambda, \end{aligned}$$

using (D.2). Writing $\zeta_m(\lambda) = \xi_m + (\zeta_m(\lambda) - \xi_m)$ with $\xi_m \overset{\text{def}}{=} \zeta_m(\tau_m) \neq 0$ and noting that, by a simple computation,

$$\frac{1}{2\pi} \int_{\Gamma_m} \frac{\bar{\sigma}_m - \lambda}{\sqrt[s]{(\lambda_{2m} - \lambda)(\lambda - \lambda_{2m-1})}} \, \mathrm{d}\lambda = \bar{\sigma}_m - \tau_m,$$

we obtain

$$(\bar{\sigma}_m - \tau_m)\xi_m = \frac{1}{2\pi} \int_{\Gamma_m} \frac{(\lambda - \bar{\sigma}_m)(\zeta_m(\lambda) - \xi_m)}{\sqrt[s]{(\lambda_{2m} - \lambda)(\lambda - \lambda_{2m-1})}} \, \mathrm{d}\lambda. \tag{D.8}$$

If $\gamma_m = 0$, then the right hand side vanishes, and there is nothing to do. So assume that $\gamma_m \neq 0$. Choosing the contour Γ_m close of order γ_m to G_m,

$$|\zeta_m(\lambda) - \xi_m| = |\zeta_m(\lambda) - \zeta_m(\tau_m)| \leq M_m \gamma_m$$

along Γ_m, where M_m denotes the supremum of $|\zeta_m'|$ over the convex hull of Γ_m.

In view of (D.3), (D.7) and Lemma L.2 we have for all large m the asymptotic estimate $w_m \zeta_m(\lambda) = O(1)$ in a neighbourhood of size m around Γ_m. Hence,

$$\begin{aligned} w_m M_m &= O(1/m), \\ w_m \xi_m &= 1 + O(\log m/m) \end{aligned}$$

by Cauchy's estimate and (D.5), respectively. Moreover, for any $m \geq 1$, the eigenvalues $\lambda_{2m}(q), \lambda_{2m-1}(q)$ as well as $\bar{\sigma}_m(q)$ are contained in an isolating neighbourhood U_m locally uniformly in q, whence also

$$|w_m \xi_m| \geq c > 0$$

for *all* m, n locally uniformly in q.

From all this and (D.8) we thus obtain, in view of Lemma M.1,

$$|\bar{\sigma}_m - \tau_m| \, |\xi_m| = \sup_{\Gamma_m} |\lambda - \bar{\sigma}_m| \, O(M_m \gamma_m),$$

and subsequently

$$|\bar{\sigma}_m - \tau_m| = \sup_{\Gamma_m} |\lambda - \bar{\sigma}_m| \, O\Big(\frac{\gamma_m}{m}\Big).$$

Together with (D.7) this gives $|\bar{\sigma}_m - \tau_m| = O(\gamma_m/m)$. But this in turn implies

$$\sup_{\Gamma_m} |\lambda - \bar{\sigma}_m| \le |\bar{\sigma}_m - \tau_m| + \sup_{\Gamma_m} |\lambda - \tau_m| = O(\gamma_m),$$

which finally gives the claimed estimate.

The same arguments show that these estimates hold on some complex neighbourhood of L^2_0. Since only the asymptotic behavior of the periodic eigenvalues of q and the initial estimate (D.7) enter, but not n, they hold locally uniformly on this neighbourhood and uniformly in n. $\square$

Normalization

Consider now the not yet normalized entire functions $\tilde{\varphi}_n = \phi_n(\sigma^n)$. Clearly, we have $A_m\tilde{\varphi}_n = 0$ for any $m \ne n$, and $A_n\tilde{\varphi}_n \ne 0$ by Lemma D.3. Indeed, writing

$$\frac{\tilde{\varphi}_n}{\sqrt[c]{\Delta^2(\lambda) - 4}} = \frac{n^2\pi^2}{\sqrt[s]{(\lambda_{2n} - \lambda)(\lambda - \lambda_{2n-1})}} \cdot \frac{(-1)^{n+1}}{2\sqrt[+]{\lambda - \lambda_0}} \prod_{l \ne n} \frac{\bar{\sigma}_l - \lambda}{\sqrt[+]{(\lambda_{2l} - \lambda)(\lambda_{2l-1} - \lambda)}} \tag{D.9}$$

for λ near Γ_n in analogy to (D.2), Lemmas L.2 and M.1 lead to

$$A_n\tilde{\varphi}_n = \frac{\pi n}{2}\left(1 + O\Big(\frac{\log n}{n}\Big)\right).$$

Thus, we may set

$$\psi_n = \frac{c_n}{\pi n}\,\phi_n(\sigma^n)$$

with a constant $c_n = 2 + O(\log n/n)$ depending real analytically on q. This weaker result is completely sufficient to develop the theory of Birkhoff coordinates in chapter III.

The proof of Theorem D.1 is complete once we show the following result.

Proposition D.10. *For any* $n \ge 1$,

$$A_n\tilde{\varphi}_n = \frac{\pi n}{2}.$$

We prove this, maybe surprising, fact in the next appendix with the help of Riemann bilinear relations. It is connected with the rigidity of periodic spectra, illustrated for example by the fact that the positive gap lengths alone together with λ_0 already determine the entire spectrum of a potential.

A Sampling Formula

Finally we establish a *sampling formula* for ψ_n which is used in section 10. For more general formulas of this type see [90].

Proposition D.11. *For each $n \geq 1$,*

$$\sum_{k\geq 1} \frac{\psi_n(\mu_k)}{\dot{m}_2(\mu_k)} \frac{m_2(\lambda)}{\lambda - \mu_k} = \psi_n(\lambda)$$

everywhere on L_0^2, where $m_2(\lambda) = y_2(1, \lambda)$ and the μ_k are the Dirichlet eigenvalues of q.

Proof. Fix $n \geq 1$. By the product expansions for ψ_n in the theorem above and m_2 in appendix B one has the crude estimate

$$\frac{\psi_n(\mu_k)}{\dot{m}_2(\mu_k)} = O(1) \quad \text{for} \quad k \neq n$$

locally uniformly on L_0^2, by using Lemma L.2 and the asymptotic estimates of the μ_k and σ_m^n. Hence the sum converges to a real analytic function of λ and q. By the continuity of both sides on L_0^2 it then suffices to verify the identity on the dense subset of finite gap potentials.

So fix such a finite gap potential. Then $\lambda_{2k-1} = \mu_k = \sigma_k^n = \tau_k = \lambda_{2k}$ and $\psi_n(\mu_k) = 0$ for all $k > K$, with K sufficiently large. It remains to show that

$$\sum_{1\leq k\leq K} \frac{\psi_n(\mu_k)}{\dot{m}_2(\mu_k)} \frac{m_2(\lambda)}{\lambda - \mu_k} = \psi_n(\lambda).$$

By the product representations for ψ_n and m_2,

$$\psi_n(\lambda) = \frac{2}{\pi n} \prod_{\substack{1\leq m\leq K \\ m\neq n}} \frac{\sigma_m^n - \lambda}{m^2\pi^2} \cdot P(\lambda)$$

and, for $1 \leq k \leq K$,

$$\frac{m_2(\lambda)}{\lambda - \mu_k} = -\frac{1}{k^2\pi^2} \prod_{\substack{1\leq m\leq K \\ m\neq k}} \frac{\mu_m - \lambda}{m^2\pi^2} \cdot P(\lambda),$$

with

$$P(\lambda) = \prod_{m>K} \frac{\tau_m - \lambda}{m^2\pi^2}.$$

Factoring out $P(\lambda)$, both sides of the sampling formula reduce to polynomials in λ of degree at most $K - 1$. One also checks with l'Hospital's rule, that both sides agree at the points $\mu_1, \ldots, \mu_K$, while P does not vanish there. Consequently, both sides must be equal. $\square$

E A Trace Formula

In this appendix we prove the following trace formula which is used in the proof of Lemma 11.6 as well as Proposition D.10.

Theorem E.1. *For any $q \in L_0^2$,*

$$\sum_{n \geq 1} 2\pi n I_n = \frac{1}{2} \int_0^1 q(x)^2 \,\mathrm{d}x.$$

Note that the right hand side is the zero-th Hamiltonian in the KdV hierarchy which corresponds to translation. Hence the theorem says that in the angle-action coordinates introduced in chapter III the zero-th KdV Hamiltonian is just

$$H^0 = \sum_{n \geq 1} 2\pi n I_n,$$

and its frequencies are just

$$\frac{\partial H^0}{\partial I_n} = 2\pi n, \qquad n \geq 1.$$

This result will be extended to higher KdV Hamiltonians in the next appendix.

Riemann Bilinear Relations

The proof of the trace formula is based on the Riemann bilinear relations for meromorphic differentials on a Riemann surface. Specifically, we consider the Riemann surface

$$\Sigma(q) = \left\{ (\lambda, z) \colon z^2 = \Delta^2(\lambda, q) - 4 \right\} \subset \mathbb{C}^2$$

associated with a potential q. In the case of an N-gap potential this is a hyperelliptic surface of genus N, which may be viewed as two copies of the complex plane slit open along the $N+1$ intervals $(-\infty, \lambda_0), (\lambda_1, \lambda_2), \ldots, (\lambda_{2N-1}, \lambda_{2N})$ and then glued together crosswise along the slits.

On $\Sigma(q)$ consider the canonical basis (a_k) and (b_k) of cycles known as a- and b-cycles, as indicated in Figure 8. Each of the a-cycles is homotopic to the corresponding a'-cycle in Figure 9, while the b-cycles are the same. The cycles in Figure 9 only intersect in λ_0, as required by the 2nd reciprocity law – see [52, chapter XX], for example. The configuration given first, however, is more convenient for computations.

By convention, the root $\sqrt{\Delta^2(\lambda) - 4}$ on the sheet containing the a-cycles is given by the c-root

$$\mathrm{i} \sqrt[c]{\Delta^2(\lambda) - 4} > 0 \quad \text{for} \quad \lambda \in (\lambda_0, \lambda_1),$$

as defined in section 6 on page 62.

We will use the Riemann bilinear relations on $\Sigma(q)$ in the form of the 2nd reciprocity law.

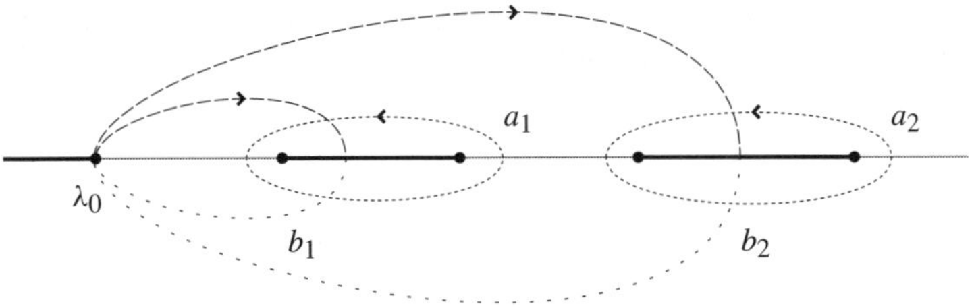

Figure 8 a- and b-cycles for $N = 2$

Proposition E.2. *Let q be an N-gap potential. If η and χ are two meromorphic differentials on $\Sigma(q)$ with a single pole at infinity with vanishing residuum, then*

$$\frac{1}{2\pi \mathrm{i}} \sum_{k=1}^{N} \left(\int_{a_k} \eta \int_{b_k} \chi - \int_{a_k} \chi \int_{b_k} \eta \right) = \mathrm{Res}_{\lambda=\infty} \left(\chi \int_{\lambda_0}^{\lambda} \eta \right), \tag{E.1}$$

where the integral is taken above the negative real axis on that sheet of $\Sigma(q)$ containing the a-cycles.

We also need an asymptotic representation of the Δ-function. Note that the following is a weaker form of Theorem C.3.

Proposition E.3. *For $q \in L_0^2$ and $\lambda < \lambda_0(q)$,*

$$\operatorname{arcosh} \frac{\Delta(\lambda)}{2} = \sqrt{-\lambda} - \frac{1}{4} \frac{H^0}{\sqrt{-\lambda}^{\,3}} + O\left(\frac{1}{\lambda^2} \right),$$

where $H^0 = \frac{1}{2} \int_0^1 q(x)^2 \, \mathrm{d}x$ is the zero-th Hamiltonian in the KdV hierarchy.

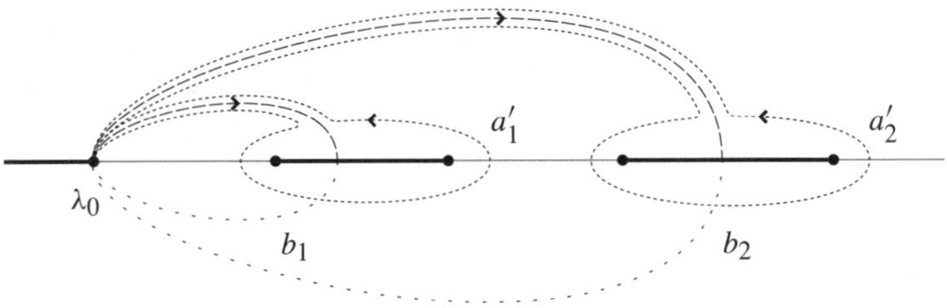

Figure 9 a'- and b-cycles with basepoint λ_0 for $N = 2$

Proof. By Proposition B.2,

$$\Delta(\lambda) = 2\cos\sqrt{\lambda} + 2Q\frac{\sin\sqrt{\lambda}}{\lambda^{3/2}} + O\left(\frac{\mathrm{e}^{|\mathrm{Im}\sqrt{\lambda}|}}{\lambda^2}\right)$$

with $4Q = H^0$. For $\lambda = -\mu^2$ with $\mu > 0$, this gives $\sqrt{\lambda} = \mathrm{i}\mu$ and

$$\frac{\Delta(\lambda)}{2} = \cosh\mu - \frac{Q}{\mu^3}\sinh\mu + O\left(\frac{\mathrm{e}^{\mu}}{\mu^4}\right).$$

Expanding arcosh at $\cosh\mu$ and noting that

$$\operatorname{arcosh}'(\cosh\mu) = \frac{1}{\sinh\mu}, \qquad \operatorname{arcosh}''(\cosh\mu) = -\frac{\cosh\mu}{\sinh^3\mu},$$

we obtain

$$\begin{aligned}\operatorname{arcosh}\frac{\Delta(\lambda)}{2} &= \operatorname{arcosh}(\cosh\mu)\\ &\quad + \frac{1}{\sinh\mu}\left(-\frac{Q}{\mu^3}\sinh\mu + O\left(\frac{\mathrm{e}^{\mu}}{\mu^4}\right)\right)\\ &\quad - \frac{1}{2}\frac{\cosh\tilde{\mu}}{\sinh^3\tilde{\mu}}\left(-\frac{Q}{\mu^3}\sinh\mu + O\left(\frac{\mathrm{e}^{\mu}}{\mu^4}\right)\right)^2\\ &= \mu - \frac{Q}{\mu^3} + O\left(\frac{1}{\mu^4}\right),\end{aligned}$$

where $\tilde{\mu} = \mu + O(1/\mu^3)$. With $\mu = \sqrt{-\lambda}$ the result follows. □

Proof of Theorem E.1

It suffices to prove the trace formula for q in the dense union of N-gap potentials, N arbitrary, as both sides of the formula are continuous in q.

We apply the Riemann bilinear relations to the differentials

$$\eta = \frac{\dot{\Delta}(\lambda)}{\sqrt{\Delta^2(\lambda) - 4}}\,\mathrm{d}\lambda, \qquad \chi = \frac{\lambda\dot{\Delta}(\lambda)}{\sqrt{\Delta^2(\lambda) - 4}}\,\mathrm{d}\lambda.$$

Clearly, they are meromorphic on $\Sigma(q)$ with a single pole at infinity. To determine its residue, note that $\eta = \mathrm{d}\operatorname{arcosh}\Delta(\lambda)/2$, hence

$$\eta\Big|_{\lambda=-z^{-2}} = \left(-\frac{1}{z^2} - 3Qz^2 + \dots\right)\mathrm{d}z \tag{E.2}$$

by the preceding proposition with $4Q = H^0$. Thus, the residues of η and χ at infinity vanish, and we can apply the Riemann bilinear relations.

We have $\int_{a_k} \eta = 0$, since η is exact near a_k, and $\int_{a_k} \chi = \pi I_k$ by the definition of the actions I_k in section 7. Shrinking b_k to the real axis in the usual way,

$$\begin{aligned}\int_{b_k} \eta = 2 \int_{\lambda_0}^{\lambda_1} \frac{\dot{\Delta}(\lambda)}{\sqrt[c]{\Delta^2(\lambda) - 4}} \, d\lambda \\ + \cdots + 2 \int_{\lambda_{2k-2}}^{\lambda_{2k-1}} \frac{\dot{\Delta}(\lambda)}{\sqrt[c]{\Delta^2(\lambda) - 4}} \, d\lambda,\end{aligned}$$

since the integrals along the slits cancel each other. With

$$\begin{aligned}\int_{\lambda_{2l}}^{\lambda_{2l+1}} \frac{\dot{\Delta}(\lambda)}{\sqrt[c]{\Delta^2(\lambda) - 4}} \, d\lambda &= \frac{(-1)^{l+1}}{i} \int_{\lambda_{2l}}^{\lambda_{2l+1}} \frac{\dot{\Delta}(\lambda)}{\sqrt[+]{4 - \Delta^2(\lambda)}} \, d\lambda \\ &= (-1)^l i \arcsin \frac{\Delta(\lambda)}{2} \bigg|_{\lambda_{2l}}^{\lambda_{2l+1}} \\ &= -\pi i,\end{aligned}$$

we obtain $\int_{b_k} \eta = -2\pi i k$. Hence, for the left hand side of the bilinear relation we obtain

$$\frac{1}{2\pi i} \sum_{k=1}^{N} \left(\int_{a_k} \eta \int_{b_k} \chi - \int_{a_k} \chi \int_{b_k} \eta \right) = \sum_{k=1}^{N} \pi k I_k .$$

On the other hand, with $\lambda = -z^{-2}$ and (E.2),

$$\begin{aligned}\chi \int_{\lambda_0}^{\lambda} \eta &= (z^{-4} + 3Q + \ldots)(z^{-1} - Qz^3 + \ldots) \, dz \\ &= (\cdots + 2Qz^{-1} + \ldots) \, dz,\end{aligned}$$

hence

$$\operatorname{Res}_{\lambda=\infty} \left(\chi \int_{\lambda_0}^{\lambda} \eta \right) = 2Q.$$

This proves the theorem.

Proof of Proposition D.10

We end this section by proving Proposition D.10 stated at the end of appendix D. It again suffices to consider N-gap potentials with N sufficiently large. We apply the Riemann bilinear relations to the meromorphic differentials

$$\eta = \frac{\dot{\Delta}(\lambda)}{\sqrt{\Delta^2(\lambda) - 4}} \, d\lambda, \qquad \chi = \frac{\tilde{\varphi}_n(\lambda)}{\sqrt{\Delta^2(\lambda) - 4}} \, d\lambda.$$

Note that χ has no poles at τ_k for $k > N$, since also $\tilde{\varphi}_n$ vanishes there by Proposition D.7.

As before, $\int_{a_k} \eta = 0$ and $\int_{b_k} \eta = -2\pi\mathrm{i}k$ for $1 \leq k \leq N$. By construction of the function $\tilde{\varphi}_n$, also $\int_{a_k} \chi = 0$ for $k \neq n$. The left hand side of the bilinear relation thus amounts to

$$\frac{1}{2\pi\mathrm{i}} \sum_{k=1}^{N} \left(\int_{a_k} \eta \int_{b_k} \chi - \int_{a_k} \chi \int_{b_k} \eta \right) = n \int_{a_n} \chi = 2\pi n A_n \tilde{\varphi}_n,$$

with A_n as on page 212.

On the other hand, a straightforward calculation using a representation analogous to (D.9) with $\lambda = -z^{-2}$ gives

$$\begin{aligned} \chi &= n^2\pi^2 z^3 \left(1 + O(z^2)\right) \frac{\mathrm{d}z}{z^3} \\ &= n^2\pi^2 \left(1 + O(z^2)\right) \mathrm{d}z. \end{aligned} \tag{E.3}$$

Together with $\int_{\lambda_0}^{\lambda} \eta = z^{-1} - Qz^3 + \ldots$ for $\lambda = -z^{-2}$ by (E.2) we obtain

$$\operatorname{Res}_{\lambda=\infty} \left(\chi \int_{\lambda_0}^{\lambda} \eta \right) = n^2\pi^2.$$

Combining this with the left hand side of the bilinear relation above we arrive at

$$2\pi n \, A_n \tilde{\varphi}_n = n^2\pi^2.$$

This proves the proposition.

F Frequencies

By the results of chapter III, and in particular Theorem 11.13, every Hamiltonian H^m in the KdV hierarchy, when expressed in terms of the Birkhoff coordinates in a suitable subspace of $\mathit{h}_{1/2}$, is a function of the actions $I_n = (x_n^2 + y_n^2)/2$ alone and thus classically integrable. Hence their associated frequencies ω_n^m are

$$\omega_n^m = \frac{\partial H^m}{\partial I_n}, \qquad n \geq 1.$$

For example, the zero-th KdV Hamiltonian is $H^0 = \frac{1}{2} \int_0^1 q(x)^2 \,\mathrm{d}x$, and

$$\omega_n^0 = \frac{\partial H^0}{\partial I_n} = 2\pi n, \qquad n \geq 1,$$

by Theorem E.1.

To obtain an asymptotic estimate of the frequencies of higher KdV Hamiltonians H^m, we use a procedure first developed in [50]. To start we recall that in Theorem C.3 we obtained the Hamiltonians in the KdV hierarchy as the coefficients in the formal expansion of the Δ-function at infinity. More precisely, we have the following result, which extends Theorem E.3.

Theorem F.1. *For a smooth, 1-periodic potential q with mean value zero,*

$$\operatorname{arcosh}\frac{\Delta(\lambda)}{2} \sim \sqrt{-\lambda} - \sum_{m\geq 0}\frac{(-1)^m}{4^{m+1}}\frac{H^m}{\sqrt{-\lambda}^{2m+3}},$$

for $\lambda \to -\infty$.

Expressing a potential q in terms of the Birkhoff coordinates $(\boldsymbol{x}, \boldsymbol{y})$ of Theorem 6.1, we may view the Δ-function as an analytic function of λ and $(\boldsymbol{x}, \boldsymbol{y})$. As Δ is a spectral invariant, it is indeed an analytic function of λ and the actions $\boldsymbol{I}$ alone. Thus it makes sense to consider its gradient with respect to I_n for each $n \geq 1$, and to introduce the one-forms

$$\eta_n = \frac{\partial\Delta/\partial I_n}{\sqrt{\Delta^2(\lambda)-4}}\,\mathrm{d}\lambda.$$

These are holomorphic one-forms on $\Sigma(q)$ except possibly at infinity.

Proposition F.2. *For an N-gap potential q, each* η_n *with* $1 \leq n \leq N$ *is a holomorphic one-form on* $\Sigma(q)$ *with*

$$\eta_n \sim -\sum_{m\geq 0}\frac{1}{4^{m+1}}\frac{\omega_n^m}{\sqrt{-\lambda}^{2m+3}}\,\mathrm{d}\lambda$$

for $\lambda \to -\infty$.

Proof. On the proper sheet of $\Sigma(q)$ we can write

$$\eta_n = \left(\frac{\partial}{\partial I_n}\operatorname{arcosh}\frac{\Delta(\lambda)}{2}\right)\mathrm{d}\lambda$$

and apply Theorem F.1 and the definition of the frequencies ω_m^n. □

It turns out that these one-forms η_n can be identified with the one-forms

$$\chi_n = \frac{\psi_n}{\sqrt{\Delta^2(\lambda)-4}}\,\mathrm{d}\lambda,$$

with ψ_n as in Theorem 8.1 or D.1.

Proposition F.3. *At an N-gap potential,* $\chi_n = -2\eta_n$ *for* $1 \leq n \leq N$.

Proof. Consider an N-gap potential q and its Riemann surface $\Sigma(q)$. By the preceding proposition, η_n with $1 \leq n \leq N$ is a holomorphic one-form on $\Sigma(q)$. Further, χ_n is holomorphic at infinity in view of the expansion (E.3), hence on all of $\Sigma(q)$ by its definition. To prove the claim, it therefore suffices to compare the a-periods of η_n and χ_n.

On one hand, by the construction of the ψ-functions,

$$\int_{a_m}\chi_n = 2\pi\delta_{mn}.$$

On the other hand, recall from the proof of Theorem 7.1 on page 64 that for a circuit Γ_m sufficiently close to $[\lambda_{2m-1}, \lambda_{2m}]$, the principal branch of the logarithm

$$\phi(\lambda) = \log(-1)^n \left(\Delta(\lambda) - \sqrt[c]{\Delta^2(\lambda) - 4}\right)$$

is well defined along Γ_m and depends analytically on I_n. Further,

$$\frac{\partial}{\partial I_n}\phi(\lambda) = -\frac{\partial \Delta / \partial I_n}{\sqrt[c]{\Delta^2(\lambda) - 4}}.$$

Therefore we get

$$\int_{a_m} \eta_n = \int_{\Gamma_m} \frac{\partial \Delta / \partial I_n}{\sqrt[c]{\Delta^2(\lambda) - 4}} \,\mathrm{d}\lambda = -\frac{\partial}{\partial I_n} \int_{\Gamma_m} \phi(\lambda)\,\mathrm{d}\lambda = -\pi\delta_{mn}$$

by equation (7.2). Hence

$$\int_{a_m} \chi_n = -2 \int_{a_m} \eta_n$$

for all $m \geq 1$, and the result follows. □

Comparing the expansions of χ_n and η_n in $\lambda = -z^{-2}$ we now obtain identities for the frequencies of the first and second KdV Hamiltonian.

Theorem F.4. *The frequencies of the first and second KdV Hamiltonian, defined for potentials in the appropriate Sobolev spaces, are*

$$\omega_n^1 = 8\pi n\,(\tau_n - r_n),$$
$$\omega_n^2 = 32\pi n\,(\tau_n^2 - r_n\tau_n + s_n),$$

with

$$r_n = -\frac{\lambda_0}{2} + \sum_{m\geq 1} (\sigma_m^n - \tau_m)$$

and

$$s_n = \frac{3}{8}\lambda_0^2 - \frac{\lambda_0}{2} \sum_{m\geq 1} (\sigma_m^n - \tau_m) + \sum_{m\geq 1} \left(\tau_m^2 - \tau_m\sigma_m^n + \gamma_m^2/8\right)$$
$$+\frac{1}{2} \sum_{m\neq r} (\sigma_m^n - \tau_m)\,(\sigma_r^n - \tau_r).$$

Remark 1. One verifies that the coefficients r_n and s_n are locally uniformly bounded on L_0^2 and $\mathcal{H}_0^{1/2}$, respectively.

Remark 2. One can show that $\lambda_0/2 = \sum_{m\geq 1} (\dot{\lambda}_m - \tau_m)$ and hence also

$$r_n = \sum_{m\geq 1} (\sigma_m^n - \dot{\lambda}_m).$$

Proof of Theorem F.4. The frequencies ω_n^1 and ω_n^2 are analytic on the appropriate Sobolev spaces, since the corresponding KdV Hamiltonians are. By Propositions B.11, D.7 and D.9, the functions τ_n, γ_n^2 and σ_m^n are all real analytic on L_0^2 and satisfy

$$\sigma_m^n - \tau_m = O(\gamma_m^2/m), \quad \tau_n = n^2\pi^2 + O(1), \quad \gamma_n = \ell^2(n),$$

locally uniformly on L_0^2. By continuity, it therefore suffices to prove the identities on a dense subset of finite gap potentials.

Choose an N-gap potential q. Then, for λ near $-\infty$,

$$\frac{\psi_n}{\sqrt[+]{\Delta^2(\lambda)-4}} = \frac{\pi n}{\tau_n - \lambda}\frac{1}{\sqrt[+]{\lambda_0 - \lambda}}\prod_{1\le m\le N}\frac{\sigma_m - \lambda}{\sqrt[+]{(\lambda_{2m}-\lambda)(\lambda_{2m-1}-\lambda)}},$$

where $\sigma_m = \sigma_m^n$, and in particular $\sigma_n = \tau_n$. Hence, on the canonical sheet near $-\infty$ above the real axis,

$$\begin{aligned}\chi_n\Big|_{\lambda=-z^{-2}} &= \frac{2\pi n}{\tau_n + z^{-2}}\frac{\Pi(z^2)}{\sqrt[+]{\lambda_0 + z^{-2}}}\frac{\mathrm{d}z}{z^3}\\ &= \frac{2\pi n}{1+\tau_n z^2}\frac{\Pi(z^2)}{\sqrt[+]{1+\lambda_0 z^2}}\,\mathrm{d}z\end{aligned}$$

with

$$\Pi(z^2) = \prod_{1\le m\le N}\frac{1+\sigma_m z^2}{\sqrt[+]{(1+\lambda_{2m-1}z^2)(1+\lambda_{2m}z^2)}}.$$

Using the identity

$$(1+\lambda_{2m-1}z^2)\,(1+\lambda_{2m}z^2) = 1 + 2\tau_m z^2 + (\tau_m^2 - \gamma_m^2/4)z^4,$$

one obtains

$$\frac{1+\sigma_m z^2}{\sqrt[+]{(1+\lambda_{2m-1}z^2)(1+\lambda_{2m}z^2)}} = \\ 1 + (\sigma_m - \tau_m)z^2 + (\tau_m^2 - \tau_m\sigma_m + \gamma_m^2/8)z^4 + O(z^6).$$

Combining this with the standard expansion of $1/\sqrt[+]{1+\lambda_0 z^2}$ one finds that

$$\frac{\Pi(z^2)}{\sqrt[+]{1+\lambda_0 z^2}} = 1 + \tilde{r}_n z^2 + \tilde{s}_n z^4 + O(z^6)$$

with

$$\tilde{r}_n = -\frac{\lambda_0}{2} + \sum_{1\le m\le N}(\sigma_m^n - \tau_m)$$

and an expression for $\tilde{s}_n$, which is analogous to s_n, but with all summation indices restricted to $\{1, 2, \ldots, N\}$.

As $\gamma_m = 0$ and $\tau_m = \sigma_m^n$ for all $m > N$, these coefficients agree at an N-gap potential with the coefficients r_n and s_n given above. Together with

$$\frac{1}{1+\tau_n z^2} = 1 - \tau_n z^2 + \tau_n^2 z^4 \mp \ldots,$$

we thus have

$$\chi_n \Big|_{\lambda=-z^{-2}} = 2\pi n \left(1 - (\tau_n - r_n) z^2 + (\tau_n^2 - r_n \tau_n + s_n) z^4 \mp \ldots\right) \mathrm{d}z.$$

On the other hand, by Proposition F.2,

$$\eta_n \Big|_{\lambda=-z^{-2}} = \frac{1}{2} \sum_{m \geq 0} \frac{\omega_n^m}{4^m} z^{2m} \, \mathrm{d}z.$$

In view of Proposition F.3 we thus obtain, for an N-gap potential,

$$\omega_n^1 = 8\pi n\,(\tau_n - r_n),$$
$$\omega_n^2 = 32\pi n \left(\tau_n^2 - r_n \tau_n + s_n\right),$$

for $1 \leq n \leq N$. □

Corollary F.5. *The frequencies of the first and second KdV Hamiltonian satisfy the asymptotic identities*

$$\omega_n^1 = (2\pi n)^3 + O(n),$$
$$\omega_n^2 = (2\pi n)^5 + O(n^3),$$

locally uniformly on the appropriate Sobolev spaces.

Proof. Just observe that

$$\tau_n = n^2 \pi^2 + O(1)$$

locally uniformly on L_0^2, and use the identities stated in the preceding theorem together with the result that on the appropriate Sobolev spaces the r_n and s_n are bounded in n locally uniformly. □

Remark. In fact, Theorem F.4 allows to derive more precise asymptotic estimate than the ones presented in the preceding corollary. But the latter are sufficient for our purposes, and were obtained by different methods in [7], see also [42].

IX

Birkhoff Normal Forms

G Two Results on Birkhoff Normal Forms

Generalized Normal Forms

Consider a Hamiltonian on the space $h_r^\bullet$ introduced in section 14 of the form

$$H = H_2 + H_3 + \ldots,$$

where the H_k are homogeneous of degree k in $v \in h_r^\bullet$. Assume that the quadratic term is already in *normal form*:

$$H_2 = \sum_{n \geq 1} \lambda_n \, |v_n|^2$$

with certain frequencies $\lambda_1, \lambda_2, \ldots$. Assume further that these frequencies are *non-resonant up to order* $m \geq 3$:

$$\sum_{n \geq 1} k_n \lambda_n \neq 0 \qquad \text{whenever} \qquad 1 \leq \sum_{n \geq 1} |k_n| \leq m, \tag{G.1}$$

where $k_1, k_2, \ldots$ are arbitrary integers.

Then there exists a symplectic transformation $v = \Phi(w) = w + \ldots$, where the dots stand for higher order terms, which takes H into its *Birkhoff normal form up to order* m:

$$H \circ \Phi = H_2 + N_4 + \cdots + N_m + \ldots,$$

where the $N_k, 4 \leq k \leq m$, are homogeneous terms of order k, which are actually functions of $|w_n|^2$, $n \geq 1$, and where the trailing dots stand for arbitrary terms of order strictly greater than m [26, 98]. In particular, the normalized part of the Hamiltonian contains no monomials of odd order. Moreover, the coefficients of the terms up to order m are *uniquely* determined by H, in the sense that for *any* normalizing transformation Φ of the form $\Phi = \text{id} + \ldots$ one obtains $H \circ \Phi = H_2 + N_4 + \cdots + N_m + \ldots$.

The normalization process may be taken to any order beyond m and may include resonant terms corresponding to resonant frequencies, leading to a so called *resonant Birkhoff normal form*. Its coefficients, however, are no longer uniquely determined, *except* for coefficients of the terms of order $m+1$ and $m+2$.

To formulate this result, denote by $\mathcal{P}_k$ the space of all homogeneous functions of degree k. The nullspace $\mathcal{N}_k$ of the operator $\{H_2, \cdot\}$ in $\mathcal{P}_k$ together with its range $\mathcal{R}_k$ in $\mathcal{P}_k$ provide an invariant splitting $\mathcal{P}_k = \mathcal{N}_k \oplus \mathcal{R}_k$. Any term in an expansion $H = H_2 + H_3 + \ldots$ can be uniquely written as

$$H_k = N_k + R_k, \qquad N_k \in \mathcal{N}_k, \quad R_k \in \mathcal{R}_k,$$

and such a Hamiltonian H is said to be in generalized normal form up to order m, if $R_k = 0$ for $3 \le k \le m$.

Theorem G.1. *Suppose H_2 is nonresonant up to order $m \ge 3$, and H is in Birkhoff normal form up to order m:*

$$H = H_2 + N_4 + \cdots + N_m + H_{m+1} + H_{m+2} + \ldots .$$

Then the generalized Birkhoff normal form is uniquely determined up to order $m+2$ with respect to transformations Φ of the form $\Phi = \mathrm{id} + \ldots$, and is obtained by projecting H_{m+1} and H_{m+2} onto $\mathcal{N}_{m+1}$ and $\mathcal{N}_{m+2}$, respectively.

Proof. The last statement is evident by considering the normalization process: as $m+1 \ge 4$, removing from H_{m+1} all nonresonant terms will generate only terms of order $\ge m+3$, hence it will not affect the normal form terms of H_{m+2}. So it remains to prove uniqueness.

Suppose the Hamiltonian is transformed into the normal form

$$H = H_2 + N_4 + \cdots + N_m + N_{m+1} + N_{m+2} + \ldots,$$

where N_{m+1}, N_{m+2} are resonant normal form terms in $\mathcal{N}_{m+1}$ and $\mathcal{N}_{m+2}$, respectively. Consider a symplectic transformation $\Phi = \mathrm{id} + \ldots$ so that

$$H \circ \Phi = H_2 + N_4' + \cdots + N_m' + N_{m+1}' + N_{m+2}' + \ldots$$

is in similar normal form. Φ can be written as

$$\Phi = F_3^* \circ F_4^* \circ \cdots \circ F_{m+2}^* \circ \Phi_{m+3},$$

where F_k^* is the time-1-map of the flow of a homogeneous Hamiltonian F_k of degree k, and where Φ_{m+3} is of the form "identity + terms of order $m+3$ or more".

Suppose that for some k with $3 \le k \le m$ the effect of $F_3^* \circ \cdots \circ F_{k-1}^*$ is already known to be nil, where the composition is understood to be the identity for $k = 3$:

$$H \circ F_3^* \circ \cdots \circ F_{k-1}^* = H_2 + N_4 + \cdots + N_m + N_{m+1} + N_{m+2} + \ldots .$$

Then

$$H \circ F_3^* \circ \cdots \circ F_{k-1}^* \circ F_k^* =$$
$$H_2 + N_4 + \cdots + N_m + N_{m+1} + N_{m+2} + \dots$$
$$+ \{H_2, F_k\} + \{N_4, F_k\} + \cdots + \tfrac{1}{2}\{\{H_2, F_k\}, F_k\} + \dots,$$

which by assumption on F_k^* is in normal form up to order $m+2$. Thus,

$$\{H_2, F_k\} + N_k \in \mathcal{N}_k.$$

In view of $N_k \in \mathcal{N}_k$ and the splitting $\mathcal{P}_k = \mathcal{N}_k \oplus \mathcal{R}_k$ we conclude that $\{H_2, F_k\} = 0$, hence

$$F_k \in \mathcal{N}_k.$$

As H_2 is nonresonant up to order m, all elements in $\mathcal{N}_2 \oplus \cdots \oplus \mathcal{N}_m$ commute. Hence,

$$\{N_j, F_k\} = 0, \qquad 4 \le j \le m.$$

In particular, as $\mathcal{N}_3 = 0$ by the assumption $m \ge 3$, we have $F_3 = 0$. We conclude that F_k^* does not affect the normal form terms of H up to order $m+2$ included, whereas there might be a contribution of order $m+3$ from $\{N_{m+1}, F_4\}$. By induction, this holds for $3 \le k \le m$.

Next, consider the effect of F_{m+1}^*. Since $\{H_2, F_{m+1}\} + N_{m+1} = N'_{m+1}$ is in normal form by assumption, one obtains $\{H_2, F_{m+1}\} \in \mathcal{N}_{m+1}$ and thus

$$\{H_2, F_{m+1}\} = 0, \qquad N'_{m+1} = N_{m+1}.$$

This transformation therefore affects $H \circ \Phi$ only through $\{H_4, F_{m+1}\}$ at the earliest, that is, at terms of order $m+3$. So we also have $N'_{m+2} = N_{m+2}$.

The same argument applies to F_{m+2}^*. Finally, the transformation Φ_{m+3} has no effect on terms up to order $m+2$. □

Simultaneous Normalization

We now consider a second, similar Hamiltonian $G = G_2 + G_3 + \dots$, which we assume to be in involution with H:

$$\{G, H\} = 0.$$

The finite dimensional version of this situation was investigated carefully by Ito [56]. Here we only need a simplified version for infinite dimensional systems.

Theorem G.2. *Suppose H satisfies the nonresonance condition* (G.1), *and $\Phi = \mathrm{id} + \hat{\Phi}$ is a real analytic symplectic transformation around the origin in $\hbar_r^\bullet$ that takes the Hamiltonian H into its Birkhoff normal form up to order m. Then Φ also normalizes every other Hamiltonian G up to order m, if G is in involution with H and starts with quadratic terms.*

Proof. The proof proceeds by induction. Suppose we know already that the transformed Hamiltonian

$$G \circ \Phi = G_2 + \tilde{G}_3 + \cdots + \tilde{G}_{k-1} + \tilde{G}_k + \ldots$$

is in normal form up to order $k-1$ with $2 \le k \le m$. For $k = 2$ this is true, since G is assumed to start with quadratic terms. Thus we can start the induction. Since Φ is symplectic, $H \circ \Phi$ and $G \circ \Phi$ are in involution:

$$\{H \circ \Phi, G \circ \Phi\} = \{H, G\} \circ \Phi = 0.$$

Collecting terms of order k and using analogous notation for $H \circ \Phi$, we obtain

$$\{H_2, \tilde{G}_k\} + \sum_{3 \le l \le k} \{\tilde{H}_l, \tilde{G}_{k+2-l}\} = 0,$$

where the latter sum is absent for $k = 2$. For $k > 2$, this sum vanishes, since all the functions involved are known to be in normal form and hence in involution. Hence,

$$\{H_2, \tilde{G}_k\} = 0.$$

Since H_2 is assumed to be nonresonant up to order m, $\tilde{G}_k$ is in normal form. □

For our purposes we need a slight generalization of Theorem G.2 to the effect that the nonresonance condition (G.1) need only be required for those terms which are actually present in H and G.

Addendum to Theorem G.2. *The preceding theorem remains valid, if H and G belong to some Poisson subalgebra of Hamiltonians on $\mathscr{h}_r^\bullet$, and the nonresonance conditions* (G.1) *hold for all those integer sequences $(k_n)_{n\ge1}$ which determine monomials $\prod_{n\ge1} v_n^{k_n}$ appearing in the expansion of some element in this subalgebra.*

Proof. Let us consider the case at hand. The Hamiltonians of the KdV hierarchy are integrals of polynomials in u and its derivatives. When expressed in the $\boldsymbol{v}$-coordinates via the ansatz (14.2),

$$v(x) = \mathcal{F}(\boldsymbol{v}) \stackrel{\text{def}}{=} \sum_{n \ne 0} \gamma_n v_n \mathrm{e}^{2\pi i n x}$$

with $\gamma_n = \sqrt{2\pi\,|n|}$, we obtain a Hamiltonian $H = H_2 + H_3 + \ldots$ with homogeneous terms of the form

$$H_k = \sum_{j_1 + \cdots + j_k = 0} H_{j_1 \ldots j_k} v_{j_1} \cdots v_{j_k}, \tag{G.2}$$

where $j_1, \ldots, j_k$ are arbitrary nonzero integers. Such Hamiltonians form a Poisson subalgebra, since the condition

$$j_1 + \cdots + j_k = 0$$

is preserved under Poisson brackets. For this subalgebra it suffices to pose a weaker form of the nonresonance condition (G.1), namely

$$\sum_{n\geq 1} k_n\lambda_n \neq 0$$

when

$$1 \leq \sum_{n\geq 1} |k_n| \leq m \quad \text{and} \quad \sum_{n\geq 1} nk_n = 0.$$

The reason is that for $F_k = \sum_{j_1+\cdots+j_k=0} F_{j_1\ldots j_k} v_{j_1} \cdots v_{j_k}$, we have

$$\{F_k, H_2\} = \mathrm{i} \sum_{j_1+\cdots+j_k=0} (\lambda_{j_1} + \cdots + \lambda_{j_k}) F_{j_1\ldots j_k} v_{j_1} \cdots v_{j_k},$$

where we understand that $\lambda_n = -\lambda_{-n}$ for $n < 0$. If $\lambda_{j_1} + \cdots + \lambda_{j_k} \neq 0$ for any term with $j_1 + \cdots + j_k = 0$ not already in normal form, then one can normalize H by a symplectic transformation as described in section 14. But this condition is equivalent with

$$\sum_{n\geq 1} k_n\lambda_n \neq 0, \qquad \sum_{n\geq 1} nk_n = 0,$$

by setting $k_n = \mathrm{card}\{j_l : j_l = n\}$. Moreover, if $\{G_k, H_2\} = 0$ for a G_k with the same representation as F_k above, then one concludes that G_k must be in normal form. This establishes the Addendum in our special case. The general case is handled in the same way. □

As a first application of Theorem G.2 we consider the zero-th Hamiltonian in the KdV hierarchy,

$$H^0(\boldsymbol{v}) = \frac{1}{2}\int_{S^1} v^2 \,\mathrm{d}x = \sum_{n\geq 1} \gamma_n^2 |v_n|^2,$$

for $v = \mathcal{F}(\boldsymbol{v})$. This Hamiltonian is quadratic and in involution with the KdV Hamiltonian, so the transformation Φ of Theorem 14.2 normalizes it up to order four. Recall that

$$\Phi = \Psi \circ \Xi = F_3^* \circ F_4^*,$$

where F_3, F_4 are homogeneous of order 3 and 4, respectively. As

$$H^0 \circ F_3^* = H^0 + \{H^0, F_3\} + \frac{1}{2}\{\{H^0, F_3\}, F_3\} + \ldots,$$

and as F_3^* normalizes the KdV Hamiltonian up to order three, we conclude from Theorem G.2 that $\{H^0, F_3\} = 0$. This implies that

$$H^0 \circ F_3^* = H^0.$$

Arguing similarly for F_4^* we obtain the following result.

Corollary G.3. *The zero-th Hamiltonian H^0 in the KdV hierarchy is invariant under the normalizing transformation Φ. That is, $H^0 \circ \Phi = H^0$.*

The same argument suggest an alternate way to determine the Birkhoff normal form of the KdV Hamiltonian $H_c = H^1 + 6cH^0$. It suffices to determine Φ for the Hamiltonian with $c = 0$, that is for $H_0 = H^1$, so that $H_0 \circ \Phi = H^1 \circ \Phi$ is in normal form. For general $c \neq 0$,

$$H_c \circ \Phi = (H^1 + 6cH^0) \circ \Phi = H^1 \circ \Phi + 6cH^0$$

is then also in normal form. In particular, we conclude that the higher order terms *do not depend on c*.

As a second application of Theorem G.2 and its Addendum we determine the Birkhoff normal form of the second Hamiltonian in the KdV hierarchy. This Hamiltonian, as any other in the hierarchy, is of the form

$$H = H_2 + H_3 + H_4 + \dots,$$

and it is in involution with the KdV Hamiltonian. Moreover, the Addendum applies, since all terms H_k are of the form (G.2). Hence, the transformation Φ of Theorem 14.2 also normalizes this Hamiltonian up to order 4.

To determine this normal form write again $\Phi = \Psi \circ \Xi = F_3^* \circ F_4^*$. With the usual expansions into Poisson brackets we have

$$\begin{aligned} H \circ F_3^* = H_2 + H_3 + \{H_2, F_3\} \\ + H_4 + \{H_3, F_3\} + \tfrac{1}{2}\{\{H_2, F_3\}, F_3\} + \dots \end{aligned}$$

and

$$\begin{aligned} H \circ F_3^* \circ F_4^* = H_2 + H_3 + \{H_2, F_3\} \\ + H_4 + \{H_2, F_4\} + \{H_3, F_3\} \\ + \tfrac{1}{2}\{\{H_2, F_3\}, F_3\} + \dots, \end{aligned}$$

where the dots comprise all terms of order five or more. This transformed Hamiltonian is known to be in normal form up to order four. Hence, $H_3 + \{H_2, F_3\} = 0$, and thus

$$H \circ \Phi = H_2 + H_4 + \{H_2, F_4\} + \tfrac{1}{2}\{H_3, F_3\} + \dots .$$

The term $\{H_2, F_4\}$ does not generate any contribution to the normal form, but removes those terms not belonging to it. Thus we must have

$$H \circ \Phi = H_2 + \pi_N H_4 + \tfrac{1}{2}\pi_N \{H_3, F_3\} + \dots,$$

where π_N denotes the projection onto the subspace of Hamiltonians in normal form.

We now calculate the normal form for $H = H_c^2$. We observed in section 14 that the Hamiltonian of KdV-2 is

$$\begin{aligned} H_c^2(v) &= H^2(v) + 10cH^1(v) + 30c^2H^0(v) \\ &= \int_{S^1} \left(\tfrac{1}{2}v_{xx}^2 + 5vv_x^2 + \tfrac{5}{2}v^4 + 5cv_x^2 + 10cv^3 + 15c^2v^2\right) \mathrm{d}x. \end{aligned} \tag{G.3}$$

The third order term is thus

$$H_3 = \int_{S^1} \left(5vv_x^2 + 10cv^3\right) \mathrm{d}x.$$

Using the expansion $v = \sum_{n \neq 0} \gamma_n v_n \mathrm{e}^{2\pi i n x}$ one finds that

$$H_3 = -5 \sum_{k+l+m=0} \sigma_{lm} \gamma_k \gamma_l^3 \gamma_m^3 v_k v_l v_m + 10c \sum_{k+l+m=0} \gamma_k \gamma_l \gamma_m v_k v_l v_m.$$

The coefficient in the first sum may be symmetrized with respect to k, l, m, and since $(k+l+m)^2 = 0$ one finds

$$\begin{aligned} &\sigma_{lm}\gamma_k\gamma_l^3\gamma_m^3 + \sigma_{mk}\gamma_l\gamma_m^3\gamma_k^3 + \sigma_{kl}\gamma_m\gamma_k^3\gamma_l^3 \\ &\quad = \gamma_k\gamma_l\gamma_m(\sigma_{lm}\gamma_l^2\gamma_m^2 + \sigma_{mk}\gamma_m^2\gamma_k^2 + \sigma_{kl}\gamma_k^2\gamma_l^2) \\ &\quad = 4\pi^2\gamma_k\gamma_l\gamma_m(lm + mk + kl) \\ &\quad = -2\pi^2\gamma_k\gamma_l\gamma_m(k^2 + l^2 + m^2) \end{aligned}$$

with $\sigma_{kl} = \operatorname{sgn} kl$. It follows that

$$H_3 = \frac{1}{3} \sum_{k+l+m=0} s_{klm} \gamma_k \gamma_l \gamma_m v_k v_l v_m$$

with $s_{klm} = 10\pi^2(k^2 + l^2 + m^2) + 30c$. On the other hand, in (14.4) we had

$$F_3 = -\frac{\mathrm{i}}{3} \sum_{k+l+m=0} \frac{v_k v_l v_m}{\tilde{\gamma}_k \tilde{\gamma}_l \tilde{\gamma}_m}$$

with $\tilde{\gamma}_k = \sigma_k \gamma_k$. We obtain

$$\begin{aligned} \{H_3, F_3\} &= \mathrm{i} \sum_{j \neq 0} \sigma_j \frac{\partial H_3}{\partial v_j} \frac{\partial F_3}{\partial v_{-j}} \\ &= \sum_{j \neq 0} \sigma_j \sum_{k+l=-j} s_{klj} \gamma_k \gamma_l \gamma_j v_k v_l \sum_{m+n=j} \frac{1}{\tilde{\gamma}_m \tilde{\gamma}_n \tilde{\gamma}_{-j}} v_m v_n \\ &= - \sum_{\substack{k+l+m+n=0 \\ k+l \neq 0}} s_{kl(k+l)} \frac{\gamma_k \gamma_l}{\tilde{\gamma}_m \tilde{\gamma}_n} v_k v_l v_m v_n. \end{aligned}$$

The normal form contribution of the last sum consists of those terms with $k+m=0$ (and thus also $l+n=0$) or $k+n=0$ (and thus also $l+m=0$). In precisely the same fashion as in section 14 one finds that the contribution of all terms with $k \neq l$,

$$s_{kl(k+l)}\sigma_{kl}\,|v_k|^2\,|v_l|^2,$$

to the normal form is zero, since the coefficients change sign with k and l. Hence,

$$\frac{1}{2}\pi_N\,\{H_3, F_3\} = -\frac{1}{2}\sum_{n\neq 0} s_{nn(n+n)}\,|v_n|^4 = -\sum_{n\neq 0}(15c + 30\pi^2 n^2)\,|v_n|^4.$$

It remains to consider the fourth order term. Clearly, from (G.3) we get

$$H_4 = \frac{5}{2}\int_{S^1} v^4\,\mathrm{d}x = \frac{5}{2}\sum_{k+l+m+n=0} \gamma_k\gamma_l\gamma_m\gamma_n v_k v_l v_m v_n.$$

Counting terms correctly,

$$\pi_N H_4 = \frac{5}{2}\cdot 16\pi^2 \sum_{k,l\geq 1}(3 - 2\delta_{kl})\,|kl|\,|v_k|^2\,|v_l|^2.$$

Leaving to the reader the calculation of the term H_2, this proves Theorem 14.5.

H Birkhoff Normal Form of Order 6

In section 14 we calculated the Birkhoff normal form of the KdV Hamiltonian up to order four. The KdV Hamiltonian is nonresonant up to order four at the origin, so by a first transformation we could eliminate all terms of third order, and by a second transformation put the resulting fourth order terms into a unique normal form.

According the Theorem G.1 this process even uniquely determines the Birkhoff normal form coefficients up to order six. For this we do not have to require additional nonresonance properties. On the other hand, by the existence of global angle-action coordinates for KdV we do have a classical Birkhoff normal form to any order for KdV. Therefore, by Theorem G.1, our construction even determines the normal form coefficients up to order six, and thus the Taylor series expansion of the KdV Hamiltonian up to order three in the actions.

Formally, it is not difficult to determine the sixth order coefficients. Write the KdV Hamiltonian as

$$H = \Lambda + G = H_2 + H_3,$$

dropping the subscript c from the notation, and the normalizing transformation as

$$\Phi = \Psi \circ \Xi = F_3^* \circ F_4^*$$

as in appendix G. Let π_N denote the projection onto normal form terms.

Lemma H.1. *The sixth order terms of the Birkhoff normal form of the KdV Hamiltonian are*

$$N_6 = \frac{1}{4}\pi_N \big(\{\{G_4, F_3\}, F_3\} + 2\{G_4, F_4\}\big), \qquad G_4 = \frac{1}{2}\{H_3, F_3\}.$$

To shorten notation, we let

$$GF = \{G, F\}, \qquad GF^2 = \{\{G, F\}, F\},$$

and so on in the following.

Proof. For $H = H_2 + H_3$ we obtain

$$\begin{aligned} H \circ F_3^* &= H_2 + H_2 F_3 + \frac{1}{2} H_2 F_3^2 + \frac{1}{6} H_2 F_3^3 + \frac{1}{24} H_2 F_3^4 \\ &\quad + H_3 + H_3 F_3 + \frac{1}{2} H_3 F_3^2 + \frac{1}{6} H_3 F_3^3 + \dots, \end{aligned}$$

where here and below the dots stand for terms of order 7 and more. The Hamiltonian F_3 was chosen so that $H_2 F_3 + H_3 = 0$. Hence we have

$$\begin{aligned} H \circ F_3^* &= H_2 + \frac{1}{2} H_3 F_3 + \frac{1}{3} H_3 F_3^2 + \frac{1}{8} H_3 F_3^3 + \dots \\ &= H_2 + G_4 + \frac{2}{3} G_4 F_3 + \frac{1}{4} G_4 F_3^2 + \dots, \qquad G_4 = \frac{1}{2} H_3 F_3. \end{aligned}$$

By solving $H_2 F_4 + G_4 = \pi_N G_4$ in the second step we then obtain

$$\begin{aligned} H \circ F_3^* \circ F_4^* &= H_2 + G_4 + \frac{2}{3} G_4 F_3 + \frac{1}{4} G_4 F_3^2 \\ &\quad + H_2 F_4 + G_4 F_4 + \frac{1}{2} H_2 F_4^2 + \dots \\ &= H_2 + \pi_N G_4 + \frac{2}{3} G_4 F_3 \\ &\quad + \frac{1}{4} G_4 F_3^2 + G_4 F_4 + \frac{1}{2} (\pi_N G_4 - G_4) F_4 + \dots. \end{aligned}$$

Since $\{\pi_N G_4, F_4\}$ contains no normal form terms, π_N applied to the last line gives

$$N_6 = \frac{1}{4} \pi_N \big(G_4 F_3^2 + 2 G_4 F_4\big). \qquad \square$$

In section 14 we calculated

$$G_4 = \frac{1}{2} H_3 F_3 = -B - Q,$$

where $B = 3 \sum_{k \geq 1} |v_k|^4$ is the normal form part of $-G_4$, and $\pi_N Q = 0$. Compare (14.5).

Corollary H.2. *The sixth order terms of the Birkhoff normal form of the KdV Hamiltonian are*

$$N_6 = -\frac{1}{4}\,\pi_N\left(BF_3^2 + QF_3^2 + 2QF_4\right).$$

Before calculating the Poisson brackets appearing in N_6 we collect all the ingredients. From (14.5) we have

$$B = 3\sum_{k\geq 1} |v_k|^4 = \frac{3}{2}\sum_{k\neq 0} v_k^2 v_{-k}^2$$

and

$$\begin{aligned} Q &= \frac{3}{2}\sum_{k+l+m+n=0} \frac{\gamma_k\gamma_l}{\tilde\gamma_m\tilde\gamma_n} v_k v_l v_m v_n \\ &= \frac{3}{2}\sum_{k+l+m+n=0} \frac{v_k v_l v_m v_n}{[klmn]}\, kl, \end{aligned}$$

where $\gamma_k = \sqrt{2\pi\,|k|}$ and $\tilde\gamma_k = \sigma_k\gamma_k$ with $\sigma_k = \operatorname{sgn}(k)$ as before, and

$$[k] = \sigma_k\sqrt{|k|}$$

for the rest of this appendix. The sum is restricted to $k+l, k+m, k+n \neq 0$. Symmetrizing the last series in all indices and observing that $(k+l+m+n)^2 = 0$ gives

$$k^2 + l^2 + m^2 + n^2 + 2(kl + km + kn + lm + ln + mn) = 0, \qquad \text{(H.1)}$$

we have

$$Q = -\frac{1}{8}\sum_{\substack{k+l+m+n=0\\ k+l,k+m,k+n\neq 0}} \frac{v_k v_l v_m v_n}{[klmn]}\,(k^2 + l^2 + m^2 + n^2).$$

Furthermore,

$$F_3 = \frac{1}{3\mathrm{i}}\sum_{k+l+m=0} \frac{v_k v_l v_m}{\tilde\gamma_k\tilde\gamma_l\tilde\gamma_m} = \frac{1}{3\mathrm{i}(2\pi)^{3/2}}\sum_{k+l+m=0} \frac{v_k v_l v_m}{[klm]}$$

and

$$\begin{aligned} F_4 &= \frac{3\mathrm{i}}{64\pi^5}\sum_{k+l+m+n=0} \frac{\gamma_k\gamma_l\gamma_m\gamma_n}{k^3 + l^3 + m^3 + n^3}\,\frac{v_k v_l v_m v_n}{mn} \\ &= \frac{3\mathrm{i}}{16\pi^3}\sum_{k+l+m+n=0} \frac{v_k v_l v_m v_n}{[klmn]}\,\frac{kl}{k^3 + l^3 + m^3 + n^3}, \end{aligned}$$

where the sum is restricted to the same integers as for Q. Symmetrizing in k, l, m, n

and using again (H.1),

$$F_4 = \frac{1}{8\mathrm{i}(2\pi)^3} \sum_{\substack{k+l+m+n=0 \\ k+l,k+m,k+n\neq 0}} \frac{v_k v_l v_m v_n}{[klmn]} \frac{k^2+l^2+m^2+n^2}{k^3+l^3+m^3+n^3}.$$

Finally, recall that

$$FG = \{F,G\} = \mathrm{i}\sum_{j\neq 0} \sigma_j \frac{\partial F}{\partial v_j}\frac{\partial G}{\partial v_{-j}}.$$

To shorten notation, we set $\tilde{F}_3 = (2\pi)^{3/2} F_3$ and $\tilde{F}_4 = (2\pi)^3 F_4$.

The Term BF_3^2

One finds

$$\begin{aligned} B\tilde{F}_3 &= -6 \sum_{k+l+m=0} \sigma_m \frac{v_k v_l v_m^2 v_{-m}}{[klm]} \\ &= -2 \sum_{k+l+m=0} \frac{v_k v_l v_m}{[klm]} (\sigma_k v_k v_{-k} + \sigma_l v_l v_{-l} + \sigma_m v_m v_{-m}) \end{aligned}$$

and then

$$\begin{aligned} B\tilde{F}_3^2 = -6 \sum_{\substack{k+l+u+v=0 \\ k+l\neq 0}} & \frac{v_k v_l v_u v_v}{[kluv]} \frac{1}{k+l} \\ & \cdot (\sigma_k v_k v_{-k} + \sigma_l v_l v_{-l} - 2\sigma_{k+l} v_{k+l} v_{-k-l}). \end{aligned}$$

Its contribution to the normal stems from the index matches $k+u=0, l+v=0$ or $k+v=0, l+u=0$, whereas the match $k+l=0, u+v=0$ can not arise in this sum. The two matches are different only when $k\neq l$. Hence we obtain

$$\begin{aligned} \pi_N B\tilde{F}_3^2 = -12 \sum_{k,l} & \frac{v_k v_{-k} v_l v_{-l}}{|kl|} \frac{\mu_{kl}}{k+l} \\ & \cdot (\sigma_k v_k v_{-k} + \sigma_l v_l v_{-l} - 2\sigma_{k+l} v_{k+l} v_{-k-l}) \end{aligned}$$

where $\mu_{kl} = 1$ for $k \neq l$ and $\mu_{kk} = \frac{1}{2}$.

The Term QF_3^2

One finds

$$Q\tilde{F}_3 = \frac{1}{2} \sum_{k+l+m+u+v=0} \frac{v_k v_l v_m v_u v_v}{[klmuv]} \frac{k^2+l^2+m^2+(u+v)^2}{u+v}.$$

The sum is restricted to $k+l, k+m, l+m, k+l+m \neq 0$. After some calculations, one then obtains

$$\begin{aligned} Q\tilde{F}_3^2 &= -\frac{3}{2} \sum_{k+l+p+q+u+v=0} \frac{v_k v_l v_p v_q v_u v_v}{[klpquv]} \frac{k^2+l^2+(p+q)^2+(u+v)^2}{(p+q)(u+v)} \\ &\quad + \sum_{k+l+m+p+q+u=0} \frac{v_k v_l v_m v_p v_q v_u}{[klmpqu]} \frac{k^2+l^2+m^2+(k+l+m)^2}{(k+l+m)(p+q)} \\ &= X + Y, \end{aligned}$$

where the second sum is restricted to integers with $k+l, k+m, l+m, k+l+m \neq 0$ and $p+q \neq 0$, while the first sum is restricted to integers with $k+l$, $p+q, u+v \neq 0$ and $k+p+q, l+p+q \neq 0$. The two terms are now discussed separately.

As to the term X, to contribute to the normal form the indices have to match in the following way. For each k and l, there are 8 ways to match with p, q and u, v, since they can not match both u and v, or both p and q at the same time. In each case there is only one way for the remaining indices to match. Due to the symmetry in p, q and u, v, each such match leads to the same coefficient. A prototypical match is

$$\begin{aligned} k+p &= 0 \\ l+u &= 0 \\ q+v &= 0 \end{aligned}$$

and renaming q as m, the coefficient of $-v_k v_{-k} v_l v_{-l} v_m v_{-m}/|klm|$ becomes

$$-\frac{k^2+l^2+(m-k)^2+(m+l)^2}{(m-k)(m+l)},$$

with the restriction that the denominator does not vanish and $k+l \neq 0$, $k \neq m+l$. If $k=l$, but $p \neq q$ and $u \neq v$, then the count of the terms has to be divided by 2. It is not possible, however, that all three indices k, l, m are equal. Hence we obtain

$$\pi_N X = 12 \sum_{k,l,m} \frac{v_k v_{-k} v_l v_{-l} v_m v_{-m}}{|klm|} \eta_{klm} \frac{k^2+l^2+(k-m)^2+(l+m)^2}{(k-m)(l+m)},$$

where $\eta_{klm} = 1, \frac{1}{2}, 0$, depending on whether no, exactly two except k and l, or all indices are equal. The sum is restricted to integers with $k+l, k-m, l+m \neq 0$ and $k \neq l+m$.

As to the term Y, for each choice of k, l, m with $k+l, k+m, l+m, k+l+m \neq 0$ we can find matching p, q, u to obtain a term in normal form. The possible matches are

$$\begin{aligned} k+p&=0 \\ l+q&=0 \\ m+u&=0 \end{aligned} \quad \text{or} \quad \begin{aligned} k+p&=0 \\ l+u&=0 \\ m+q&=0 \end{aligned} \quad \text{or} \quad \begin{aligned} k+u&=0 \\ l+p&=0 \\ m+q&=0 \end{aligned},$$

and the three matches we obtain from these by interchanging p and q, which lead to the same coefficient. In the case that k, l, m are pairwise different, the coefficient of $-v_k v_{-k} v_l v_{-l} v_m v_{-m}/|klm|$ is

$$-2\,\frac{k^2+l^2+m^2+(k+l+m)^2}{k+l+m}\left(\frac{1}{k+l}+\frac{1}{k+m}+\frac{1}{l+m}\right)$$
$$=-4\,\frac{k^2+l^2+m^2+kl+km+lm}{k+l+m}\cdot\frac{k^2+m^2+m^2+3(kl+km+lm)}{(k+l)(k+m)(l+m)}.$$

If exactly two of the three indices k, l, m are equal, then the count of the terms has to be divided by two, and if all three indices k, l, m are equal, it has to be divided by 6. Hence we obtain

$$\pi_N Y = 4\sum_{k,l,m}\frac{v_k v_{-k} v_l v_{-l} v_m v_{-m}}{|klm|}\,\mu_{klm}$$
$$\cdot\,\frac{k^2+l^2+m^2+kl+km+lm}{k+l+m}\cdot\frac{k^2+l^2+m^2+3(kl+km+lm)}{(k+l)(k+m)(l+m)},$$

where $\mu_{klm} = 1, \frac{1}{2}, \frac{1}{6}$, depending on whether no, exactly two, or all indices are equal.

The Term QF_4

One finds

$$Q\tilde{F}_4 = -\frac{1}{4}\sum_{k+l+m+u+v+w=0}\frac{v_k v_l v_m v_u v_v v_w}{[klmuvw]}\cdot$$
$$\cdot\,\frac{k^2+l^2+m^2+(k+l+m)^2}{(k+l+m)}\cdot\frac{u^2+v^2+w^2+(u+v+w)^2}{u^3+v^3+w^3-(u+v+w)^3},$$

where the sum extends over integers with

$$k+l, k+m, l+m, u+v, u+w, v+w \neq 0 \quad \text{and} \quad k+l+m \neq 0.$$

Using $u^3+v^3+w^3-(u+v+w)^3 = -3(u+v)(u+w)(v+w)$ from Lemma 14.4 and expanding squares leads to

$$Q\tilde{F}_4 = -\frac{1}{3}\sum_{k+l+m+u+v+w=0}\frac{v_k v_l v_m v_u v_v v_w}{[klmuvw]}\cdot$$
$$\cdot\,\frac{(k^2+l^2+m^2+kl+km+lm)(u^2+v^2+w^2+uv+uw+vw)}{(u+v)(u+w)(v+w)(u+v+w)}.$$

To contribute to the normal form, each index in $\{k, l, m\}$ must match with one index in $\{u, v, w\}$. Each match gives the same coefficient of $-v_k v_{-k} v_l v_{-l} v_m v_{-m}/|klm|$, namely

$$\frac{(k^2+l^2+m^2+kl+km+lm)^2}{(k+l)(k+m)(l+m)(k+l+m)}.$$

If all three indices are different, then there are 6 ways to match indices. If exactly two of them are equal, the number of matches is 3, and if they are all equal, there is 1 such match. Hence we obtain

$$\pi_N Q\tilde{F}_4 = 2 \sum_{k,l,m} \frac{v_k v_{-k} v_l v_{-l} v_m v_{-m}}{|klm|} \mu_{klm} \cdot \frac{(k^2+l^2+m^2+kl+km+lm)^2}{(k+l)(k+m)(l+m)(k+l+m)},$$

where μ_{klm} is defined as above.

Before proceeding we combine the last two terms to

$$\pi_N \left(Y + 2Q\tilde{F}_4\right) = 8 \sum_{k,l,m} \frac{v_k v_{-k} v_l v_{-l} v_m v_{-m}}{|klm|} \mu_{klm} \cdot \frac{(k+l+m)(k^2+l^2+m^2+kl+km+lm)}{(k+l)(k+m)(l+m)}.$$

In this sum, the restriction $k+l+m \neq 0$ can be dropped, as the factor $k+l+m$ appears in the numerator.

These normal form expansions should now be represented in terms of the action variables $I_p = v_p v_{-p}$ for $p > 0$. We will not do this for all coefficients – for our purposes it suffices to determine the coefficient of

$$I_p^2 I_q$$

for $p \neq q$. Let $\tilde{\pi}_N$ denote the projection onto just those normal form terms.

The Term $\tilde{\pi}_N(Y + 2Q\tilde{F}_4)$

To obtain the coefficient of $I_p^2 I_q$, two indices in the sum must both be equal either to p or $-p$, as they can not have opposite signs. The third index has to be equal to either q or $-q$. Adding the coefficients for $k = l = p$, $m = q$ and $k = l = p$, $m = -q$ we obtain

$$\frac{(2p+q)(3p^2+2pq+q^2)}{2p(p+q)^2} + \frac{(2p-q)(3p^2-2pq+q^2)}{2p(p-q)^2} = 2 \cdot \frac{3p^4-2p^2q^2+q^4}{(p+q)^2(p-q)^2}.$$

The same contribution is obtained, if $k = l = -p$ and $m = q$ or $m = -q$. Further, for the choice of q, there are three possibilities. Thus the above coefficient has to be multiplied by $2 \cdot 3$. As $\mu_{klm} = \frac{1}{2}$ in our situation we thus obtain

$$\tilde{\pi}_N(Y + 2Q\tilde{F}_4) = 48 \sum_{p \neq q} \frac{I_p^2 I_q}{p^2 q} \cdot \frac{3p^4-2p^2q^2+q^4}{(p+q)^2(p-q)^2}. \tag{H.2}$$

The Term $\tilde{\pi}_N X$

The contributions from $\pi_N X$ to $\tilde{\pi}_N X$ stem from the matches $k = l$, $k = -m$ and $l = m$, as the possibilities $k = -l$, $k = m$ or $l = -m$ are excluded from the sum of $\pi_N X$. Adding up the coefficients for $k = l = p$, $m = \pm q$ gives

$$\frac{2p^2 + (p-q)^2 + (p+q)^2}{(p-q)(p+q)} + \frac{2p^2 + (p+q)^2 + (p-q)^2}{(p+q)(p-q)} = 4\frac{2p^2 + q^2}{p^2 - q^2}.$$

Adding up the coefficients for $k = p$, $m = -p$, $l = \pm q$ gives

$$-\frac{5p^2 + q^2 + (p-q)^2}{2p(p-q)} - \frac{5p^2 + q^2 + (p+q)^2}{2p(p+q)} = -6\frac{p^2}{p^2 - q^2}.$$

To take account of the restriction $k \neq l + m$, however, we have to add the term $-5\delta_{q-2p}$ for the case $k = p$, $m = -p$, $l = q = 2p$, where δ_j is one for $j = 0$ and zero otherwise. The same contribution arises from $l = m = p$, $k = \pm q$. The count of each term has to be multiplied by 2, for we obtain the same terms by reversing the signs of all the integers k, l, m. The total coefficient is

$$8\frac{2p^2 + q^2}{p^2 - q^2} - 24\frac{p^2}{p^2 - q^2} - 20\delta_{q-2p} = -8 - 20\delta_{q-2p}.$$

As $\eta_{klm} = \frac{1}{2}$ in the situation at hand we obtain

$$\tilde{\pi}_N X = -48 \sum_{p \neq q} \frac{I_p^2 I_q}{p^2 q} - 60 \sum_p \frac{I_p^2 I_{2p}}{p^3}. \tag{H.3}$$

The Term $\tilde{\pi}_N B \tilde{F}_3^2$

We collect the terms for $k = p$, $l = q$ and $k = p$, $l = -q$. The same set of indices with all signs reversed gives the same contribution, so we obtain

$$\begin{aligned}
\pi_N B \tilde{F}_3^2 &= -24 \sum_{p,q} \frac{I_p I_q}{pq} \frac{\mu_{pq}}{p+q} \left(I_p + I_q - 2I_{p+q}\right) \\
&\quad - 24 \sum_{p \neq q} \frac{I_p I_q}{pq} \frac{1}{p-q} \left(I_p - I_q - 2\sigma_{p-q} I_{|p-q|}\right) \\
&= -48 \sum_{p \neq q} \frac{I_p I_q}{pq} \left(\frac{p I_p}{p^2 - q^2} + \frac{q I_q}{q^2 - p^2} - \frac{I_{p+q}}{p+q} - \frac{I_{|p-q|}}{|p-q|} \right) \\
&\quad - 12 \sum_p \frac{I_p^3}{p^3} + 12 \sum_p \frac{I_p^2 I_{2p}}{p^3}.
\end{aligned}$$

Taking into account the special cases $q = 2p$ and $p = 2q$, this leads to

$$\tilde{\pi}_N B\tilde{F}_3^2 = -96 \sum_{p \neq q} \frac{I_p^2 I_q}{p^2 q} \frac{p^2}{p^2 - q^2} + 60 \sum_p \frac{I_p^2 I_{2p}}{p^3}. \tag{H.4}$$

Collecting all terms from equations (H.2), (H.3) and (H.4) we arrive at

$$\tilde{\pi}_N \left(B\tilde{F}_3^2 + Q\tilde{F}_3^2 + 2Q\tilde{F}_4 \right) = 96 \sum_{p \neq q} \frac{I_p^2 I_q}{p^2 q} \frac{p^2 q^2}{(p^2 - q^2)^2}.$$

This has to be divided by $8\pi^3$ in view of the definitions of $\tilde{F}_3$ and $\tilde{F}_4$ and by -4 in view of Corollary H.2 to give the following result.

Proposition H.3. *For the sixth order term of the Birkhoff normal form of the KdV Hamiltonian one obtains*

$$\tilde{\pi}_N N_6 = -\frac{3}{\pi^3} \sum_{p \neq q} I_p^2 I_q \frac{q}{(p^2 - q^2)^2}.$$

Consequently, the frequencies $\boldsymbol{\omega} = (\omega_n)_{n \geq 1}$ *of the KdV Hamiltonian satisfy*

$$\left. \frac{\partial^2 \omega_q}{\partial I_p^2} \right|_{\boldsymbol{I}=0} = -\frac{6}{\pi^3} \frac{q}{(p^2 - q^2)^2}.$$

I Kramer's Lemma

In Proposition 15.5 we verified the nondegeneracy and nonresonance conditions of the KAM Theorem 16.1 for the KdV Hamiltonian H_c by looking at the first two terms of the expansion of its frequencies $\boldsymbol{\omega}$ given in (15.1),

$$\boldsymbol{\omega} = \boldsymbol{\lambda} - 6\boldsymbol{I} + \ldots,$$

where $\boldsymbol{\lambda} = (\lambda_n)_{n \geq 1}$ with $\lambda_n = 8\pi^3 n^3 + 12c\pi n$ depends also on c. This was not difficult except in the following case. If $A \subset \mathbb{N}$ is a given finite index set, we have to show that as a function of $\boldsymbol{I}_A = (I_k)_{k \in A}$, we have

$$\begin{aligned} \omega_n &\not\equiv 0, \\ \omega_m + \omega_n &\not\equiv 0, \\ \omega_m - \omega_n &\not\equiv 0, \qquad m \neq n, \end{aligned}$$

for all $m, n \notin A$. As these expressions do not depend on $\boldsymbol{I}_A$ up to first order, we had to resort to exclude certain values of c to establish these conditions. See Proposition 15.5 on page 131.

This restriction on c can be dropped by looking at the second derivative $\partial^2_{I_k}\omega_m$, which we obtained in the preceding appendix after some lengthy calculations. To verify the above three conditions, it is clearly sufficient to show that for some $k \in A$,

$$\begin{aligned} \partial^2_{I_k}\omega_m &\neq 0, \\ \partial^2_{I_k}(\omega_m + \omega_n) &\neq 0, \\ \partial^2_{I_k}(\omega_m - \omega_n) &\neq 0, \qquad m \neq n, \end{aligned}$$

for all $m, n \notin A$. Of these the first two follow immediately from Proposition H.3, as the derivatives are all strictly negative at $\boldsymbol{I} = 0$. It remains to verify the third condition.

First we record a direct consequence of Proposition H.3.

Lemma I.1. *For $k \in A$ and $m, n \notin A$,*

$$\partial^2_{I_k}(\omega_m - \omega_n)\big|_{\boldsymbol{I}=0} = -\frac{6(m-n)}{\pi^3}\,\frac{k^4 + 2mnk^2 - (m^3n + m^2n^2 + mn^3)}{(k^2 - m^2)^2(k^2 - n^2)^2}.$$

Thus, to verify the third condition we have to show that the numerator above, $k^4 + 2mnk^2 - (m^3n + m^2n^2 + mn^3)$, does not vanish for pairwise distinct, positive integers k, m, n. — We first reduce this diophantine problem to a simpler one.

Lemma I.2. *If the equation*

$$k^4 + 2mnk^2 - (m^3n + m^2n^2 + mn^3) = 0$$

has a solution in positive integers k, m, n, then up to a common factor they are of the form $k = luv$, $m = u^4$, $n = v^4$ with positive integers l, u, v, such that u and v are relatively prime and

$$l^2 = u^4 + v^4 - u^2v^2.$$

Proof. We can always divide the equation by the fourth power of (k, m, n), the greatest common divisor of these three integers. Thus we can assume without loss of generality that $(k, m, n) = 1$.

Consider then $r = (m, n)$. If $r > 1$ then $r \nmid k$, since $(k, m, n) = 1$. On the other hand, dividing the equation by r^2 we conclude that $r^2 \mid k^4$, or $r \mid k$, a contradiction. Hence, also $(m, n) = 1$.

Now, solving the quadratic equation for k^2 we have

$$k^2 = -mn + (m+n)\sqrt{mn}.$$

To give an integer solution, both m and n must be square numbers. Thus, $m = a^2$ and $n = b^2$ with $(a, b) = 1$. This then gives

$$k^2 = ab(a^2 + b^2) - a^2b^2 = ab(a-b)^2 + a^2b^2. \tag{I.1}$$

This in turn implies that a and b are also square numbers,

$$a = u^2, \qquad b = v^2,$$

for if a or b contained a prime factor p with an odd power, then the right hand sides of (I.1) would contain p with an odd power, too, while k^2 contains p with an even power. Thus we have

$$k^2 = u^4v^4 + u^2v^2(u^2 - v^2)^2,$$

where u and v are relatively prime. Setting

$$l = \frac{k}{uv} \qquad \Leftrightarrow \qquad k = luv,$$

we get

$$l^2 = (uv)^2 + (u^2 - v^2)^2 = u^4 + v^4 - u^2v^2. \qquad \square$$

Finally we show that the diophantine equation $l^2 = u^4 + v^4 - u^2v^2$ has only the obvious trivial solutions. For the following result we are indebted to Jürg Kramer.

Lemma I.3 (J. Kramer). *The set of integral solutions of the equation*

$$R^2 = S^4 + T^4 - S^2T^2 \tag{I.2}$$

is $\{(\pm n^2, n, 0), (\pm n^2, 0, n), (\pm n^2, \pm n, n) \colon n \in \mathbb{Z}\}$.

Proof. Let $\mathcal{S}_a$ denote the affine surface defined by the equation (I.2) in affine 3-space. We will not only determine the set of integral points on $\mathcal{S}_a$, but even the set of rational points $\mathcal{S}_a(\mathbb{Q})$ on $\mathcal{S}_a$. First, we observe that the intersection of $\mathcal{S}_a$ with the plane given by $T = 0$ contains precisely the rational points (r, s, t) of the form

$$(r, s, t) = (\pm n^2, n, 0)$$

with arbitrary $n \in \mathbb{Q}$. Let now $\mathcal{U}_a$ denote the open subset of $\mathcal{S}_a$ given by the set of points having non-zero T-coordinate. Dividing (I.2) by T^4 and setting

$$U := S/T, \qquad V := R/T^2, \tag{I.3}$$

we obtain the morphism $\varphi\colon \mathcal{U}_a \to \mathcal{C}_a$ from the open subset $\mathcal{U}_a$ to the affine curve $\mathcal{C}_a$ defined by the equation

$$V^2 = U^4 - U^2 + 1.$$

Since we have an inclusion $\varphi(\mathcal{U}_a(\mathbb{Q})) \subseteq \mathcal{C}_a(\mathbb{Q})$, hence $\mathcal{U}_a(\mathbb{Q}) \subseteq \varphi^{-1}(\mathcal{C}_a(\mathbb{Q}))$, it suffices to find the rational points on the quartic $\mathcal{C}_a$. To do this, we set

$$X := U^2, \qquad Y := UV, \tag{I.4}$$

which gives rise to the morphism $\psi\colon \mathcal{C}_a \to \mathcal{E}_a$, $\mathcal{E}_a$ being given as the affine cubic

$$Y^2 = X^3 - X^2 + X.$$

As before, we have an inclusion $\psi(\mathcal{C}_a(\mathbb{Q})) \subseteq \mathcal{E}_a(\mathbb{Q})$, that is, $\mathcal{C}_a(\mathbb{Q}) \subseteq \psi^{-1}(\mathcal{E}_a(\mathbb{Q}))$, and hence we are reduced to determine the rational points on the cubic $\mathcal{E}_a$. Now, $\mathcal{E}_a$ is the affine part (that is, $Z = 1$) of the elliptic curve $\mathcal{E}$ defined by the equation

$$Y^2Z = X^3 - X^2Z + XZ^2.$$

This curve is well-known and listed as curve 24A in the tables given in [127, p. 83]; it is modular of conductor 24, and the rank of its Mordell-Weil group is zero. Hence the set of rational points on $\mathcal{E}$ consists of finitely many points, the so-called torsion points of $\mathcal{E}(\mathbb{Q})$, which correspond to the elements of $\mathcal{E}_a(\mathbb{Q})$ with the exception of the point at infinity.

To determine these torsion points we use the theorem of Nagell-Lutz – see [125, p. 221]: changing the coordinates X, Y, Z to

$$\tilde{X} := 9X - 3, \qquad \tilde{Y} := 27Y, \qquad \tilde{Z} := Z,$$

the elliptic curve $\mathcal{E}$ is isomorphically mapped onto the elliptic curve $\tilde{\mathcal{E}}$ given by the equation

$$\tilde{Y}^2\tilde{Z} = \tilde{X}^3 + A\tilde{X}\tilde{Z}^2 + B\tilde{Z}^3$$

with $A = 54$, $B = 189$ and discriminant $4A^3 + 27B^2 = 3^{13}$. The torsion points $(\tilde{x} : \tilde{y} : 1)$ (different from the point $(0 : 1 : 0)$ at infinity) of $\tilde{\mathcal{E}}(\mathbb{Q})$ are then characterized by those pairs $(\tilde{x}, \tilde{y})$ satisfying $\tilde{x}, \tilde{y} \in \mathbb{Z}$ and $\tilde{y} = 0$ or $\tilde{y}^2 \mid 3^{13}$. A straightforward calculation shows that these conditions are only satisfied for $\tilde{y} = 0$ and $\tilde{y} = \pm 27$. This leads, up to the point at infinity, to the three torsion points

$$(-3 : 0 : 1), \quad (+6 : +27 : 1), \quad (+6 : -27 : 1)$$

of $\tilde{\mathcal{E}}(\mathbb{Q})$, and hence proves

$$\mathcal{E}_a(\mathbb{Q}) = \{(0, 0), (+1, +1), (+1, -1)\}. \tag{I.5}$$

By means of the formulae (I.4), we then immediately find the following six rational points on the quartic $\mathcal{C}_a$ (observing that each one of the three rational points of $\mathcal{E}_a$ given above has exactly two preimages in $\mathcal{C}_a(\mathbb{Q})$ for the morphism ψ)

$$\mathcal{C}_a(\mathbb{Q}) = \{(0, +1), (0, -1), (+1, +1), (-1, -1), (+1, -1), (-1, +1)\}.$$

Finally, using the formulae (I.3), it is now a simple task to determine the rational points in the fibers of φ over the six points of $\mathcal{C}_a(\mathbb{Q})$ given in (I.5); in this way we find

$$\mathcal{U}_a(\mathbb{Q}) = \left\{(\pm n^2, 0, n), (\pm n^2, \pm n, n) : n \in \mathbb{Q}^*\right\},$$

which completes the proof of the lemma. □

J Nondegeneracy of the Second KdV Hamiltonian

We establish the nondegeneracy properties of the second KdV Hamiltonian as stated in Lemma J.2. First we consider the Jacobian of the frequency map $I_A \to \omega_A^2$ at $I_A = 0$. This Jacobian is represented by the matrix

$$C_A = (C_{kl})_{k,l\in A} = Q_A^2\big|_{I_A=0},$$

whose coefficients are given in Theorem 15.3. Recall that they are linear functions of the parameter $c \in \mathbb{R}$, which we do not indicate for the sake of the simplicity of the notation. Also, we do not use bold face symbols to denote infinite dimensional vectors, since there is no danger of confusion.

Lemma J.1. *For every finite set $A \subset \mathbb{N}$ there exists an $|A|$-point set $\mathcal{C}_A \subset \mathbb{R}$ not containing 0 such that*

$$\det C_A = 0 \quad \Leftrightarrow \quad c \in \mathcal{C}_A.$$

In particular, if $A = \{i\}$, then $\mathcal{C}_A = \{-\frac{2}{3}\pi^2 i^2\}$, while for $A = \{i_1 < \cdots < i_n\}$ one has

$$\mathcal{C}_A = \left\{c_A^n < \cdots < c_A^2 < 0 < c_A^1\right\}$$

with

$$-\frac{14}{3}\pi^2 i_\nu^2 < c_A^\nu < -\frac{14}{3}\pi^2 i_{\nu-1}^2, \qquad 2 \le \nu \le n,$$

and $c_A^1 \to \infty$ as $|A| \to \infty$.

Proof. In view of Theorem 15.3 we can write the matrix under consideration as $C = D - B$, where $D = \operatorname{diag}(D_i)_{i\in A}$ and $B = (B_{ij})_{i,j\in A}$ have coefficients

$$D_i = 280\pi^2 i^2 + 60c, \qquad B_{ij} = 240\pi^2 ij.$$

The matrix B has rank 1, so by the multi-linearity of the determinant we have

$$\det C = \det D - \sum_{i\in A} B_{ii} \prod_{j\in A,\, j\ne i} D_j.$$

If one of the D_j vanishes, say $D_k = 0$, then all other D_j do not vanish, and we have $\det C = -B_{kk}\prod_{j\ne k} D_j \ne 0$. Otherwise, $\det C = (\det D)(1 - \sum_{i\in A} B_{ii}/D_i)$, and the determinant vanishes if and only if

$$1 = \sum_{i\in A} \frac{B_{ii}}{D_i} = \sum_{i\in A} \frac{12}{14 + cf_i}, \qquad f_i = \frac{3}{\pi^2 i^2}.$$

Each summand is a hyperbola in c. It has a single pole at $c = -\frac{14}{3}\pi^2 i^2$, value $\frac{6}{7}$ at $c = 0$, and asymptotic value 0 as $c \to \pm\infty$. Also, both branches of the hyperbola are monotonically decreasing. From these considerations it follows that the above equation has exactly n solutions, which for $n \geq 2$ are located as

$$c_A^n < c_A^{n-1} < \cdots < c_A^2 < 0 < c_A^1,$$

as described in the lemma. For $n = 1$ the result is also immediately read off. □

The lemma shows that for any given $A \subset \mathbb{N}$ the Jacobian of the frequency map $I_A \mapsto \omega_A^2$ of the second KdV Hamiltonian *does* become singular, at least at $I_A = 0$ for $c \in \mathcal{C}_A$. This is in contrast to the first KdV Hamiltonian, where the Jacobian is always regular at $I_A = 0$. Moreover, Krichever [68] and Bobenko & Kuksin [11] have shown that for $c = 0$ it is indeed regular everywhere. See the second reference for a complete proof.

We now fix a finite set $A \subset \mathbb{N}$ and consider the frequency combinations $k \cdot \omega^2$ as functions of I_A on $\mathbb{P}\ell_A$. In view of $\omega^2 = \lambda^2 - CI + \ldots$ and the symmetry of the matrix C we have

$$k \cdot \omega^2 = k \cdot \lambda^2 - (Ck)_A I_A + \ldots$$

on $\mathbb{P}\ell_A$. To prove that $k \cdot \omega^2 \not\equiv 0$ on $\mathbb{P}\ell_A$ it is thus sufficient to show that

$$k \cdot \lambda^2 \neq 0 \qquad \text{or} \qquad (Ck)_A \neq 0. \tag{J.1}$$

We first prove a general statement to this effect. Recall that C depends on the parameter $c \in \mathbb{R}$.

Lemma J.2. *For each $k \in \mathbb{Z}^\infty$ with $1 \leq |k_Z| \leq 2$ there exists at most one $c_k \in \mathbb{R}$ such that the alternative* (J.1) *does not hold. This c_k is a rational multiple of π^2. Moreover, within every compact subset of $\mathbb{R} - \mathcal{C}_A$ there are only finitely many such c_k.*

Proof. We have

$$(Ck)_A = C_A k_A + C_{AZ} k_Z,$$

where $C_{AZ} = (C_{ij})_{i \in A, j \in Z}$. The diagonal elements of C_A are linear functions of c, namely $60c + 40\pi^2 i^2$, while all other coefficients of both matrices are integer multiples of π^2, namely $-240\pi^2 ij$. Hence, given k the above vector $(Ck)_A$ can vanish for at most one value of c, and this value must be a rational multiple of π^2.

To prove the remaining statement suppose that $(Ck)_A = 0$, and that c belongs to some compact subset $F \subset \mathbb{R} - \mathcal{C}_A$. Then C_A is invertible,

$$k_A = -C_A^{-1} C_{AZ} k_Z,$$

and we can bound C_A^{-1} uniformly for $c \in F$. Moreover, the vector $C_{AZ} k_Z$ has coefficients $-240\pi^2 ip$, $i \in A$, where

$$p = k_Z \cdot \lambda_Z^{\circ}, \qquad \lambda_Z^{\circ} = (j)_{j \in Z}.$$

It follows that for any $c \in F$,

$$|k_A| \leq \left|C_A^{-1}\right| |C_{AZ} k_Z| \leq K \left|k_Z \cdot \lambda_Z^{\circ}\right|,$$

where here and below, K stands for various constants bigger than 1 that depend only on A and the compact set F.

Now suppose that also

$$k \cdot \lambda^2 = k_A \cdot \lambda_A^2 + k_Z \cdot \lambda_Z^2 = 0.$$

In view of $\lambda_n^2 = \tilde{n}^5 + 10c\tilde{n}^3 + 30c^2\tilde{n}$ from Theorem 15.3 it is a routine estimate to show that for $1 \leq |k_Z| \leq 2$ one has

$$\left|k_Z \cdot \lambda_Z^2\right| \geq K^{-1} \left|k_Z \cdot \lambda_Z^{\mathrm{o}}\right|^5 - K \left|k_Z \cdot \lambda_Z^{\mathrm{o}}\right|^3.$$

Thus we obtain

$$K \left|k_Z \cdot \lambda_Z^{\mathrm{o}}\right| \left|\lambda_A^2\right| \geq \left|k_A \cdot \lambda_A^2\right| = \left|k_Z \cdot \lambda_Z^2\right| \geq K^{-1} \left|k_Z \cdot \lambda_Z^{\mathrm{o}}\right|^5 - K \left|k_Z \cdot \lambda_Z^{\mathrm{o}}\right|^3.$$

Therefore, $\left|k_Z \cdot \lambda_Z^{\mathrm{o}}\right| \leq K$ with a different constant K. Combining this estimate with the estimate for $|k_A|$ we find

$$|k_A| \leq K, \qquad \left|k_Z \cdot \lambda_Z^2\right| \leq K.$$

Thus, for $c \in F$ there can be only *finitely* many $k \in \mathbb{Z}^\infty$ with $1 \leq |k_Z| \leq 2$ for which the alternative (J.1) does not hold. Consequently, there can be at most finitely many exceptional values c_k in F. This proves the last statement. □

The preceding lemma does not make explicit the value of c_k. So it may happen that the mean value zero case $c = 0$ is excluded. We now show that this is not the case.

Lemma J.3. *For every finite set $A \subset \mathbb{N}$ the second KdV Hamiltonian at $c = 0$ satisfies* (J.1) *and is thus nondegenerate.*

Proof. As in the previous proof suppose that $(Ck)_A = 0$. At $c = 0$ the coefficients of C are, up to a common multiplicative factor, $(7\delta_{ij} - 6)ij$. Hence the coefficients of k satisfy

$$7ik_i = 6 \sum_{j \geq 1} jk_j, \qquad i \in A.$$

It follows that $ik_i = r$ is independent of i for $i \in A$. Substituting these identities in the above sum we obtain

$$7r = 6\left(\sum_{i \in A} r + \sum_{j \in Z} jk_j\right) = 6nr + 6p,$$

where $n = |A|$ and $p = \sum_{j \in Z} jk_j$. Hence,

$$ik_i = r = \frac{6p}{7 - 6n}, \qquad i \in A. \tag{J.2}$$

In particular, all k_i for $i \in A$ are *distinct* and of the *same* sign.

Now consider

$$k \cdot \lambda^2 = (2\pi)^5 \left(\sum_{i \in A} i^5 k_i + \sum_{j \in Z} j^5 k_j \right).$$

Solving equation (J.2) for i and for k_i we have

$$-\sum_{i \in A} i^5 k_i = \frac{6p}{6n-7} \sum_{i \in A} i^4 = \left(\frac{6p}{6n-7} \right)^5 \sum_{i \in A} \frac{1}{k_i^4}.$$

To show that $k \cdot \lambda^2 \neq 0$ it thus suffices to show that the two terms

$$\mathrm{I} = \sum_{i \in A} \frac{1}{k_i^4}$$

and

$$\mathrm{II} = \left(\frac{6n-7}{6p} \right)^5 \sum_{j \in Z} j^5 k_j, \qquad p = \sum_{j \in Z} j k_j > 0$$

are not equal.

In most cases, this can be done by straightforward estimates. For $n = 1$, the terms I and II have opposite sign. For $n \geq 3$, we note that with $1 \leq |k_Z| \leq 2$,

$$\sum_{j \in Z} j^5 k_j \geq \frac{1}{2^4} \left(\sum_{j \in Z} j k_j \right)^5 = \frac{1}{2^4} p^5.$$

It thus suffices to show that

$$\sum_{i \in A} \frac{1}{k_i^4} < \left(\frac{6n-7}{6} \right)^5 \frac{1}{2^4}.$$

This is obviously true for $n \geq 4$, and for $n = 3$ it follows from

$$\sum_{i \in A} \frac{1}{k_i^4} \leq 1 + \frac{1}{2^4} + \frac{1}{3^4} < \frac{11^5}{6^5} \cdot \frac{1}{2^4}.$$

For $n = 2$ the above estimates still suffice in the case where k_Z has only *one* nonzero component, so that $k_Z = l e_j$ with $1 \leq l \leq 2$. Then

$$\sum_{j \in Z} j^5 k_j = \frac{p^5}{l^4},$$

and so

$$\mathrm{II} = \mathrm{II}_l = \left(\frac{5}{6} \right)^5 \frac{1}{l^4}, \qquad l = 1, 2.$$

On the other hand, depending on whether $\min_{i\in A}(-k_i)$ is 1, 2, or ≥ 3, let

$$\mathrm{I} = \mathrm{I}_m = \sum_{i\in A} \frac{1}{k_i^4}, \qquad m = 1, 2, 3,$$

respectively. One then checks that

$$\mathrm{I}_3 < \mathrm{II}_2 < \mathrm{I}_2 < \mathrm{II}_1 < \mathrm{I}_1 .$$

It remains to discuss the case $n = 2$ and $k_Z = e_{j_1} \pm e_{j_2}$ with $j_1 > j_2$. By equation (J.2) we have

$$ik_i = -\frac{6p}{5}, \qquad i \in A.$$

In particular, $5 \mid p$. Assuming to the contrary that $k \cdot \lambda^2 = 0$, we have I = II, which is equivalent to

$$6p \sum_{i\in A} i^4 = 5 \sum_{j\in Z} j^5 k_j = 5(j_1^5 \pm j_2^5).$$

We now show that this equation has no integer solutions.

We have $p = j_1 \pm j_2$. Let $q = j_1 \mp j_2 > 0$. Solving for j_1 and j_2, taking powers and adding up one obtains

$$2^4(j_1^5 \pm j_2^5) = p(p^4 + 2 \cdot 5p^2q^2 + 5q^4).$$

So it suffices to show that

$$2^5 \cdot 3 \cdot (i_1^4 + i_2^4) = 5(p^4 + 2 \cdot 5p^2q^2 + 5q^4)$$

can not have an integer solution, if p is divisible by 5.

For every integer n, n^4 is congruent to 0 or 1 modulo 5. Therefore,

$$5 \mid i_1^4 + i_2^4 \quad \Leftrightarrow \quad 5 \mid i_1 \quad \text{and} \quad 5 \mid i_2 .$$

If the above equation holds, then $5 \mid i_1^4 + i_2^4$, hence $5 \mid i_1$ and $5 \mid i_2$, and so the left hand side is divisible by 5^4. On the right hand side the first two terms are divisible by 5^4, since $5 \mid p$. Hence also $5^4 \mid 5^2q^4$, and thus $5 \mid q$, too. Thus, *every integer* in the equation is divisible by 5, the whole equation is divisible by 5^4, and so we get

$$2^5 \cdot 3 \cdot (\tilde{i}_1^4 + \tilde{i}_2^4) = 5(\tilde{p}^4 + 2 \cdot 5 \cdot \tilde{p}^2\tilde{q}^2 + 5\tilde{q}^4),$$

where $\tilde{p} = p/5$ and so on.

The preceding argument can now be repeated *ad infinitum*. But this is absurd. So also in the remaining cases we have $k \cdot \lambda^2 \neq 0$, and the lemma is proven. □

X

Some Technicalities

K Symplectic Formalism

We describe some basic notions of symplectic geometry in infinite dimensions to the extent as they are needed here. We follow Kuksin's exposition in [72], with a few simplifications and modifications.

A *scale of Hilbert spaces*, or a *Hilbert scale*, is a sequence $\boldsymbol{E} = (E_p)_{p=0,1,\dots}$ of Hilbert spaces E_p with inner products $\langle\cdot\,,\cdot\rangle_p$ and norms $\|\cdot\|_p$ such that

$$E_0 \supset E_1 \supset E_2 \supset \dots, \qquad \|\cdot\|_0 \le \|\cdot\|_1 \le \dots,$$

and such that all E_p are dense in E_0. A *linear isomorphism of order d* between two Hilbert scales $\boldsymbol{E}$ and $\boldsymbol{F}$, where d is an integer, is a linear isomorphism

$$L\colon\ E_q \to F_{q-d}, \qquad q = \max(0, d),$$

whose restrictions $L|E_p$ give rise to linear isomorphisms $E_p \to F_{p-d}$ for each $p \ge q$.

Recall that a constant symplectic form on a finite dimensional Hilbert space E with inner product $\langle\cdot\,,\cdot\rangle$ is a bilinear, anti-symmetric, nondegenerate form α on E. It can be represented as

$$\alpha(\xi, \eta) = \langle \xi, K\eta\rangle, \qquad \xi, \eta \in E,$$

with an anti-symmetric isomorphism K of E. In view of our applications to the KdV equations we extend this notion to infinite dimensional spaces as follows. A constant *symplectic form* α on a Hilbert scale $\boldsymbol{E}$ is given by a linear anti-symmetric isomorphism K of the scale $\boldsymbol{E}$ of some order $-d_\alpha \le 0$ such that

$$\alpha(\xi, \eta) = \langle \xi, K\eta\rangle_0, \qquad \xi, \eta \in E_0.$$

The pair $(\boldsymbol{E}, \alpha)$ is called a *symplectic Hilbert scale*. Note that α is anti-symmetric, nondegenerate and *closed*, since it is constant.

Consider an open domain $U_p \subset E_p$ and a function $H \in C^1(U_p; \mathbb{C})$. A continuous map

$$X_H\colon \; U_p \to E_0$$

is called the *Hamiltonian vector field of the Hamiltonian* H on U_p, if

$$\alpha(X_H, \,\cdot\,) = \mathrm{d}H$$

at every point of U_p. That is, $\alpha(X_H(x), \xi) = d_x H(\xi)$ for all $x \in U_p$ and $\xi \in E_p$.

The vector field is said to be *of order* d, where $d \le p$ is an integer, if in fact X_H maps into $E_{p-d} \subset E_0$:

$$X_H\colon \; U_p \to E_{p-d}.$$

This definition requires some explanations. First, X_H is not a vector field in the strict sense, since it is not required to be a section of the tangent bundle over U_p. This allows us to deal with unbounded operators. On the other hand, not every differentiable Hamiltonian H admits a Hamiltonian vector field X_H in this sense, since in general $d_x H$ is an element of the dual space E_p^* only and thus lacks the required regularity. But if it exists, then it is unique.

To be more precise, given $H \in C^1(U_p; \mathbb{C})$ we always have a representation

$$\mathrm{d}H(\xi) = \langle \nabla H, \xi \rangle_0, \qquad \xi \in E_p,$$

with a continuous map $\nabla H\colon U_p \to E_p^*$, if we identify the dual space E_p^* with the completion of the space E_0 with respect to the norm

$$\| \cdot \|_{-p} = \sup_{\|\xi\|_p \le 1} |\langle \,\cdot\, , \xi \rangle_p|.$$

Then the equation $\alpha(X_H, \,\cdot\,) = \mathrm{d}H$ is equivalent to

$$-\langle K X_H, \cdot \rangle_0 = \langle \nabla H, \cdot \rangle_0.$$

This equation can be solved for X_H in E_0, if ∇H is in the domain of K^{-1}, that is, if

$$\nabla H\colon \; U_p \to E_{d_\alpha} \subset E_0,$$

since K is an anti-symmetric isomorphism of the scale $\boldsymbol{E}$ of order $-d_\alpha \le 0$. In this case we have

$$X_H = J \nabla H, \qquad J = -K^{-1},$$

where J is an anti-symmetric isomorphism of the Hilbert scale $\boldsymbol{E}$ of order $d_\alpha \ge 0$. This is the situation we encounter in our applications.

For the sake of completeness we mention the notion of a strong solution. Let I be an interval. A differentiable map $u\colon I \to U_p$ is called a *strong solution* of the Hamiltonian vector field X_H on U_p, or more precisely of the evolution equation $\dot{u} = X_H(u)$, if $\dot{u} = \mathrm{d}u/\mathrm{d}t$ satisfies this equation in E_0 for every $t \in I$.

We are interested in constructing strong solutions of perturbed KdV equations. However, we will not do this directly, nor will we construct local flows or semi-flows. Rather, we *transform* the vector fields into a form such that the existence of certain strong solutions is evident.

We consider the relevant transformation rule. Let $(\boldsymbol{E}, \alpha_E)$ and $(\boldsymbol{F}, \alpha_F)$ be two symplectic Hilbert scales. Fix p. A differentiable map

$$\Phi\colon \quad F_p \supset V_p \to E_p$$

is *symplectic*, if

$$\Phi^* \alpha_E = \alpha_F$$

on V_p. That is, in every point of V_p one has

$$\alpha_E(\Phi_* \xi, \Phi_* \eta) = \alpha_F(\xi, \eta), \qquad \xi, \eta \in F_p,$$

where Φ_* denotes the push forward map between tangent spaces induced by Φ. Using representations of α_E, α_F by isomorphisms K_E, K_F, respectively, this is equivalent to $\langle \Phi_* \xi, K_E \Phi_* \eta \rangle_{E_0} = \langle \xi, K_F \eta \rangle_{F_0}$, or

$$\Phi^* K_E \Phi_* = K_F,$$

where Φ^* is the adjoint of Φ_*.

In a finite dimensional setting a symplectic transformation takes Hamiltonian vector fields into Hamiltonian vector fields – see section 2. Our definition of Hamiltonian vector fields guarantees that this is also true in infinite dimensions.

Transformation Rule. *Suppose H admits a Hamiltonian vector field X_H on a domain $U_p \subset E_p$ of order $d \le p$, and*

$$\Phi\colon \quad F_p \supset V_p \to U_p \subset E_p$$

is a symplectic transformation, such that at every point in V_p,

$$D\Phi\colon F_p \to E_p$$

extends to a linear isomorphism $F_{p-d} \to E_{p-d}$. Then the transformed vector field $\Phi^* X_H = D\Phi^{-1} X_H \circ \Phi$ is a Hamiltonian vector field on V_p of order d with Hamiltonian $H \circ \Phi$. That is,

$$\Phi^* X_H = X_{H \circ \Phi}\colon \quad V_p \to F_{p-d} \subset F_0.$$

Proof. Clearly,

$$\Phi^* X_H = D\Phi^{-1} X_H \circ \Phi\colon \quad V_p \to F_{p-d} \subset F_0,$$

since by assumption,

$$V_p \xrightarrow{\Phi} U_p \xrightarrow{X_H} E_{p-d} \xrightarrow{D\Phi^{-1}} F_{p-d}.$$

This vector field is Hamiltonian, since

$$\begin{aligned} d(H \circ \Phi)(\xi) &= (\mathrm{d}H \circ \Phi)(D\Phi \cdot \xi) \\ &= \alpha_E(X_H \circ \Phi, D\Phi \cdot \xi) \\ &= \alpha_E(D\Phi \cdot \Phi^* X_H, D\Phi \cdot \xi) \\ &= \Phi^* \alpha_E(\Phi^* X_H, \xi) \\ &= \alpha_F(\Phi^* X_H, \xi) \end{aligned}$$

for all $\xi \in F_p$. Hence, $\Phi^* X_H$ is the Hamiltonian vector field for $H \circ \Phi$ as defined above. □

L Infinite Products

We collect some estimates concerning infinite products of complex numbers that are quite elementary but used repeatedly in the construction of the Birkhoff coordinates.

Lemma L.1. *Suppose a_{mn}, $m, n \geq 1$, are complex numbers satisfying*

$$|a_{mn}| = O\left(\frac{1}{|m^2 - n^2|}\right), \qquad m \neq n.$$

Then

$$\prod_{\substack{m \geq 1 \\ m \neq n}} (1 + a_{mn}) = 1 + O\left(\frac{\log n}{n}\right), \qquad n \geq 1,$$

where the implicit constant in the conclusion depends only on the implicit constant in the assumption.

Proof. By assumption,

$$\sum_{\substack{m \geq 1 \\ m \neq n}} |a_{mn}| \leq C \sum_{\substack{m \geq 1 \\ m \neq n}} \frac{1}{|m^2 - n^2|}$$

with some positive constant C. The sum on the right can be estimated by

$$\begin{aligned} \sum_{\substack{m \geq 1 \\ m \neq n}} \frac{1}{|m^2 - n^2|} &= \sum_{\substack{1 \leq m \leq 2n \\ m \neq n}} \frac{1}{|m - n|} \frac{1}{m + n} + \sum_{m > 2n} \frac{1}{m^2 - n^2} \\ &\leq \frac{2}{n} \sum_{1 \leq k \leq n} \frac{1}{k} + \sum_{k > n} \frac{1}{k^2} \\ &\leq \frac{2}{n}(1 + \log n) + \frac{1}{n}. \end{aligned}$$

Hence we obtain

$$\begin{aligned}\left|\prod_{\substack{m\geq 1\\ m\neq n}}(1+a_{mn})-1\right| &\leq \prod_{\substack{m\geq 1\\ m\neq n}}(1+|a_{mn}|)-1\\ &\leq \exp\left(\sum |a_{mn}|\right)-1\\ &\leq \exp\left(\frac{C'\log n}{n}\right)-1 = O\left(\frac{\log n}{n}\right)\end{aligned}$$

with a different constant C' depending only on C. □

For the sake of reference we separately state an important special case of the preceding lemma.

Lemma L.2. *Suppose ξ_m and ζ_m are complex numbers satisfying*

$$\xi_m, \zeta_m = m^2\pi^2 + O(1)$$

for $m \geq 1$. Then

$$\prod_{m\neq n}\frac{\zeta_m-\lambda}{\xi_m-\lambda} = 1 + O\left(\frac{\log n}{n}\right)$$

uniformly for $(n-\frac{1}{2})^2\pi^2 \leq |\lambda| \leq (n+\frac{1}{2})^2\pi^2$ with n sufficiently large, where the implicit constant in the conclusion depends only on the implicit constant in the assumption.

Proof. For $\lambda = n^2\pi^2 + O(n)$ and $m \neq n$,

$$\left|\frac{\zeta_m-\lambda}{\xi_m-\lambda}-1\right| = \left|\frac{\zeta_m-\xi_m}{\xi_m-\lambda}\right| = O\left(\frac{1}{|m^2-n^2|}\right),$$

and the preceding lemma applies. □

Lemma L.3. *Suppose ξ_m are complex numbers with $\xi_m = m^2\pi^2 + O(1)$. Then the infinite product*

$$\prod_{m\geq 1}\frac{\xi_m-\lambda}{m^2\pi^2}$$

defines an entire function of λ, whose roots are precisely ξ_m, $m \geq 1$, and which satisfies

$$\prod_{m\geq 1}\frac{\xi_m-\lambda}{m^2\pi^2} = \frac{\sin\sqrt{\lambda}}{\sqrt{\lambda}}\left(1+O\left(\frac{\log n}{n}\right)\right)$$

uniformly on the circles $|\lambda| = (n+\frac{1}{2})^2\pi^2$.

Proof. We just prove the last statement. To this end recall that

$$\frac{\sin\sqrt{\lambda}}{\sqrt{\lambda}} = \prod_{m\geq 1} \frac{m^2\pi^2 - \lambda}{m^2\pi^2}.$$

Hence the quotient of the given product and $\sin\sqrt{\lambda}/\sqrt{\lambda}$ is

$$\prod_{m\geq 1} \frac{\xi_m - \lambda}{m^2\pi^2 - \lambda} = \frac{\xi_n - \lambda}{n^2\pi^2 - \lambda} \prod_{m\neq n} \frac{\xi_m - \lambda}{m^2\pi^2 - \lambda}.$$

By the preceding lemma, the infinite product on the right is $1+O(\log n/n)$ uniformly on the circle $|\lambda| = (n+\frac{1}{2})^2\pi^2$, while on the same circle,

$$\frac{\xi_n - \lambda}{n^2\pi^2 - \lambda} = 1 + O\Big(\frac{1}{n}\Big).$$

From this the result follows. □

M Auxiliary Results

Here we collect a few frequently used estimates for analytic maps and a result about extensions of Lipschitz functions.

Lemma M.1. *Let Γ be a circuit around the interval $[a,b]$ within some complex neighbourhood U of it. If f is analytic on U, then*

$$\left| \frac{1}{2\pi} \int_\Gamma \frac{f(\lambda)}{\sqrt{(b-\lambda)(\lambda-a)}} \,\mathrm{d}\lambda \right| \leq \max_{a\leq t\leq b} |f(t)|.$$

Proof. If $a = b$, the given integral turns into a Cauchy integral, and the result follows immediately. If $a < b$, then we may shrink the contour of integration to the interval $[a,b]$ to obtain

$$\begin{aligned} \left| \frac{1}{2\pi} \int_\Gamma \frac{f(\lambda)}{\sqrt{(b-\lambda)(\lambda-a)}} \,\mathrm{d}\lambda \right| &\leq \frac{1}{\pi} \int_a^b \frac{|f(\lambda)|}{\sqrt[+]{(b-\lambda)(\lambda-a)}} \,\mathrm{d}\lambda \\ &\leq \frac{1}{\pi} \int_a^b \frac{\mathrm{d}\lambda}{\sqrt[+]{(b-\lambda)(\lambda-a)}} \max_{a\leq t\leq b} |f(t)|. \end{aligned}$$

The last integral is 1, giving the result. □

Lemma M.2. *Let u_j, $j \geq 1$, be complex functions on $\mathbb{T}^n$ that are real analytic on $D(s) = \{|\mathrm{Im}\, x| < s\}$. Then*

$$\Big(\sum_{j\geq 1} \sup_{x\in D(s-\sigma)} |u_j(x)|^2 \Big)^{1/2} \leq \frac{4^n}{\sigma^n} \sup_{x\in D(s)} \Big(\sum_{j\geq 1} |u_j(x)|^2 \Big)^{1/2}$$

for $0 < \sigma < s \leq 1$.

Proof. First consider the case $n = 1$. For each $j \geq 1$ there exists a point x_j in the rectangle $Q = \{x\colon\ |\operatorname{Re} x| \leq \pi,\ |\operatorname{Im} x| \leq s - \sigma\}$ such that

$$\sup_{x \in D(s-\sigma)} \left|u_j(x)\right| \leq \left|u_j(x_j)\right|.$$

By the Cauchy integral formula,

$$u_j(x_j) = \frac{1}{2\pi \mathrm{i}} \int_\Gamma \frac{u_j(\zeta)}{\zeta - x_j}\,\mathrm{d}\zeta,$$

where Γ describes a rectangle with distance $0 < \rho < \sigma$ around Q, independent of j. By the triangle inequality we then get

$$\begin{aligned}
\Bigg(\sum_{j\geq 1} \sup_{x\in D(s-\sigma)} \left|u_j(x)\right|^2\Bigg)^{1/2} &\leq \Bigg(\sum_{j\geq 1} \left|\frac{1}{2\pi\mathrm{i}} \int_\Gamma \frac{u_j(\zeta)}{\zeta - x_j}\,\mathrm{d}\zeta\right|^2\Bigg)^{1/2} \\
&\leq \frac{1}{2\pi} \int_\Gamma \Bigg(\sum_{j\geq 1} \left|\frac{u_j(\zeta)}{\zeta - x_j}\right|^2\Bigg)^{1/2} |\mathrm{d}\zeta| \\
&\leq \frac{4}{\rho} \sup_{x\in D(s)} \Bigg(\sum_{j\geq 1} \left|u_j(x)\right|^2\Bigg)^{1/2}.
\end{aligned}$$

Letting $\rho \to \sigma$ the estimate follows for $n = 1$. For $n > 1$ the single Cauchy integral is replaced by an n-fold Cauchy integral. □

We also need a version of this lemma for bounded operators on ℓ^2.

Lemma M.3. *Let $A = (A_{ij})_{i,j\geq 1}$ be a bounded operator on ℓ^2 which depends on $x \in \mathbb{T}^n$ such that all coefficients are analytic on $D(s) = \{|\operatorname{Im} x| < s\}$. Suppose $B = (B_{ij})_{i,j\geq 1}$ is another operator on ℓ^2 depending on x whose coefficients satisfy*

$$\sup_{x\in D(s)} \left|B_{ij}(x)\right| \leq \frac{1}{|i-j|} \sup_{x\in D(s)} \left|A_{ij}(x)\right|, \qquad i \neq j,$$

and $B_{jj} = 0$ for $j \geq 1$. Then B is a bounded operator on ℓ^2 for every $x \in D(s)$, and

$$\sup_{x\in D(s-\sigma)} \|B(x)\| \leq \frac{4^{n+1}}{\sigma^n} \sup_{x\in D(s)} \|A(x)\|$$

for $0 < \sigma \leq s \leq 1$.

Proof. For $x \in D(s-\sigma)$ we have by the Schwarz inequality and the preceding lemma

$$\begin{aligned}
\sum_{j\geq 1} \left|B_{ij}(x)\right| &\leq \sum_{j\geq 1} \sup_{x\in D(s-\sigma)} \left|B_{ij}(x)\right| \\
&\leq \Bigg(\sum_{j\geq 1} \sup_{x\in D(s)} \left|A_{ij}(x)\right|^2\Bigg)^{1/2} \Bigg(\sum_{j\neq i} \frac{1}{|i-j|^2}\Bigg)^{1/2}
\end{aligned}$$

$$\leq \frac{4^{n+1}}{\sigma^n} \sup_{x \in D(s)} \Bigl(\sum_{j \geq 1} \bigl| A_{ij}(x) \bigr|^2 \Bigr)^{1/2}$$

$$\leq \frac{4^{n+1}}{\sigma^n} \sup_{x \in D(s)} \| A(x) \| .$$

The same estimate applies to $\sum_{i \geq 1} \bigl| B_{ij}(x) \bigr|$. Hence, for $x \in D(s - \sigma)$,

$$\begin{aligned} \| B(x) v \|^2 &\leq \sum_{i \geq 1} \Bigl(\sum_{j \geq 1} \bigl| B_{ij}(x) \bigr| \bigl| v_j \bigr| \Bigr)^2 \\ &\leq \sum_{i \geq 1} \Bigl(\sum_{j \geq 1} \bigl| B_{ij}(x) \bigr| \Bigr) \Bigl(\sum_{j \geq 1} \bigl| B_{ij(x)} \bigr| \bigl| v_j \bigr|^2 \Bigr) \\ &\leq \Bigl(\sup_i \sum_{j \geq 1} \bigl| B_{ij}(x) \bigr| \Bigr) \Bigl(\sup_j \sum_{i \geq 1} \bigl| B_{ij}(x) \bigr| \Bigr) \Bigl(\sum_{j \geq 1} \bigl| v_j \bigr|^2 \Bigr) \\ &\leq \Bigl(\frac{4^{n+1}}{\sigma^n} \sup_{x \in D(s)} \| A(x) \| \Bigr)^2 \| v \|^2 . \end{aligned}$$

From this the final estimate follows. □

Let V be an open domain in a complex Banach space E with norm $\| \cdot \|$, Π an arbitrary subset of parameters in $\mathbb{R}^n$, and

$$X \colon \quad V \times \Pi \to E$$

a parameter dependent vector field on V, which is analytic on some neighbourhood $W \supset V$ and Lipschitz on Π. Let Φ^t denote the flow of X, and assume that there is a subdomain $U \subset V$ such that

$$\Phi^t \colon \quad U \times \Pi \to V, \qquad -1 \leq t \leq 1.$$

Lemma M.4. *If, with the preceding assumptions,* $\| DX \|^{\sup}_V \leq 1$, *then*

$$\begin{aligned} \bigl\| \Phi^t - \mathrm{id} \bigr\|^{\sup}_{U \times \Pi} &\leq \| X \|^{\sup}_{V \times \Pi}, \\ \bigl\| \Phi^t - \mathrm{id} \bigr\|^{\mathrm{lip}}_{U \times \Pi} &\leq 3 \, \| X \|^{\mathrm{lip}}_{V \times \Pi}, \end{aligned}$$

for $-1 \leq t \leq 1$. *Moreover, if* $| \cdot |$ *is any operator norm on* E *with* $| DX |^{\sup}_{V \times \Pi} \leq 1$, *then*

$$\begin{aligned} \bigl| D\Phi^t - I \bigr|^{\sup}_{U \times \Pi} &\leq 3 \, |I| \, | DX |^{\sup}_{V \times \Pi}, \\ \bigl| D\Phi^t - I \bigr|^{\mathrm{lip}}_{U \times \Pi} &\leq 9 \, |I| \, | DX |^{\mathrm{lip}}_{V \times \Pi} + 27 K \, |I| \, | DX |^{\sup}_{W \times \Pi}, \end{aligned}$$

where $K = \| X \|^{\mathrm{lip}}_{V \times \Pi} / \operatorname{dist}(V, \partial W)$, *and* I *is the identity operator.*

Proof. In the following we suppress Π from the notation. Fix $0 \le t \le 1$. The first estimate follows from $\Phi^t - \mathrm{id} = \int_0^t X \circ \Phi^s \, \mathrm{d}s$. To prove the second one, let $\Delta\Phi^t = \Phi^t_\xi - \Phi^t_\zeta$ for $\xi, \zeta \in \Pi$, where $\Phi^t_\xi = \Phi^t(\,\cdot\,, \xi)$. Then

$$\Delta\Phi^t = \int_0^t \Delta(X \circ \Phi^s)\,\mathrm{d}s = \int_0^t \Delta X \circ \Phi^s_\xi \,\mathrm{d}s + \int_0^t (X \circ \Phi^s_\xi - X \circ \Phi^s_\zeta)\,\mathrm{d}s,$$

hence, as $\|DX\|_V^{\sup} \le 1$ by assumption,

$$\|\Delta\Phi^t\|_U^{\sup} \le \int_0^t \|\Delta X\|_V^{\sup} \mathrm{d}s + \int_0^t \|\Delta\Phi^s\|_U^{\sup} \mathrm{d}s.$$

With Gronwall's inequality we get $\|\Delta\Phi^t\|_U^{\sup} \le 3\,\|\Delta X\|_V^{\sup}$. Dividing by $|\xi - \zeta|$ and taking the supremum over Π the Lipschitz estimate follows.

Now let $|\cdot|$ be any operator norm on E, and assume that $|DX|_V^{\sup} \le 1$. We have

$$D\Phi^t = I + \int_0^t DX \circ \Phi^s \cdot D\Phi^s \,\mathrm{d}s,$$

from which $|D\Phi^t|_U^{\sup} \le 3\,|I|$ and the third estimate of the lemma follow with Gronwall's inequality. For the Lipschitz estimate we write

$$\Delta D\Phi^t = \int_0^t \Delta(DX \circ \Phi^s) \cdot D\Phi^s_\xi \,\mathrm{d}s + \int_0^t DX \circ \Phi^s_\xi \cdot \Delta D\Phi^s \,\mathrm{d}s.$$

With $|D\Phi^s|_U^{\sup} \le 3\,|I|$ from the preceding estimate and $|DX|_V^{\sup} \le 1$ this gives

$$|\Delta D\Phi^t|_U^{\sup} \le 3\,|I| \int_0^t |\Delta(DX \circ \Phi^s)|_U^{\sup} \mathrm{d}s + \int_0^t |\Delta D\Phi^s|_U^{\sup} \mathrm{d}s.$$

Hence, by Gronwall,

$$|\Delta D\Phi^t|_U^{\sup} \le 9\,|I| \int_0^t |\Delta(DX \circ \Phi^s)|_U^{\sup} \mathrm{d}s.$$

Again, we write

$$\Delta(DX \circ \Phi^s) = (\Delta DX) \circ \Phi^s_\xi + DX \circ \Phi^s_\xi - DX \circ \Phi^s_\zeta.$$

To estimate the second term, we use the Cauchy inequality for the operator norm of $D(DX)$ with the norms $\|\cdot\|$ and $|\cdot|$ in the source and target space, respectively. With $\rho = \mathrm{dist}(V, \partial W)$ we get

$$\begin{aligned} |DX \circ \Phi^s_\xi - DX \circ \Phi^s_\zeta|_U^{\sup} &\le |D(DX)|_V^{\sup} \, \|\Delta\Phi^s\|_U^{\sup} \\ &\le \frac{|\xi - \zeta|}{\rho} \, |DX|_W^{\sup} \, \|\Phi^s - \mathrm{id}\|_U^{\mathrm{lip}}. \end{aligned}$$

It follows with $\|\Phi^s - \mathrm{id}\|_U^{\mathrm{lip}} \le 3\,\|X\|_V^{\mathrm{lip}}$ that

$$\left|\Delta D\Phi^t\right|_U^{\mathrm{sup}} \le 9\,|I|\,|\Delta DX|_V^{\mathrm{sup}} + 27\,|I|\frac{|\xi-\zeta|}{\rho}\,|DX|_W^{\mathrm{sup}}\,\|X\|_V^{\mathrm{lip}}.$$

From this the estimate for $\left|D\Phi^t - I\right|^{\mathrm{lip}}$ follows as stated. □

Lemma M.5 (Lipschitz Extension). *Let $F \subset \mathbb{R}^n$ be a closed set and $u\colon F \to \mathbb{R}$ a bounded Lipschitz continuous function. Then there exists an extension $U\colon \mathbb{R}^n \to \mathbb{R}$ of u, which preserves minimum, maximum and Lipschitz semi-norm of u.*

Proof. Let $\lambda = |u|_F^{\mathrm{lip}}$, and define

$$\tilde{u}(x) = \sup_{\xi\in F}\,(u(\xi) - \lambda\,|x-\xi|)$$

for $x \in \mathbb{R}^n$. This is an extension of u to all of $\mathbb{R}^n$. By the triangle inequality,

$$\tilde{u}(x) \ge u(\xi) - \lambda\left|x'-\xi\right| - \lambda\left|x-x'\right|$$

for all $\xi \in F$ and hence $\tilde{u}(x) \ge \tilde{u}(x') - \lambda\left|x-x'\right|$. Interchanging x and x', we get

$$\frac{\left|\tilde{u}(x)-\tilde{u}(x')\right|}{|x-x'|} \le \lambda.$$

It follows that $|\tilde{u}|_{\mathbb{R}^n}^{\mathrm{lip}} = |u|_F^{\mathrm{lip}}$. Setting

$$U = (\tilde{u} \wedge \max_F u) \vee \min_F u$$

we note that $|U|_{\mathbb{R}^n}^{\mathrm{lip}} = |\tilde{u}|_{\mathbb{R}^n}^{\mathrm{lip}}$. Thus the function U has all the required properties. □

References

1. L. AHLFORS, *Complex Analysis*. McGraw-Hill, New York, 1966.
2. V. I. ARNOLD, Proof of a theorem by A. N. Kolmogorov on the invariance of quasi-periodic motions under small perturbations of the Hamiltonian. *Russian Math. Surveys* **18** (1963), 9–36.
3. V. I. ARNOLD, *Mathematical Methods of Classical Mechanics*. Springer, New York, 1978.
4. V. I. ARNOLD & A. AVEZ, *Ergodic Problems of Classical Mechanics*. W. A. Benjamin, New York, 1968.
5. D. BÄTTIG, A. M. BLOCH, J.-C. GUILLOT & T. KAPPELER, On the symplectic structure of the phase space for periodic KdV, Toda, and defocusing NLS. *Duke Math. J.* **79** (1995), 549–604.
6. D. BÄTTIG, T. KAPPELER & B. MITYAGIN, On the Korteweg-de Vries equation: Convergent Birkhoff normal form. *J. Funct. Anal.* **140** (1996), 335–358.
7. D. BÄTTIG, T. KAPPELER & B. MITYAGIN, On the Korteweg-de Vries equation: Frequencies and initial value problem. *Pacific J. Math.* **181** (1997), 1–55.
8. D. BERNSTEIN & A. KATOK, Birkhoff periodic orbits for small perturbations of completely integrable Hamiltonian systems with convex Hamiltonian. *Invent. Math.* **88** (1987), 225–241.
9. R. F. BIKBAEV & S. B. KUKSIN, On the parametrization of finite gap solutions by frequency vector and wave number vector and a theorem of I. Krichever. *Lett. Math. Phys.* **28** (1993), 115–122.
10. R. F. BIKBAEV & S. B. KUKSIN, A periodic boundary value problem for the sine-Gordon equation, small Hamiltonian perturbations of it, and KAM-deformations of finite gap tori. *Algebra and Analysis* **4** (1992) [Russian]; English translation in *St. Petersburg Math. J.* **4** (1993), 439–468.
11. A. BOBENKO & S. B. KUKSIN, Finite-gap periodic solutions of the KdV equation are non-degenerate. *Physics Letters A* **161** (1991), 274–276.
12. A. I. BOBENKO & S. B. KUKSIN, The nonlinear Klein-Gordon equation on an interval as a perturbed sine-Gordon equation. *Comm. Math. Helv.* **70** (1995), 63–112.
13. J.-B. BOST, Tores invariants des systèmes dynamiques hamiltoniens (d'après Kolmogorov, Arnol'd, Moser, Rüssmann, Zehnder, Herman, Pöschel, ...). *Astérisque* **133–134** (1986), 113–157.

14. J. BOURGAIN, Fourier transform restriction phenomena for certain lattice subsets and applications to nonlinear evolution equations. II. The KdV equation. *Geom. Funct. Anal.* **3** (1993), 209–262.
15. J. BOURGAIN, On the Cauchy problem for periodic KdV-type equations. Proceedings of the Conference in Honor of Jean-Pierre Kahane (Orsay, 1993). *J. Fourier Anal. Appl.* (1995), Special Issue, 17–86.
16. J. BOURGAIN, Construction of approximative and almost periodic solutions of perturbed linear Schrödinger and wave equations. *Geom. Funct. Anal.* **6** (1996), 201–230.
17. J. BOURGAIN, *Global Solutions of Nonlinear Schrödinger Equations.* Colloquium Publications, American Mathematical Society, Providence, 1999.
18. J. BOUSSINESQ, Théorie de l'intumescence liquid appelée onde solitaire ou de translation, se propageant dans un canal rectangulaire. *Comptes Rend. Acad. Sci. (Paris)* **72** (1871), 755–759.
19. A. D. BRJUNO, Convergence of transformations of differential equations to normal form. *Dokl. Akad. Nauk SSSR* **165** (1965), 987–989.
20. H. W. BROER, G. B. HUITEMA & M. B. SEVRYUK, *Quasi-Periodic Motions in Families of Dynamical Systems.* Lecture Notes in Mathematics 1645, Springer, Berlin, 1996.
21. C. Q. CHENG, Lower dimensional invariant tori in the regions of instability for nearly integrable Hamiltonian systems. *Comm. Math. Phys.* **203** (1999), 385–419.
22. C. Q. CHENG & Y. S. SUN, Existence of KAM tori in degenerate Hamiltonian systems. *J. Diff. Equ.* **114** (1994), 288–335.
23. C. Q. CHENG & S. WANG, The surviving of lower dimensional tori from a resonant torus of Hamiltonian systems. *J. Diff. Equ.* **155** (1999), 311–326.
24. L. CHIERCHIA & P. PERFETTI, Maximal almost-periodic solutions for Lagrangian equations on infinite-dimensional tori. In: *Seminar on Dynamical Systems,* S. Kuksin, V. Lazutkin, J. Pöschel (Eds), Birkhäuser, Basel, 1994, 203–212.
25. L. CHIERCHIA & J. YOU, KAM tori for 1D nonlinear wave equations with periodic boundary conditions. *Comm. Math. Phys.* **211** (2000), 497–525.
26. R. C. CHURCHILL, M. KUMMER & D. L. ROD, On averaging, reduction and symmetry in Hamiltonian systems. *J. Diff. Equations* **49** (1983), 359–414.
27. Y. COLIN DE VERDIÈRE & T. KAPPELER, On double eigenvalues of Hill's operator. *J. Funct. Anal.* **86** (1989), 127–135.
28. W. CRAIG & C. E. WAYNE, Newton's method and periodic solutions of nonlinear wave equations. *Comm. Pure Appl. Math.* **46** (1993), 1409–1498.
29. W. CRAIG, *Problèmes de petits diviseurs dans les équations aux dérivées partielles.* Panoramas et Synthèses 9, Société Mathématique de France, Paris, 2000.
30. P. DEIFT, F. LUND & E. TRUBOWITZ, Nonlinear wave equations and constrained harmonic motion. *Comm. Math. Phys.* **74** (1980), 141–188.
31. L. A. DICKEY, *Soliton equations and Hamiltonian systems.* Advanced Series in Mathematical Physics 12, World Scientific, Singapore, 1991.
32. B. A. DUBROVIN, Periodic problems for the Korteweg-de Vries equation in the class of finite band potentials. *Funct. Anal. Appl.* **9** (1975), 215–223.
33. B. A. DUBROVIN, I. M. KRICHEVER & S. P. NOVIKOV, The Schrödinger equation in a periodic field and Riemann surfaces. *Sov. Math. Dokl.* **17** (1976), 947–951.

34. B. A. DUBROVIN, I. M. KRICHEVER & S. P. NOVIKOV, Integrable Systems I. In: *Dynamical Systems IV*, Encyclopedia of Mathematical Sciences vol. 4, V. I. ARNOLD & S. P. NOVIKOV (eds.). Springer, 1990, 173–280.

35. B. A. DUBROVIN, V. B. MATVEEV & S. P. NOVIKOV, Nonlinear equations of Korteweg-de Vries type, finite-zone linear operators and Abelian varieties. *Russ. Math. Surv.* **31** (1976), 59–146.

36. B. A. DUBROVIN & S. P. NOVIKOV, Periodic and conditionally periodic analogues of the many-soliton solutions of the Korteweg-de Vries equation. *Sov. Phys.-JETP* **40** (1974), 1058–1063.

37. J. J. DUISTERMAAT, On global action-angle coordinates. *Comm. Pure Appl. Math.* **33** (1980), 687–706.

38. L. H. ELIASSON, Perturbations of stable invariant tori for Hamiltonian systems. *Ann. Sc. Norm. Sup. Pisa* **15** (1988), 115–147.

39. L. D. FADDEEV & L. A. TAKHTADZHYAN, *Hamiltonian methods in the theory of solitons*. Springer Series in Soviet Mathematics, Springer, Berlin, 1987.

40. L. D. FADDEEV & V. E. ZAKHAROV, Kortweg-de Vries equation: a completely integrable Hamiltonian system. *Funct. Anal. Appl.* **5** (1971), 280–287.

41. H. FLASCHKA, On the inverse problem for Hill's operator. *Arch. Ration. Mech. Anal.* **59** (1975), 293–309.

42. H. FLASCHKA, M. G. FOREST & D. MCLAUGHLIN, Multiphase averaging and the inverse spectral solution of the Korteweg-de Vries equation. *Comm. Pure Appl. Math.* **33** (1980), 739–784.

43. H. FLASCHKA & D. MCLAUGHLIN, Canonically conjugate variables for the Korteweg-de Vries equation and Toda lattices with periodic boundary conditions. *Progress Theor. Phys.* **55** (1976), 438–456.

44. J. P. FRANÇOISE, The Arnol'd formula for algebraically completely integrable systems. *Bull. Amer. Math. Soc. (N.S.)* **17** (1987), 301–303.

45. J. FRÖHLICH, T. SPENCER & C. E. WAYNE, Localization in disordered, nonlinear dynamical systems. *J. Stat. Phys.* **42** (1986), 247–274.

46. C. S. GARDNER, Korteweg-de Vries equation and generalizations. IV. The Korteweg-de Vries equation as a Hamiltonian system. *J. Math. Phys.* **12** (1971), 1548–1551.

47. C. S. GARDNER, J. M. GREENE, M. D. KRUSKAL & R. M. MIURA, Method for solving the Korteweg-de Vries equation. *Phys. Rev. Lett.* **19** (1967), 1095–1097.

48. J. GARNETT & E. T. TRUBOWITZ, Gaps and bands of one dimensional periodic Schrödinger operators. *Comment. Math. Helv.* **59** (1984), 258–312.

49. J. GARNETT & E. T. TRUBOWITZ, Gaps and bands of one dimensional periodic Schrödinger operators II. *Comment. Math. Helv.* **62** (1987), 18–37.

50. B. GRÉBERT & T. KAPPELER, Perturbations of the defocusing NLS equation with periodic boundary conditions. Preprint.

51. B. GRÉBERT & T. KAPPELER, Normal form theory for the NLS equation. Preprint.

52. PH. GRIFFITHS & J. HARRIS, *Principles of Algebraic Geometry*. John Wiley & Sons, New York, 1978.

53. M. R. HERMAN, *Sur les courbes invariantes par les difféomorphismes de l'anneau. Vol. 1. Complété par un appendice au chapitre 1 de Albert Fathi*. Astérisque, 103–104. Société Mathématique de France, Paris, 1983.

54. M. R. HERMAN, *Sur les courbes invariantes par les difféomorphismes de l'anneau. Vol. 2.* Astérisque, 144. Société Mathématique de France, Paris, 1986.

55. H. HOFER & E. ZEHNDER, *Symplectic Invariants and Hamiltonian Dynamics.* Birkhäuser, Basel, 1994.

56. H. ITO, Convergence of Birkhoff normal forms for integrable systems. *Comment. Math. Helvetici* **64** (1989), 412–461.

57. A. R. ITS & V. B. MATVEEV, A class of solutions of the Korteweg-de Vries equation. *Probl. Mat. Fiz. 8,* Leningrad State University, Leningrad, 1976, 70–92.

58. R. JOST, Winkel- und Wirkungsvariable für allgemeine mechanische Systeme. *Helvetica Physica Acta* **41** (1968), 965–968.

59. T. KAPPELER, Fibration of the phase space for the Korteweg-de Vries equation. *Ann. Inst. Fourier* **41** (1991), 539–575.

60. T. KAPPELER & M. MAKAROV, On Birkhoff coordinates for KdV. *Ann. Henri Poincaré* **2** (2001), 807–856.

61. T. KAPPELER & B. MITYAGIN, Estimates for periodic and Dirichlet eigenvalues of the Schrödinger operator. *SIAM J. Math. Anal.* **33** (2001), 101–136.

62. A. B. KATOK, Ergodic properties of degenerate integrable Hamiltonian systems. *Math. USSR Izv.* **7** (1973), 185–214.

63. Y. KATZNELSON, *An Introduction to Harmonic Analysis.* Dover, New York, 1976.

64. C. KENIG, G. PONCE & L. VEGA, A bilinear estimate with applications to the KdV equation. *J. Amer. Math. Soc.* **9** (1996), 573–603.

65. A. N. KOLMOGOROV, On the conservation of conditionally periodic motions for a small change in Hamilton's function. *Dokl. Akad. Nauk SSSR* **98** (1954), 527–530 [Russian]. English translation in: *Lectures Notes in Physics* **93,** Springer, 1979.

66. D. J. KORTEWEG & G. DE VRIES, On the change of form of long waves advancing in a rectangular canal, and on a new type of long stationary waves. *Phil. Mag. Ser. 5* **39** (1895), 422–443.

67. I. KRICHEVER, Integration of nonlinear equations by methods of algebraic geometry. *Funkt. Anal. Appl.* **11** (1977), 15–31.

68. I. KRICHEVER, "Hessians" of integrals of the Korteweg-de Vries equation and perturbations of finite-gap solutions. *Dokl. Akad. Nauk SSSR* **270** (1983), 1312–1317 [Russian]. English translation in *Sov. Math. Dokl.* **27** (1983), 757–761.

69. M. D. KRUSKAL & N. J. ZABUSKY, Interaction of "solitons" in a collisionless plasma and the recurrence of initial states. *Phys. Rev. Lett.* **15** (1965), 240–243.

70. S. B. KUKSIN, Hamiltonian perturbations of infinite-dimensional linear systems with an imaginary spectrum. *Funts. Anal. Prilozh.* **21** (1987), 22–37 [Russian]. English translation in *Funct. Anal. Appl.* **21** (1987), 192–205.

71. S. B. KUKSIN, Perturbation theory for quasiperiodic solutions of infinite-dimensional Hamiltonian systems, and its application to the Korteweg-de Vries equation. *Matem. Sbornik* **136** (1988) [Russian]. English translation in *Math. USSR Sbornik* **64** (1989), 397–413.

72. S. B. KUKSIN, *Nearly integrable infinite-dimensional Hamiltonian systems.* Lecture Notes in Mathematics 1556, Springer, 1993.

73. S. B. KUKSIN, On small-denominator equations with large variable coefficients. Preprint, 1995.

74. S. B. KUKSIN, A KAM-theorem for equations of the Korteweg-de Vries type. *Rev. Math. Phys.* **10** (1998), 1–64.

75. S. B. KUKSIN, *Analysis of Hamiltonian PDEs*. Oxford University Press, Oxford, 2000.

76. S. B. KUKSIN & J. PÖSCHEL, Invariant Cantor manifolds of quasi-periodic oscillations for a nonlinear Schrödinger equation. *Ann. Math.* **143** (1996), 149–179.

77. P. LAX, Integrals of nonlinear equations of evolution and solitary waves. *Comm. Pure Appl. Math.* **21** (1968), 467–490.

78. P. LAX, Periodic solutions of the KdV equation. *Comm. Pure Appl. Math.* **28** (1975), 141–188.

79. V. F. LAZUTKIN, *KAM Theory and Semiclassical Approximations of Eigenfunctions*. Springer, Berlin, 1993.

80. B. M. LEVITAN & I. S. SARGSJAN, *Introduction to Spectral Theory: Selfadjoint Ordinary Differential Operators*. American Mathematical Society, Providence, 1975.

81. R. DE LA LLAVE, A tutorial on KAM theory. *Proc. Symp. Pure Math.* **69** (2001), 175–292.

82. W. MAGNUS & S. WINKLER, *Hill's Equation*. Second edition, Dover, New York, 1979.

83. F. MAGRI, A simple model of the integrable Hamiltonian equation. *J. Math. Phys.* **19** (1978), 1156–1162.

84. V. A. MARČHENKO, *Sturm-Liouville Operators and Applications*. Birkhäuser, Basel, 1986.

85. L. MARKUS & K. MEYER, Generic Hamiltonian systems are neither integrable nor ergodic. *Mem. Am. Math. Soc.* **144,** 1974.

86. D. MCDUFF & D. SALAMON, *Introduction to Symplectic Topology*. Clarendon Press, Oxford, 1995.

87. H. P. MCKEAN, Compatible brackets in Hamiltonian mechanics. In: A. S. FOKAS ET AL., *Important Developments in Soliton Theory,* Springer, Berlin, 1993, 344–354.

88. H. P. MCKEAN & P. VAN MOERBECKE, The spectrum of Hill's equation. *Invent. Math.* **30** (1975), 217–274.

89. H. P. MCKEAN & E. TRUBOWITZ, Hill's operator and hyperelliptic function theory in the presence of infinitely many branch points. *Comm. Pure Appl. Math.* **29** (1976), 143–226.

90. H. P. MCKEAN & E. TRUBOWITZ, Hill's surfaces and their theta functions. *Bull. Am. Math. Soc.* **84** (1978), 1042–1085.

91. H. P. MCKEAN & K. L. VANINSKY, Action-angle variables for the cubic Schroedinger equation. *Comm. Pure Appl. Math.* **50** (1997), 489–562.

92. H. P. MCKEAN & K. L. VANINSKY, Cubic Schroedinger: The petit canonical ensemble in action-angle variables. *Comm. Pure Appl. Math.* **50** (1997), 594–622.

93. V. K. MELNIKOV, On some cases of conservation of conditionally periodic motions under a small change of the Hamilton function. *Soviet Math. Doklady* **6** (1965), 1592–1596.

94. R. M. MIURA, Korteweg-de Vries equation and generalizations. I. A remarkable explicit nonlinear transformation. *J. Math. Phys.* **9** (1968), 1202–1204.

95. R. M. MIURA, C. S. GARDNER & M. D. KRUSKAL, Korteweg-de Vries equation and generalizations. II. Existence of conservation laws and constants of motion. *J. Math. Phys.* **9** (1968), 1204–1209.

96. J. MOSER, On invariant curves of area preserving mappings of an annulus. *Nachr. Akad. Wiss. Gött., Math. Phys. Kl.* (1962), 1–20.

97. J. MOSER, Convergent series expansions for quasi-periodic motions. *Math. Ann.* **169** (1967), 136–176.

98. J. MOSER, Lectures on Hamiltonian systems. *Mem. Amer. Math. Soc.* **81** (1968), 1–60.

99. J. MOSER, On the continuation of almost periodic solutions for ordinary differential equations. In: *Proc. Int. Conf. Func. Anal. and Rel. Topics,* Tokyo, 1969, 60–67.

100. J. MOSER, *Stable and Random Motions in Dynamical Systems.* Princeton University Press, 1973.

101. R. NARASIMHAN, *Compact Riemann Surfaces.* Lectures in Mathematics ETH Zürich, Birkhäuser, Basel 1992.

102. N. V. NIKOLENKO, Complete integrability of the nonlinear Schrödinger equation. *Sov. Math. Dokl.* **17** (1976), 398–402.

103. N. V. NIKOLENKO, The method of Poincaré normal form in problems of integrability of equations of evolution type. *Russ. Math. Surveys* **41** (1986), 63–114.

104. S. NOVIKOV, S. V. MANAKOV, L. P. PITAEVSKIJ & V. E. ZAKHAROV, *Theory of solitons.* Translated from the Russian. Plenum Publishing Corporation, Consultants Bureau, New York, 1984.

105. J. PÖSCHEL, Über invariante Tori in differenzierbaren Hamiltonschen Systemen. *Bonn. Math. Schr.* **120** (1980), 1–103.

106. J. PÖSCHEL, Integrability of Hamiltonian systems on Cantor sets. *Comm. Pure Appl. Math.* **35** (1982), 653–695.

107. J. PÖSCHEL, On elliptic lower dimensional tori in Hamiltonian systems. *Math. Z.* **202** (1989), 559–608.

108. J. PÖSCHEL, Small divisors with spatial structure in infinite dimensional Hamiltonian systems. *Comm. Math. Phys.* **127** (1990), 351–393.

109. J. PÖSCHEL, A KAM-theorem for some nonlinear partial differential equations. *Ann. Sc. Norm. Sup. Pisa* **23** (1996), 119–148.

110. J. PÖSCHEL, Quasi-periodic solutions for a nonlinear wave equation. *Comment. Math. Helv.* **71** (1996), 269–296.

111. J. PÖSCHEL, On the construction of almost periodic solutions for a nonlinear Schrödinger equation. *Ergodic Theory Dyn. Syst.* **22** (2002), 1–22.

112. J. PÖSCHEL & E. TRUBOWITZ, *Inverse Spectral Theory.* Academic Press, Boston, 1987.

113. LORD RAYLEIGH, On waves. *Phil. Mag. Ser. 5* **1** (1876), 257–279.

114. J. S. RUSSELL, Report on waves. In: *Report of the Fourteenth Meeting of the British Association for the Advancement of Sciences,* John Murray, London, 1844, 311–390.

115. H. RÜSSMANN, Über das Verhalten analytischer Hamiltonscher Differentialgleichungen in der Nähe einer Gleichgewichtslösung. *Math. Ann.* **154** (1964), 285–300.

116. H. RÜSSMANN, Über die Iteration analytischer Funktionen. *J. Math. Mech.* **17** (1967), 523–532.

117. H. RÜSSMANN, On the one-dimensional Schrödinger equation with a quasi-periodic potential. *Ann. N.Y. Acad. Sci.* **357** (1980) 90–107 .

118. H. RÜSSMANN, Non-degeneracy in the perturbation theory of integrable dynamical systems. In: M. M. DODSON, J. A. G. VICKERS (EDS.), *Number Theory and Dynamical Systems,* London Math. Soc. Lecture Note Series 134, Cambridge University Press, 1989, 5–18.

119. H. RÜSSMANN, Invariant tori in non-degenerate nearly integrable Hamiltonian systems. *Regul. Chaotic Dyn.* **6** (2001), 119–204.

120. D. SALAMON, The Kolmogorov-Arnold-Moser theorem. Preprint, ETH-Zürich, 1986.

121. J. SAUT & R. TEMAM, Remarks on the Korteweg-de Vries equation. *Israel J. Math.* **24** (1976), 78–87.

122. M. SEVRYUK, KAM-stable Hamiltonians. *J. Dyn. Control Syst.* **1** (1995), 351–366.

123. C. L. SIEGEL, Über die Existenz einer Normalform analytischer Hamiltonscher Differentialgleichungen in der Nähe einer Gleichgewichtslösung. *Math. Ann.* **128** (1954), 144–170.

124. C. L. SIEGEL, J. K. MOSER, *Lectures On Celestial Mechanics.* Springer, Berlin, 1971.

125. J. H. SILVERMAN, *The arithmetic of elliptic curves.* Graduate Texts in Mathematics 106, Springer, Berlin, 1986.

126. A. SJÖBERG, On the Korteweg-de Vries equation: existence and uniqueness. *J. Math. Anal. Appl.* **29** (1970), 569–579.

127. H. P. F. SWINNERTON-DYER, Numerical tables on elliptic curves. In: *Modular functions of one variable IV,* W. KUYK et al (eds), Lecture Notes in Mathematics 476, Springer, 1975, 75–144.

128. R. TEMAM, Sur un problème non linéaire. *J. Math. Pures Appl.* **48** (1969), 159–172.

129. D. V. TRESHCHEV, The mechanism of destruction of resonance tori of Hamiltonian systems. *Math. USSR Sb.* **68** (1990), 181–203.

130. P. VAN MOERBEKE, The spectrum of Jacobi matrices. *Invent. Math.* **37** (1976), 45–81.

131. A. P. VESELOV & S. P. NOVIKOV, Poisson brackets and complex tori. *Proc. Steklov Inst. Math.* **165** (1985), 53–65.

132. J. VEY, Sur certains systèmes dynamiques séparables. *Amer. J. Math.* **100** (1978), 591–614.

133. M. VITTOT & J. BELLISSARD, Invariant tori for an infinite lattice of coupled classical rotators. Preprint, CPT-Marseille, 1985.

134. B. WARE, Infinite-dimensional versions of two theorems by Carl Siegel. *Bull. Am. Math. Soc.* **82** (1976), 613–615.

135. C. E. WAYNE, Periodic and quasi-periodic solutions of nonlinear wave equation via KAM theory. *Comm. Math. Phys.* **127** (1990), 479–528.

136. G. B. WITHAM, *Linear and Nonlinear Waves.* Wiley, New York, 1974.

137. J. XU, J. YOU & Q. QIU, Invariant tori for nearly integrable Hamiltonian systems with degeneracy. *Math. Z.* **226** (1997), 375–387.

138. J. YOU, Perturbations of lower-dimensional tori for Hamiltonian systems. *J. Diff. Equ.* **152** (1999), 1–29.

139. E. ZEHNDER, Siegel's linearizaton theorem in infinite dimensions. *manus. math.* **23** (1978), 363–371.

140. N. T. ZUNG, Convergence versus integrability in Birkhoff normal form. Preprint, 2001.

Index